LINEAR PROGRAMMING

LINEAR PROGRAMMING

Methods and Applications

THIRD EDITION

SAUL I. GASS

World Systems Laboratories, Inc.
The American University

McGRAW-HILL BOOK COMPANY

New York St. Louis San Francisco Dusseldorf London
Mexico Panama Sydney Toronto

LINEAR PROGRAMMING

 Library of Congress Catalog Card Number 68-55267

22967

2 3 4 5 6 7 8 9 0 M A M M 7 5 4 3 2 1 0

TO TRUDY

For her encouragement,
patience,
and lost weekends

PREFACE

The material in this book was originally prepared for an introductory course in linear programming given at the Graduate School, U.S. Department of Agriculture, Washington, D.C. In developing and expanding the notes into a suitable text, I have attempted to pursue the same objectives that guided the presentation of the course material. These basic aims were to instill in the student an ability to recognize potential linear-programming problems, to formulate such problems as linear-programming models, to employ the proper computational techniques to solve these problems, and to understand the mathematical aspects that tie together these elements of linear programming.

It is very convenient to divide the subject matter of linear programming into three separate, but not distinct, areas: *theoretical, computational,* and *applied.* In teaching the course, I found it appropriate, instructive, and beneficial to the students to interlace material from all three areas as much as possible. Hence, after an introductory lecture on applications and the mathematical model of linear programming (Chap. 1), the mathematics of convex sets and linear inequalities were developed and followed by the computational aspects of the elimination method for solving linear equations (Chap. 2). The mathematical properties of a solution to the general linear-programming problem were next evolved. Then, in an attempt to explain fully the fundamentals of the simplex computational procedure, the ability to generate extreme-point solutions was shown to be a simple variation of the elimination technique of Jordan and Gauss (Chap. 3). The next set of lectures developed the theoretical and computational elements of the simplex method of G. B. Dantzig (Chap. 4). A discussion on the duality problems of linear programming (Chap. 5) was followed by lectures on the formulation of certain illustrative applications (Chaps. 10 and 11). The final lectures of the course[1]

[1] This one-semester course consisted of 16 evening lectures of 2½ hours' duration.

described the relationship between linear programming and the zero-sum two-person game (Chap. 12).

The basic-course notes have been revised to include full discussion of the revised simplex method (Chap. 6), degeneracy procedures (Chap. 7), parametric programming (Chap. 8), further computational techniques (Chap. 9), and other topics and applications. As a result of a suggestion to the author, all the material has been gathered into three parts: an introduction, methods, both theoretical and computational, and applications.[1] It is felt that this arrangement enhances the usefulness of the book for reference, and presents the three areas of linear programming in a related but separate manner. Consequently, the reader will find some relatively advanced topics, such as the revised simplex method and parametric linear programming, appearing before the basic discussions of the transportation problem and general applications. It is suggested that, in order to motivate one's study of linear programming, the chapters should not be studied in numerical sequence. Instead, one should, as soon as possible (probably after Chap. 4 or 5), become acquainted with material in the applications sections.[2]

It is felt that the material covered in this text is appropriate for use in mathematics courses at the senior or first-year-graduate level. However, because of the interest in linear-programming methods outside the academic field, it seemed advisable to include sufficient material on matrices and vectors to make the work complete for all readers (Chap. 2). It should be noted that much of the mathematical notation used in subsequent chapters is developed in Chap. 2.[3]

The search for the best, the maximum, the minimum, or, in general, the optimum solutions to a variety of problems has entertained and intrigued man throughout the ages. Euclid, in Book III, was concerned with finding the greatest and least straight lines that can be drawn from a point to the circumference of a circle, and in Book IV he described how to find the parallelogram of maximum area with a given perimeter. However, the rigorous approach to these and more sophisticated problems had to wait until the great mathematicians of the seventeenth and eighteenth centuries developed the powerful methods of the calculus and the calculus of variations. With these techniques we can find the maximum and minimum solutions to a wide range of optimization problems. These

[1] This edition includes an additional part on nonlinear programming, Chap. 13.

[2] In a course that meets three times a week, one lecture could be devoted to applications and/or reports of case studies cited in the Bibliography.

[3] The student will soon find that one of the main difficulties in understanding the mathematics of linear programming arises from the diverse and often intricate notation used in many of the source papers of this field. Wherever possible, I have employed "standard," consistent, and, I hope, explicit notation.

and other mathematical optimization procedures were mainly concerned with the solutions to problems of a geometric, dynamic, or physical nature. Such problems as finding the minimum curves of revolution and the curve of quickest descent are resolved by these classical optimization methods.

Recently, a new class of optimization problems has originated out of the complex organizational structures that permeate modern society. Here we are concerned with such matters as the most efficient manner in which to run an economy, or the optimum deployment of aircraft that maximizes a country's chances of winning a war, or with such mundane tasks as mixing the ingredients of a fertilizer to meet agricultural specifications at a minimum cost. Research on how to formulate and solve such problems has led to the development of new and important optimization techniques. Among these we find the subject of this book—*linear programming*. The linear-programming model, i.e., the optimization of a linear function subject to linear constraints, is simple in its mathematical structure but powerful in its adaptability to a wide range of applications.

Historically, the general problem of linear programming was first developed and applied in 1947 by George B. Dantzig, Marshall Wood, and their associates of the U.S. Department of the Air Force. At that time, this group was called on to investigate the feasibility of applying mathematical and related techniques to military programming and planning problems. This inquiry led Dantzig to propose "that interrelations between activities of a large organization be viewed as a linear programming type model and the optimizing program determined by minimizing a linear objective function." In order to develop and extend these ideas further, the Air Force organized a research group under the title of Project SCOOP (Scientific Computation of Optimum Programs). Besides putting the Air Force programming and budgeting problems on a more scientific basis, Project SCOOP's major contribution was the formal development and application of the linear-programming model. These early applications of linear-programming methods fell into three major categories: military applications generated by Project SCOOP, interindustry economics based on the Leontief input-output model, and problems involving the relationship between zero-sum two-person games and linear programming. In the past years these areas of applications have been extended and developed, but the main emphasis in linear-programming applications has shifted to the general industrial area.

The initial mathematical statement of the general problem of linear programming was made by Dantzig in 1947 along with the *simplex method*, a systematic procedure for solving the problem. Prior to this a number of problems (some unsolved) were recognized as being of the type that dealt with the optimization of a linear function subject to linear constraints. The more important examples include the transportation problem posed by Hitchcock (1941) and independently by Koopmans (1947) and the diet problem of Stigler (1945). The first successful solution of a linear-programming problem on a high-speed electronic computer occurred in January, 1952, on the National Bureau of Standards SEAC machine. Since that time, the simplex algorithm, or variations of this procedure, has been coded for most intermediate and large general-purpose electronic computers.

Linear programming has become an important tool of modern theoretical and applied mathematics. This remarkable growth can be traced to the pioneering efforts of many individuals and research organizations. Specifically, I would like to make special mention of George B. Dantzig, Murray A. Geisler, Leon Goldstein, Julian L. Holley, Walter W. Jacobs, Alex Orden, Emil D. Schell, and Marshall K. Wood, all formerly with the U.S. Department of the Air Force; Leon Gainen, Alan J. Hoffman, and Solomon Pollack, formerly with the National Bureau of Standards; and the research groups of the Graduate School of Industrial Administration of the Carnegie Institute of Technology, The RAND Corporation, the Department of Mathematics of Princeton University, and the Cowles Commission for Research in Economics. I would like to thank the above named individuals and groups, other authors, and their publishers for their kind permission to use certain basic material contained in what might be considered "source documents" of linear programming. Appropriate references are given in the text.

This edition includes new sections on network problems and the bounded-variable problem, a new chapter on nonlinear programming (Part IV), and a revised section on the decomposition algorithm. Additional material and exercises have been included in a number of sections and many new publications have been included in the bibliography and references.

I wish to acknowledge the initial encouragement to write this text by my former associates at the Directorate of Management Analysis of the Department of the Air Force and to thank Harold Fassberg, Walter W. Jacobs, Robert R. Meyer, Paul Rech, Thomas L. Saaty, and Kenneth Webb for their many valuable suggestions. Special appreciation is due Mrs. Thelma Chesley and Mrs. Anne Bache for their typing of the past editions and Miss Joanne Wagner for her excellent typing of the current-edition manuscript.

Saul I. Gass

CONTENTS

LINEAR PROGRAMMING

PART ONE / **INTRODUCTION**

CHAPTER 1 / GENERAL DISCUSSION

1. LINEAR-PROGRAMMING PROBLEMS

Programming problems are concerned with the efficient use or allocation of limited resources to meet desired objectives. These problems are characterized by the large number of solutions that satisfy the basic conditions of each problem. The selection of a particular solution as the best solution to a problem depends on some aim or over-all objective that is implied in the statement of the problem. A solution that satisfies both the conditions of the problem and the given objective is termed an *optimum solution.* A typical example is that of the manufacturer who must determine what combination of his available resources will enable him to manufacture his products in a way which not only satisfies his production schedule, but also maximizes his profit. This problem has as its basic conditions the limitations of the available resources and the requirements of the production schedule, and as its objective the desire of the manufacturer to maximize his gain.

We shall mainly consider only a very special subclass of programming problems called *linear-programming problems.* A linear-programming problem differs from the general variety in that a *mathematical model* or description of the problem can be stated, using relationships which are called "straight-line," or linear. Mathematically, these relationships are of the form

$$a_1x_1 + a_2x_2 + \cdots + a_jx_j + \cdots + a_nx_n = b\dagger$$

where the a_j's and b are known coefficients and the x_j's are unknown variables. The complete mathematical statement of a linear-programming problem includes a set of simultaneous linear equations which represent

† Geometrically, these relationships are equivalent to straight lines in two dimensions, planes in three dimensions, and hyperplanes in higher dimensions.

the conditions of the problem and a linear function which expresses the objective of the problem. In Sec. 2 we shall state a number of programming problems and formulate them as linear-programming problems.

To solve a linear-programming problem, we must initially concern ourselves with the solution of the associated set of linear equations. There are various criteria which can be applied to a set of linear equations to reveal whether a solution or solutions to the problem exist (see Dickson [34]).[1] The set of two equations in two variables

$$2x_1 + 3x_2 = 8$$
$$x_1 + 2x_2 = 5$$

has the *unique* solution $x_2 = 1$ and $x_2 = 2$, while the single equation

$$x_1 + 2x_2 = 8 \tag{1.1}$$

has an *infinite* number of solutions. From (1.1) we have

$$x_1 = 8 - 2x_2 \qquad \text{or} \qquad x_2 = 4 - \tfrac{1}{2}x_1$$

For every value of x_2 (or x_1) there is a corresponding value of x_1 (or x_2). If we further restrict the variables to be nonnegative, that is, $x_1 \geq 0$ and $x_2 \geq 0$, we limit the range of the variables, since

$$x_1 = 8 - 2x_2 \geq 0 \qquad \text{implies} \qquad 0 \leq x_2 \leq 4$$

and

$$x_2 = 4 - \tfrac{1}{2}x_1 \geq 0 \qquad \text{implies} \qquad 0 \leq x_1 \leq 8$$

We still have an infinite number of solutions, but the addition of further restrictions or constraints to (1.1) has resulted in less freedom of action. As we shall show, the condition of nonnegativity of the variables is an important requirement of linear-programming problems. Systems like (1.1) in which there are more variables than equations are called underdetermined. In general, underdetermined systems of linear equations have either no solution or an infinite number of solutions.

One important method of determining solutions to underdetermined systems of equations is to reduce the system to a set containing just as many variables as equations, i.e., a determined set. This can be accomplished by letting the appropriate number of variables equal zero. For example, the underdetermined system

$$\begin{aligned} 2x_1 + 3x_2 + x_3 &= 8 \\ x_1 + 2x_2 + 2x_3 &= 5 \end{aligned} \tag{1.2}$$

[1] Numbers in brackets refer to the publications listed in the References at the back of the book.

has three such solutions:

$$x_1 = 0 \qquad x_2 = 11/4 \qquad x_3 = -1/4$$
$$x_1 = 11/3 \qquad x_2 = 0 \qquad x_3 = 2/3$$

and

$$x_1 = 1 \qquad x_2 = 2 \qquad x_3 = 0\dagger$$

Mathematically, linear programming deals with *nonnegative solutions* to underdetermined systems of linear equations. As we shall show in the succeeding chapters, the only solutions we have to be concerned with are those corresponding to determined subsets of equations that have been obtained in the manner described above. If, for example, we let Eqs. (1.2) represent the conditions of a linear-programming problem, we need only to consider the two nonnegative solutions $x_1 = 1$, $x_2 = 2$, $x_3 = 0$ and $x_1 = 11/3$, $x_2 = 0$, $x_3 = 2/3$. The remaining solutions fail to satisfy either the nonnegativity requirements or other criteria to be discussed.

Just as the general programming problem has some objective that guides the selection of the solution to be used, the linear-programming problem has a linear function of the variables to aid in choosing a solution to the problem. This linear combination of the variables, called the *objective function*, must be optimized by the selected solution. If for (1.2) we wished to maximize the objective function $x_1 + x_2 + x_3$, then, of the two nonnegative solutions, the solution $x_1 = 11/3$, $x_2 = 0$, $x_3 = 2/3$ is the optimum, as it yields a value of $13/3$ for the objective function compared with a value of 3 for the other nonnegative solution. If we wanted to minimize the objective function $x_1 - x_2$, then the solution $x_1 = 1$, $x_2 = 2$, $x_3 = 0$ would be the optimum, with a value of -1. As we have implied, the optimum solution either maximizes or minimizes some linear combination of the variables. Since the maximum of a linear function is equal to minus the minimum of the negative of the linear function, we lose no generality by considering only the minimization problem.

With the added condition of optimizing an objective function, we are now able to select a single solution that satisfies all the conditions of the problem. There might be multiple solutions in that more than one nonnegative solution to the equations gives the same optimum value of the objective function. Generally speaking, combining the linear constraints of the programming problem with the optimization of a linear objective function transforms an underdetermined system of linear equations that

† There are, of course, an infinite number of other solutions to (1.2), which can be obtained by arbitrarily setting one of the variables equal to a constant; e.g., with $x_1 = a$ we have $x_2 = (11 - 3a)/4$ and $x_3 = (-1 + a)/4$.

describes a programming problem with many possible solutions to a system that can be solved for a solution that yields the unique optimum value of the objective function.

We next give the general mathematical statement of the linear-programming problem:

Minimize the objective function

$$c_1x_1 + c_2x_2 + \cdots + c_jx_j + \cdots + c_nx_n$$

subject to the conditions

$$\begin{aligned} x_1 \quad & \geq 0 \\ x_2 \quad & \geq 0 \\ & \vdots \\ x_j \quad & \geq 0 \\ & \vdots \\ x_n & \geq 0 \end{aligned}$$

and

$$\begin{array}{l} a_{11}x_1 + a_{12}x_2 + \cdots + a_{1j}x_j + \cdots + a_{1n}x_n = b_1 \\ a_{21}x_1 + a_{22}x_2 + \cdots + a_{2j}x_j + \cdots + a_{2n}x_n = b_2 \\ \cdots\cdots\cdots\cdots\cdots\cdots\cdots\cdots\cdots\cdots \\ a_{i1}x_1 + a_{i2}x_2 + \cdots + a_{ij}x_j + \cdots + a_{in}x_n = b_i \\ \cdots\cdots\cdots\cdots\cdots\cdots\cdots\cdots\cdots\cdots \\ a_{m1}x_1 + a_{m2}x_2 + \cdots + a_{mj}x_j + \cdots + a_{mn}x_n = b_m \end{array}$$

where the c_j for $j = 1, 2, \ldots, n$; b_i for $i = 1, 2, \ldots, m$; and a_{ij} are all constants, and $m < n$. The c_j are called *cost coefficients.*

As will be discussed in the following chapters, every linear-programming problem has either:

1. No solution in terms of nonnegative values of the variables
2. A nonnegative solution that yields an infinite value to the objective function
3. A nonnegative solution that yields a finite value to the objective function

A linear-programming problem that describes a valid, practical programming problem usually has a nonnegative solution with a corresponding finite value of the objective function.

2. EXAMPLES OF LINEAR-PROGRAMMING PROBLEMS

To illustrate the application of the above mathematical description of the linear-programming model, we shall next discuss the linear-programming formulation of three problems. A more detailed discussion of these and other problems is given in Part 3.

The transportation problem. A manufacturer wishes to ship a number of units of an item from several warehouses to a number of retail stores. Each store requires a certain number of units of the item, while each warehouse can supply up to a certain amount. Let us define the following:

m = the number of warehouses
n = the number of stores
a_i = the total amount of the item available for shipment at warehouse i
b_j = the total requirement of the item by store j
x_{ij} = the amount of the item shipped from warehouse i to store j

We shall assume that the total amount available is equal to the total required, that is, $\sum_i a_i = \sum_j b_j$. As will be shown later, this assumption is not a restrictive one.

The x_{ij} are the unknown shipments to be determined. If we form the array (for $m = 2$ and $n = 3$)

		Stores 1	2	3	
Warehouses	1	x_{11}	x_{12}	x_{13}	a_1
	2	x_{21}	x_{22}	x_{23}	a_2
		b_1	b_2	b_3	

we see that the total amount shipped from warehouse 1 can be expressed by the linear equation

$$x_{11} + x_{12} + x_{13} = a_1 \tag{2.1}$$

For warehouse 2, we have

$$x_{21} + x_{22} + x_{23} = a_2' \tag{2.2}$$

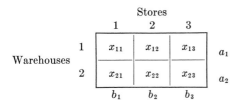

We also note that the total amounts shipped to the three stores are expressed by the equations

$$\begin{aligned} x_{11} + x_{21} &= b_1 \\ x_{12} + x_{22} &= b_2 \\ x_{13} + x_{23} &= b_3 \end{aligned} \tag{2.3}$$

The manufacturer knows the cost c_{ij} of shipping one unit of the item from warehouse i to store j. We have the additional assumption that the cost relationship is linear; i.e., the cost of shipping x_{ij} units is $c_{ij}x_{ij}$.

The manufacturer wishes to determine how many units should be sent from each warehouse to each store so that the total shipping cost is a minimum. This objective of minimizing the cost is achieved by minimizing the linear cost function

$$c_{11}x_{11} + c_{12}x_{12} + c_{13}x_{13} + c_{21}x_{21} + c_{22}x_{22} + c_{23}x_{23} \tag{2.4}$$

Since a negative x_{ij} would represent a shipment from store j to warehouse i, we require that all the variables $x_{ij} \geq 0$.

By combining Eqs. (2.1) to (2.3), the objective function (2.4), and the condition of nonnegativity of the variables, the transportation problem for $m = 2$ and $n = 3$ can be formulated in terms of the following linear-programming problem:

Minimize the cost function

$$c_{11}x_{11} + c_{12}x_{12} + c_{13}x_{13} + c_{21}x_{21} + c_{22}x_{22} + c_{23}x_{23}$$

subject to the conditions

$$\begin{array}{llllllll} x_{11} & & & & & & \geq 0 \\ & x_{12} & & & & & \geq 0 \\ & & x_{13} & & & & \geq 0 \\ & & & x_{21} & & & \geq 0 \\ & & & & x_{22} & & \geq 0 \\ & & & & & x_{23} & \geq 0 \end{array}$$

and

$$\begin{array}{lllllll} x_{11} + x_{12} + x_{13} & & & & & & = a_1 \\ & & & x_{21} + x_{22} + x_{23} & & & = a_2 \\ x_{11} & & & + x_{21} & & & = b_1 \\ & x_{12} & & & + x_{22} & & = b_2 \\ & & x_{13} & & & + x_{23} & = b_3 \end{array}$$

Activity-analysis problem. A manufacturer has at his disposal fixed amounts of a number of different resources. These resources, such as raw material, labor, and equipment, can be combined to produce any one of several different commodities or combinations of commodities. The manufacturer knows how much of resource i it takes to produce one unit of commodity j. He also knows how much profit he makes for each unit of commodity j produced. The manufacturer desires to produce that combination of commodities which will maximize the total profit. For this problem, we define the following:

m = the number of resources
n = the number of commodities
a_{ij} = the number of units of resource i required to produce one unit of commodity j
b_i = the maximum number of units of resource i available
c_j = profit per unit of commodity j produced
x_j = the level of activity (the amount produced) of the jth commodity

The a_{ij} are sometimes called input-output coefficients.

The total amount of the ith resource that is used is given by the linear expression

$$a_{i1}x_1 + a_{i2}x_2 + \cdots + a_{in}x_n$$

Since this total amount must be less than or equal to the maximum number of units of the ith resource available, we then have, for each i, a linear inequality of the form

$$a_{i1}x_1 + a_{i2}x_2 + \cdots + a_{in}x_n \leq b_i$$

As a negative x_j has no appropriate interpretation, we require that all $x_j \geq 0$. The profit derived from producing x_j units of the jth commodity is given by c_jx_j. Our formulation has yielded the problem of maximizing the profit function

$$c_1x_1 + c_2x_2 + \cdots + c_nx_n$$

subject to the conditions

$$\begin{array}{llll} x_1 & & & \geq 0 \\ & x_2 & & \geq 0 \\ & \cdot & & \cdot \\ & & \cdot & \cdot \\ & & \quad\cdot & \cdot \\ & & x_n & \geq 0 \end{array}$$

and

$$\begin{aligned} a_{11}x_1 + a_{12}x_2 + \cdots + a_{1n}x_n &\le b_1 \\ a_{21}x_1 + a_{22}x_2 + \cdots + a_{2n}x_n &\le b_2 \\ \cdots\cdots\cdots\cdots\cdots\cdots\cdots \\ a_{m1}x_1 + a_{m2}x_2 + \cdots + a_{mn}x_n &\le b_m \end{aligned}$$

Since, as will be discussed later, an inequality has an equivalent representation as an equality in nonnegative variables, the above problem is another statement of the general linear-programming problem.

The diet problem. Here we are given the nutrient content of a number of different foods. For example, we might know how many milligrams of phosphorus or iron are contained in 1 oz of each food being considered. We are also given the minimum daily requirement for each nutrient. Since we know the cost per ounce of food, the problem is to determine the diet that satisfies the minimum daily requirements and is also the minimum-cost diet. Define

m = the number of nutrients
n = the number of foods
a_{ij} = the number of milligrams of the ith nutrient in 1 oz of the jth food
b_i = the minimum number of milligrams of the ith nutrient required
c_j = the cost per ounce of the jth food
x_j = the number of ounces of the jth food to be purchased ($x_j \ge 0$)

The total amount of the ith nutrient contained in all the purchased food is given by

$$a_{i1}x_1 + a_{i2}x_2 + \cdots + a_{in}x_n$$

Since this total amount must be greater than or equal to the minimum daily requirement of the ith nutrient, this linear-programming problem involves minimizing the cost function

$$c_1x_1 + c_2x_2 + \cdots + c_nx_n$$

subject to the conditions

$$\begin{aligned} x_1 \qquad\qquad &\ge 0 \\ x_2 \qquad &\ge 0 \\ \ddots \quad &\vdots \\ x_n &\ge 0 \end{aligned}$$

and

$$\begin{aligned} a_{11}x_1 + a_{12}x_2 + \cdots + a_{1n}x_n &\geq b_1 \\ a_{21}x_1 + a_{22}x_2 + \cdots + a_{2n}x_n &\geq b_2 \\ \cdots\cdots\cdots\cdots\cdots\cdots\cdots\cdots\cdots \\ a_{m1}x_1 + a_{m2}x_2 + \cdots + a_{mn}x_n &\geq b_m \end{aligned}$$

3. SURVEY OF LINEAR-PROGRAMMING APPLICATIONS

The early applications of linear-programming methods fell into three major categories: military applications generated by the Air Force's Project SCOOP, interindustry economics based on the Leontief input-output model, and problems involving the relationship between zero-sum two-person games and linear programming.[1] In the past few years these areas of application have been extended and developed, but the emphasis in linear-programming applications has shifted to the industrial area. In this section we shall describe in general terms many of these applications, so as to give the reader an idea of the adaptability and success of the linear-programming model. Although the following material was taken from case studies, we shall not cite specific references in the body of the text. Instead we have grouped in a more readable fashion the corresponding selected set of references in the Bibliography. Additional references are cited in appropriate places in the text. For a more complete list of applications the reader is referred to the bibliography by Riley and Gass [87]. In this bibliography, the authors have compiled linear-programming applications into a number of classifications. We shall follow their arrangement in our discussion.

a. Agricultural applications. These applications fall into two categories, farm economics and farm management. The former deals with agricultural economics in the large, i.e., as related to the economy of a nation or region, while the latter is concerned with the problems of the individual farm.

One study in farm economics deals with interregional competition and the optimum spatial allocation of crop production in the United States. Efficient production patterns were specified by a linear-programming model constrained by regional land resources and national demands. The models used were based on 122 producing regions and included as many as 500 constraints, including upper bounds on each crop category within regions. Production patterns were indicated to allow minimum national food costs and alternatives in livestock feed substitution. Three

[1] For purposes of exposition we have kept the discussion of linear programming and the theory of games for Chap. 12.

models were used and a set of national supply prices for crops was derived for each. The quantitative analysis suggests the amount of land to be withdrawn from crop production and shifted to less intensive uses.

An application of linear-programming techniques to a typical farm-management problem is that of allocating limited resources such as acreage, labor, water supply, working capital, etc., in such a way as to maximize net revenue. The problem is to choose simultaneously the particular crop or crops to be grown in the following period, the number of acres of land to allocate to each of these crops, and the particular method to use in the production of each of these crops so that net cash return will be maximized. Another linear model of a more general nature considers the problem of the selection of a crop-rotation plan by an individual farmer. This application has been developed for both the static and dynamic situations.

As described in Chap. 11, one of the first applications to farm-management problems was the minimum-cost feed problem.

b. Contract awards. In addition to the procurement problem described in Chap. 10, the linear-programming model has been applied to competitive bond bidding. Competitive bidding for serial bonds issued by governments and other public authorities is based on the net-interest-cost method. The bidder presenting the lowest net-interest cost to the issuing authority wins the issue. The linear-programming model considers the factors which enter into the net-interest cost and provides a method for adjusting those variables most subject to the control of the bond bidders. The model which arises from the minimization requirement admits an explicit solution.

c. Industrial applications. CHEMICAL INDUSTRY. The applications to the chemical industry have been mostly of the production and inventory-control type. One study was made to find the optimal scheduling of 25 machines of varying capacities used in electrochemical formation of aluminum oxide films on etched aluminum foil. The foil comes in about 45 combinations of voltage capacity and width, each with its corresponding current requirements. The total current load is restricted by limitations of the company's power-distribution facilities. This scheduling problem turns out to be a straightforward linear-programming problem which involves the capacities, power limitations, and other restrictions.

THE COAL INDUSTRY. A model for the coal industry has been formulated which is composed of two interrelated linear-programming problems. The data of the model are spatially distributed demands for coal and the unit costs of deliveries from the deposits to the demand locations. The levels of the deliveries are the variables of the first programming problem. They are selected to minimize the cost of meeting the demands subject to the capacity restrictions for the coal deposits. The variables of the second programming problem are the delivered prices of coal at the demand locations and the unit royalties earned by the various deposits. The values of these variables are selected to maximize total revenue net of royalty payments.

COMMERCIAL AIRLINES. The work done in this area has dealt with the problems of routing aircraft and of airline management. One study used linear-programming techniques in determining the pattern and timing of flights in the Scottish region of British European Airways. Only a small number of aircraft and crews operate there, but relatively large losses occur. Optimal operating policy was first determined, and then a static model was formulated describing point-to-point flying as activities with linear input and output coefficients. Inputs are aircraft, aircrews, and money, while outputs are aircraft, aircrews, and services. Time of day may be added for a dynamic extension of the model.

COMMUNICATIONS INDUSTRY. The main work has been in the optimal design and utilization of communication networks. Problems involving facilities for transmission, switching, relaying, etc., have been solved by linear-programming methods. These methods provide a general system approach for resolving the complex interactions among system capacities, users' demands, and economic factors.

IRON AND STEEL INDUSTRY. A number of models dealing with production planning in the iron and steel industry have been formulated. One such study derives a minimum-cost steel production program from a linear-programming model which is used to determine the optimum monthly plan of the open-hearth shop, the rate of hot-metal production by the blast-furnace department, and the amount and type of steel scrap to be purchased. The monthly production plan becomes a function of the demand for steel by major types of steel, the amount and type of the available mill scrap, and the price and availability of several types of scrap on the open market.

A report of a linear-programming study made in the tube mill of a large manufacturer showed that scheduling for maximized profit would increase company profits by close to $350,000 over the previous year. On the other hand, scheduling for maximized output would increase the quantity by 22 per cent, but profits would decrease 23 per cent, or about $300,000. The increase in profit was the direct result of a selection of products using linear programming. Factors considered are plant and machine capacity, sales forecasts by various warehouses, shipping rates by items to the various distribution centers, and present and proposed company policies.

PAPER INDUSTRY. Two applications of linear programming in the pulp and paper industry have been in the transportation problem and in trim scheduling. Transportation scheduling deals with the problem confronting a company having several mills. The problem is how to assign the various orders to the mills so as to reduce the total company freight bill to a minimum. Trim scheduling deals with the matter of reducing trim waste on the paper-cutting machine. The manufacturer makes rolls of newsprint paper to meet customers' specifications as to width and diameter. In cutting these customer rolls from larger rolls of paper, trim losses are incurred. Here the manufacturer must determine on which machines and in what combinations the orders should be cut to result in a minimum over-all trim loss. An approach of this kind was successful in showing the way to an increase in trim efficiency of over 1.5 per cent. This is equivalent to an increase in production of over 15 tons per day.

PETROLEUM INDUSTRY. This industrial area has furnished a great many important and interesting linear-programming applications. The earliest of these was the problem of blending gasolines into required products for maximum profit. The final products must meet a variety of specifications, e.g., octane number and volatility, in such a manner as to maximize the net receipts. Other studies include the problems of optimum crude allocation to several refineries, and the optimum inventory and production rate for a seasonal product. Mathematical models of refinery operations and of the petroleum industry in general have led to the study and solution of many nonlinear programming problems.

RAILROAD INDUSTRY. A linear-programming model for optimal programming of railway freight-train movements has been formulated to handle scheduling problems as found at a large terminal-switching railroad. The constraints of the model were based on the methods of hiring and paying the trainmen, the scheduling of shipments, and the capacity limitations of the railroad, the objective being to minimize the total crew and engine expense.

Other railroad applications have dealt with the distribution of freight cars and the distribution of classification effort between yards.

d. Economic analysis. The use of linear-programming techniques in the field of economics has not been limited to the Leontief interindustry model. Another important application has been the linear-programming interpretation of the theory of the firm. Here we deal with the problem of finding the production program which will make the profits of the firm as great as possible, with the limitation that the program must not require more than the total available supply of any resource.

The problem of selecting an investment portfolio has also been treated by programming techniques. Here we start with the relevant beliefs about the securities involved and end with the selection of a portfolio.

In addition to diet, many problems in the general area of marketing analysis have been formulated as linear-programming problems. One such model was constructed to describe the behavior of customers in a food market, based upon experimental data resulting from a survey by the British Ministry of Foods of approximately one thousand urban working-class households. These data served to provide average distribution of food purchases and expenditures for the sample as a whole, and of the quantities of nutritional factors derived from the diet. For purposes of the analyses, the foods were classified into 15 groups and the number of nutritional factors considered was 12. The expenditures for each of four diets meeting a specified set of requirements were computed and compared with the results of the survey.

Special cases of the theory of plant location, i.e., the selection of sites for factories or branch stores so as to maximize profits, have been investigated by linear-programming techniques and the procedures of activity analysis. One example considers the reaction of competitors and formulates the problem as an equivalent two-person zero-sum game.

An unusual experiment involving the methods of linear programming was designed to measure the cardinal utility of nonmonetary outcomes and to use the computed utilities to predict further choices. Here seven music students were used as subjects with long-playing records as the outcomes. The students individually attended two sessions, and a corresponding joint utility curve was developed. The joint curve was used to predict choices that would be made among new combinations of records, and these predictions were tested in a third session. It was found that the predictions made by the linear-programming model were somewhat more accurate than the predictions made by the usual method.

e. Military applications. One of the earliest linear models was that of the air-lift problem. Here the constraints involved the supplies to Berlin, number of runways available, number of crews and aircraft, and money available. The objective was either to deliver a specified number of tons at a minimum cost or to maximize tonnage supplied with a given supply of aircraft and money.

Another Air Force application, the aircraft-deployment problem, is concerned with the efficient allocation of limited resources such as combat aircraft and trained crews. Here we are given the numbers of aircraft to be produced in the successive months of a program. These aircraft are either deployed (sent into a combat area) or else used to train crews. Aircraft are deployed only when there are crews available to man them, and once deployed, they remain at the assigned station for the duration of the program. Aircraft sent to training may be deployed after a given number of months. Each aircraft in training produces a specified number of new crews each month. The object is to allocate the new aircraft between the combat and training missions so as to maximize the total number of aircraft-months of deployed aircraft.

The caterer problem (see Exercise 2 of Chap. 11) is a paraphrased version of a military problem which arose in connection with the estimation of aircraft spare-engine (i.e., napkins) requirements.

Other military examples include the problem of selecting an air weapon system against guerillas so as to keep them pinned down and at the same time minimize the amount of aviation gasoline used; a variation of the transportation problem that maximizes the total tonnage of bombs dropped on a set of targets; and the problem of community defense against disaster, the solution of which yields the number of defense units that should be used in a given attack in order to provide the required level of protection at the lowest possible cost.

f. Personnel assignment. The general statement of the personnel-assignment problem is given in Chap. 10. A special dynamic assignment problem considers the assignment of toll collectors so as to man the required number of toll booths with the minimum number of men for given time periods.

g. Production scheduling and inventory control. A basic formulation of the production-scheduling and inventory-control problem is discussed in Chap. 11. There we consider the problem of smoothing production to meet stipulated requirements in a manner that minimizes storage costs. An earlier linear-programming treatment of a similar problem considers the warehouse problem. Here we wish to determine the optimum use of a storage warehouse for commodities subject to price fluctuations. This problem has since been generalized to include multiple products and warehouses, and varying prices.

A problem that has been investigated in many forms by linear-programming techniques is that of assembly-line balancing. The basic problem is as follows: The commodity to be assembled is composed of many different components. It is necessary to assemble these component parts in some specified sequence or set of sequences. Each assembly operator must be assigned a number and combination of parts to assemble such that the sum of the times required to carry out his assigned task is equal to or less than the cycle time, i.e., the amount of time elapsing between successive units as they move along the assembly line. An operator has idle time if the work time assigned is less than the cycle time. The assignment of jobs to each operator is to be done so as to minimize his idle time. A variation of this problem is the multistage production line. Here we have a number of items which must go through one production stage or machine and then a second, etc. At most, one item can be on a machine at a given time. For each item we know the setup time plus work time at each stage. The problem is to schedule these items through the production line so as to minimize the total elapsed time.

Other applications include the problem of determining the number of each type of an item to be produced by each routing in a machine-shop production line so that the total production cost is minimum and delivery time and requirements are satisfied using existing facilities; and the problem of procuring and assigning new transport aircraft to the various transport jobs in order to minimize the cumulative operations-plus-procurement costs.

h. Structural design. Problems in this area involve the linearization of engineering principles related to the theory of plastic collapse and structural design. The critical value of the loading parameter at which plastic collapse occurs in structures for certain types of loading has been characterized by means of a maximum or minimum principle. The maximum principle can be reduced in various ways to the problem of maximizing a linear function subject to linear inequalities.

The problem of designing plane frames in such a way that the material consumption is a minimum can also be formulated as a linear model.

i. Traffic analysis. This application deals with the problem of scheduling traffic signals. The mathematical formulation of a street-network system assumes knowledge of the following parameters: total traffic-signal cycle (red plus green), the fraction of the cycle that is red at each intersection, and the maximum number of vehicles that can move through the intersection in each direction. The model can handle such phenomena as variations in average speed along different portions of the route, turn-ons and turnoffs, variation in traffic capacity with intersection and flow direction, the capacity of blocks for holding stopped vehicles, three-way lights or other special light schedules, delays in starting up after the light turns green, and the appearance of random delays due to causes other than lights. The criterion for obtaining an optimum time phasing of the lights is that the number of delays be minimized.

j. Transportation problems and network theory. Discussions of the transportation problem and some of its variations are given in Chap. 10. The mathematical model of the problem has proved extremely versatile and susceptible to a number of computational procedures.

An associated problem is that of maximal flows in networks (see Chap. 11). The problem is the following: Consider a network (e.g., rail, road, communications) connecting two given points by way of a number of intermediate points, where each link of the network has a number assigned to it representing its capacity. Assuming a steady-state condition, find a maximal flow from one given point to the other. A simple computational method, based on the simplex method, has been developed for solving this problem.

An extension of the basic transportation problem is that of the transshipment problem. Here any shipping or receiving point is also permitted to act as an intermediate point in seeking an optimum solution.

k. Traveling-salesman problem. The problem is to find the shortest route for a salesman starting from a given city, visiting each city in a specified group, and then returning to the original point of departure. A computational procedure for the general problem has not been devised, but specific examples have been successfully solved by linear-programming techniques. In particular, the problem of touring 49 cities, one in each of the 48 mainland states other than Alaska and one in the District of Columbia, has been formulated as a linear-programming model and solved.

l. Other applications. Additional applications of linear programming include the efficient operation of a system of dams and the design of optical filters. The first problem is to determine variations in the storages in six Missouri River dams so as to maximize the energy obtained from the entire system and thereby increase the efficiency of operation. Physical limitations on the storages and releases appear in the form of inequalities. An approximation to a change in energy is given as a linear function of the variations.

In the second problem, products of exponentials are converted to sums of the logarithms to provide a linear-programming approach to the synthesis of filter combinations for optimum fit of combination filters to a desired spectral-transmission curve. The unknowns to be determined could be either the concentrations of various possible filter constituents, which are to be mixed to form a single filter, or the thicknesses of various filters which are to be placed in series.

In addition to their wide application to physical situations, linear-programming techniques have been firmly related to a number of theoretical areas of mathematics. Some of these disciplines have lent their power to the solution of special types of linear-programming problems; e.g., the theory of statistics has been applied in solving problems whose elements are subject to random errors. What is more important, however, has been the use of linear-programming methods in the proof of theorems and solution of problems in what appear to be totally unrelated fields of investigation. The theoretical and computational aspects of linear programming have been applied with much success to the areas of combinatorial analysis, partially ordered sets, network and graph theory, and systems of distinct representatives. Since descriptions of selected applications from these areas would entail introducing material which is beyond the scope and purposes of this book, we leave to the interested reader the investigation of the specific details of these mathematical applications of linear programming. Appropriate references are given in Sec. 12 of the Bibliography.

REMARKS

For additional introductory material, the reader is referred to Dorfman [38], Cooper and Charnes [16], Henderson and Schlaifer [55], and Directorate of Management Analysis [36].

EXERCISES

1. Write out the complete linear-programming formulation of the following transportation problem and attempt to determine the optimum solution by trial-and-error methods:

A manufacturer has distribution centers located at Atlanta, Chicago, and New York City. These centers have available 40, 20, and 40 units of his product, respectively. His retail outlets require the following number of units: Cleveland, 25; Louisville, 10; Memphis, 20; Pittsburgh, 30; and Richmond, 15. The shipping cost per unit in dollars between each center and outlet is given in the following table:

	Cleveland	Louisville	Memphis	Pittsburgh	Richmond
Atlanta	55	30	40	50	40
Chicago	35	30	100	45	60
New York	40	60	95	35	30

2. A furniture manufacturer wishes to determine how many tables, chairs, desks, or bookcases he should make in order to optimize the use of his available resources. These products utilize two different types of lumber, and he has on hand 1,500 board feet of the first type and 1,000 board feet of the second. He has 800 man-hours available for the total job. His sales forecast plus his back orders require him to make at least 40 tables, 130 chairs, 30 desks, and no more than 10 bookcases. Each table, chair, desk, and bookcase requires 5, 1, 9, and 12 board feet, respectively, of the first type of lumber and 2, 3, 4, and 1 board feet of the second type. A table requires 3 man-hours to make, a chair 2, a desk 5, and a bookcase 10. The manufacturer makes a total of $12 profit on a table, $5 on a chair, $15 on a desk, and $10 on a bookcase. Write out the complete linear-programming formulation of this problem in terms of maximizing the profit.

3. A baker starts the day with a certain supply of flour, shortening, eggs, sugar, milk, and yeast. He specializes in making bread, cakes, English muffins, and cookies. He wishes to determine how much of each product he should make so as to maximize his profit. The recipes are given in the following table (we ignore such plentiful supplies as salt, water, etc.).

	Bread	Cake	Muffins	Cookies	Available resources
Flour	12 C	3 C	$\frac{9}{2}$ C	$\frac{3}{2}$ C	b_1 C
Shortening	2 tbsp	12 tbsp	3 tbsp	4 tbsp	b_2 tbsp
Eggs	0	3	1	1	b_3
Sugar	$\frac{1}{4}$ C	$\frac{3}{2}$ C	$\frac{1}{8}$ C	1 C	b_4 C
Milk	2 C	$\frac{3}{4}$ C	1 C	0	b_5 C
Yeast	1 cake	0	1 cake	0	b_6 cakes
Profit	c_1	c_2	c_3	c_4	

Write the corresponding linear-programming formulation and discuss the applicability of such a model to a real-life situation.

4. Describe the model of a transportation problem in which the total availabilities are greater than the total requirements. From the mathematical point of view, why must the sum of the availabilities and the sum of the requirements be equal in the standard transportation model? Discuss the linearity assumption of the transportation-problem cost relationship.

5. Formulate the following diet problem. A mother wishes her children to obtain certain amounts of nutrients from their breakfast cereals. The children have the choice of eating Krunchies or Crispies or a mixture of the two. From their breakfast they should obtain at least 1 mg of thiamine, 5 mg of niacin, and 400 cal. One ounce of Krunchies contains 0.10 mg of thiamine, 1 mg of niacin, and 110 cal; 1 oz of Crispies contains 0.25 mg of thiamine, 0.25 mg of niacin, and 120 cal. One ounce of Krunchies costs 3.8 cents and 1 oz. of Crispies 4.2 cents.

CHAPTER 2 / **MATHEMATICAL BACKGROUND**

A complete development and understanding of the theoretical and computational aspects of linear programming requires the blending of the basic concepts and techniques of a number of mathematical topics. In this chapter we shall present and discuss only those elements of these topics that facilitate the discussion, are required in the succeeding chapters, or will aid the reader in applying the material to be described.

1. MATRICES AND DETERMINANTS

A *matrix* is a rectangular array of mn numbers arranged in m rows and n columns as follows:

$$\begin{matrix} a_{11} & a_{12} & \cdots & a_{1n} \\ a_{21} & a_{22} & \cdots & a_{2n} \\ \cdots & \cdots & \cdots & \cdots \\ a_{m1} & a_{m2} & \cdots & a_{mn} \end{matrix}$$

Such an array is usually enclosed in parentheses and is called matrix **A**. The individual a_{ij} are called *elements*. We sometimes denote the matrix **A** by (a_{ij}). For any m or n and any elements a_{ij} we have

$$\mathbf{A} = \begin{pmatrix} a_{11} & a_{12} & \cdots & a_{1n} \\ a_{21} & a_{22} & \cdots & a_{2n} \\ \cdots & \cdots & \cdots & \cdots \\ a_{m1} & a_{m2} & \cdots & a_{mn} \end{pmatrix} = (a_{ij})$$

Matrix **A** is called *square* if $m = n$ and is said to be of *order* n.

A *column vector* is a matrix with only one column, and a *row vector* is a matrix with only one row. A *diagonal matrix* is a square matrix whose off-diagonal elements, a_{ij} for $i \neq j$, are all equal to zero.

A *unit matrix*, or *identity matrix*, is a diagonal matrix whose diagonal elements are 1. A unit matrix of order n is denoted by $\mathbf{I}_n$ or simply $\mathbf{I}$. For $n = 3$, we have

$$\mathbf{I}_3 = \begin{pmatrix} 1 & 0 & 0 \\ 0 & 1 & 0 \\ 0 & 0 & 1 \end{pmatrix}$$

The *transposed matrix*, $\mathbf{A}'$, of a matrix $\mathbf{A}$ is

$$\mathbf{A}' = \begin{pmatrix} a_{11} & a_{21} & \cdots & a_{m1} \\ a_{12} & a_{22} & \cdots & a_{m2} \\ \cdots & \cdots & \cdots & \cdots \\ a_{1n} & a_{2n} & \cdots & a_{mn} \end{pmatrix}$$

i.e., the rows and columns are interchanged.

Two matrices are equal if and only if their corresponding elements are equal. Hence the corresponding dimensions, m and n, must also be equal.

A square matrix $\mathbf{A}$ is *triangular* if all $a_{ij} = 0$ for $i > j$ or if all $a_{ij} = 0$ for $i < j$. The matrix

$$\begin{pmatrix} a_{11} & a_{12} & a_{13} & a_{14} \\ 0 & a_{22} & a_{23} & a_{24} \\ 0 & 0 & a_{33} & a_{34} \\ 0 & 0 & 0 & a_{44} \end{pmatrix}$$

is triangular.

A matrix $\mathbf{A}$ is *symmetric* if $\mathbf{A} = \mathbf{A}'$, that is, if $a_{ij} = a_{ji}$. A matrix is *skew-symmetric* if $\mathbf{A} = -\mathbf{A}'$, that is, if $a_{ij} = -a_{ji}$. This last condition implies that all $a_{ij} = 0$ for $i = j$.

A *null matrix* has all its elements equal to zero and is denoted by $\mathbf{O} = (0)$. In the succeeding chapters the matrix $\mathbf{O}$ will be used to denote either a null column vector or a null row vector.

We next define the operations of matrix addition and multiplication.

Given any scalar α (a real number α) and any matrix $\mathbf{A}$, the product $\alpha\mathbf{A}$ is given by

$$\alpha\mathbf{A} = \begin{pmatrix} \alpha a_{11} & \alpha a_{12} & \cdots & \alpha a_{1n} \\ \cdots & \cdots & \cdots & \cdots \\ \alpha a_{m1} & \alpha a_{m2} & \cdots & \alpha a_{mn} \end{pmatrix} = (\alpha a_{ij})$$

Given any two $m \times n$ matrices $\mathbf{A}$ and $\mathbf{B}$, the sum $\mathbf{A} + \mathbf{B} = \mathbf{C}$ is given by

$$\mathbf{C} = \begin{pmatrix} a_{11} & a_{12} & \cdots & a_{1n} \\ \cdots & \cdots & \cdots & \cdots \\ a_{m1} & a_{m2} & \cdots & a_{mn} \end{pmatrix} + \begin{pmatrix} b_{11} & b_{12} & \cdots & b_{1n} \\ \cdots & \cdots & \cdots & \cdots \\ b_{m1} & b_{m2} & \cdots & b_{mn} \end{pmatrix}$$

$$= \begin{pmatrix} a_{11} + b_{11} & a_{12} + b_{12} & \cdots & a_{1n} + b_{1n} \\ \cdots & \cdots & \cdots & \cdots \\ a_{m1} + b_{m1} & a_{m2} + b_{m2} & \cdots & a_{mn} + b_{mn} \end{pmatrix} = (a_{ij} + b_{ij})$$

The properties of scalar multiplication and matrix addition are:

(*a*) $(\mathbf{A} + \mathbf{B}) + \mathbf{C} = \mathbf{A} + (\mathbf{B} + \mathbf{C})$ (Associative law)

(*b*) $\mathbf{A} + \mathbf{B} = \mathbf{B} + \mathbf{A}$ (Commutative law)

(*c*) $(\alpha + \beta)\mathbf{A} = \alpha\mathbf{A} + \beta\mathbf{A}$
$\alpha(\mathbf{A} + \mathbf{B}) = \alpha\mathbf{A} + \alpha\mathbf{B}$ (Distributive law)

(*d*) $\mathbf{A} + \mathbf{O} = \mathbf{A}$

where **A**, **B**, and **C** are all of dimensions $m \times n$ and α and β are scalars.

The multiplication of two matrices **A** and **B** is defined only on the assumption that the number of columns of **A** is equal to the number of rows of **B**. On this assumption, the elements of the product $\mathbf{AB} = \mathbf{C}$ are defined as follows:

The element in the ith row and jth column of matrix **C** is equal to the sum of the products of the elements of the ith row of **A** multiplied by the corresponding elements of the jth column of **B**. For example,

$$\mathbf{AB} = \begin{pmatrix} a_{11} & a_{12} & a_{13} \\ a_{21} & a_{22} & a_{23} \end{pmatrix} \begin{pmatrix} b_{11} & b_{12} \\ b_{21} & b_{22} \\ b_{31} & b_{32} \end{pmatrix} = \begin{pmatrix} c_{11} & c_{12} \\ c_{21} & c_{22} \end{pmatrix} = \mathbf{C}$$

where

$$c_{ij} = a_{i1}b_{1j} + a_{i2}b_{2j} + a_{i3}b_{3j}$$

The product of a matrix **A** of dimensions $m \times n$ and a matrix **B** of dimensions $n \times q$ yields a matrix **C** of dimensions $m \times q$. The reader can easily verify that matrix multiplication is not commutative; i.e., in general $\mathbf{AB} \neq \mathbf{BA}$.

Matrix multiplication has the following properties:

(*a*) $(\mathbf{AB})\mathbf{C} = \mathbf{A}(\mathbf{BC})$ (Associative law)

(*b*) $(\mathbf{A} + \mathbf{B})\mathbf{C} = \mathbf{AC} + \mathbf{BC}$
$\mathbf{C}(\mathbf{A} + \mathbf{B}) = \mathbf{CA} + \mathbf{CB}$ (Distributive law)

(*c*) $\alpha(\mathbf{AB}) = (\alpha\mathbf{A})\mathbf{B} = (\mathbf{A})(\alpha\mathbf{B})$

(*d*) $\mathbf{AI} = \mathbf{IA} = \mathbf{A}$

where matrices **A**, **B**, **C**, and **I** are of the correct dimensions and α is a scalar. We note that $(\mathbf{AB})' = \mathbf{B}'\mathbf{A}'$.

Associated with every square matrix **A** is a number called the *determinant* of **A**. The determinant of **A**, denoted by $|\mathbf{A}|$, is obtained as the sum of all possible products in each of which there appears one and only one element from each row and each column of **A**, each product of this sum being assigned a plus or minus sign according to the following rule: Let the elements in a given product be joined in pairs by line segments (see example below). If the total number of such segments sloping upward to the right is even, prefix a plus sign to the product; otherwise prefix a negative sign. (Each determinant has $n!$ such products where n is the order of the matrix.[1])

For $n = 3$, we have

$$\mathbf{A} = \begin{pmatrix} a_{11} & a_{12} & a_{13} \\ a_{21} & a_{22} & a_{23} \\ a_{31} & a_{32} & a_{33} \end{pmatrix}$$

and

$$|\mathbf{A}| = \begin{vmatrix} a_{11} & a_{12} & a_{13} \\ a_{21} & a_{22} & a_{23} \\ a_{31} & a_{32} & a_{33} \end{vmatrix}$$

$$= a_{11}a_{22}a_{33} + a_{12}a_{23}a_{31} + a_{13}a_{21}a_{32} - a_{11}a_{23}a_{32} - a_{12}a_{21}a_{33} - a_{13}a_{22}a_{31}$$

The lines for determining the sign of the third term have been drawn in the array for the determinant.

The following properties of determinants can be verified from the above definition:

1. If every element of a column or of a row of a determinant is zero, then the value of the determinant is zero.
2. The value of a determinant is not changed if corresponding columns and rows are interchanged.
3. If $|\mathbf{B}|$ is the determinant formed by interchanging two columns or rows in $|\mathbf{A}|$, then $|\mathbf{B}| = -|\mathbf{A}|$.
4. If two columns or rows of a determinant are identical, then the determinant has a value of zero.
5. If every element of a column or of a row of a determinant is multiplied by a fixed number k, then the value of the determinant is multiplied by k.
6. The value of a determinant is not changed if, to every element of a column or row, we add k times the corresponding element of another column or row.

The *rank* of any matrix $\mathbf{A}$ is the order of the largest square array in $\mathbf{A}$ whose determinant does not vanish, i.e., does not have a value of zero.

A square matrix $\mathbf{A}$ is said to be *nonsingular* if its determinant does not equal zero. If $|\mathbf{A}| = 0$, the matrix is *singular*.

[1] The factorial notation of $n!$ is defined as $n! = n(n-1)(n-2) \cdots 2 \cdot 1$, where $0! = 1$.

The minor $\mathbf{D}_{ij}$ of the element a_{ij} is the determinant obtained from the square matrix $\mathbf{A}$ by striking out the ith row and jth column.

The cofactor $\mathbf{A}_{ij}$ of the element a_{ij} equals $(-1)^{i+j}\mathbf{D}_{ij}$. $\mathbf{A}_{ij}$ is referred to as the signed minor of the element a_{ij}.

The *adjoint* of an $n \times n$ matrix $\mathbf{A}$ is another $n \times n$ matrix $\mathbf{J} = (\mathbf{A}_{ji})$ in which the element in the ith row and jth column is the cofactor of the element in the jth row and ith column of $\mathbf{A}$. We have

$$\mathbf{J} = \begin{pmatrix} \mathbf{A}_{11} & \mathbf{A}_{21} & \cdots & \mathbf{A}_{n1} \\ \mathbf{A}_{12} & \mathbf{A}_{22} & \cdots & \mathbf{A}_{n2} \\ \cdots & \cdots & \cdots & \cdots \\ \mathbf{A}_{1n} & \mathbf{A}_{2n} & \cdots & \mathbf{A}_{nn} \end{pmatrix}$$

A matrix $\mathbf{B}$ is called the *inverse* of the square matrix $\mathbf{A}$ if $\mathbf{AB} = \mathbf{I}$. The inverse of $\mathbf{A}$ is denoted by $\mathbf{A}^{-1}$. For every nonsingular square matrix $\mathbf{A}$ there exists a unique $\mathbf{A}^{-1}$ such that

$$\mathbf{AA}^{-1} = \mathbf{A}^{-1}\mathbf{A} = \mathbf{I}$$

It can be shown that, if $|\mathbf{A}| \neq 0$,

$$\mathbf{A}^{-1} = \frac{1}{|\mathbf{A}|}\mathbf{J}$$

We see that only nonsingular square matrices have inverses. We also note that $(\mathbf{AB})^{-1} = \mathbf{B}^{-1}\mathbf{A}^{-1}$.

2. VECTORS AND VECTOR SPACES

In the familiar two-dimensional Euclidean plane, points are represented by ordered pairs of numbers $\mathbf{U} = (u_1,u_2)$. $\mathbf{U}$ can be regarded as a point with coordinates (u_1,u_2) relative to a fixed origin $\mathbf{O} = (0,0)$ or as a *vector*, i.e., a translation of the origin from $(0,0)$ by the amounts u_1 and u_2 in two fixed coordinate directions (Fig. 2.1). These two notions are used interchangeably. We shall sometimes write $\mathbf{U} = \begin{pmatrix} u_1 \\ u_2 \end{pmatrix}$.

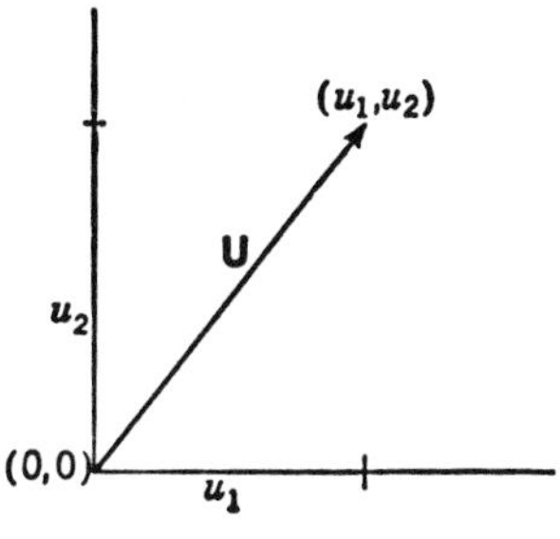

FIGURE 2.1

Let us call the two-dimensional Euclidean space $\mathbf{E}_2$ and note the important properties of vectors in $\mathbf{E}_2$.

a. Multiplication of vectors by scalars (Fig. 2.2). To every pair α and $\mathbf{U}$, where α is a scalar and $\mathbf{U}$ is a vector, there corresponds a vector called the *scalar product* of α and $\mathbf{U}$, written $\alpha\mathbf{U} = (\alpha u_1, \alpha u_2)$, and such that

(*a*) $\alpha(\beta\mathbf{U}) = (\alpha\beta)\mathbf{U}$ (Associative law)

(*b*) $\alpha(\mathbf{U} + \mathbf{V}) = \alpha\mathbf{U} + \alpha\mathbf{V}$, $(\alpha + \beta)\mathbf{U} = \alpha\mathbf{U} + \beta\mathbf{U}$ (Distributive law)

(*c*) $1\mathbf{U} = \mathbf{U}$

(*d*) $0\mathbf{U} = \mathbf{O} = (0,0)$

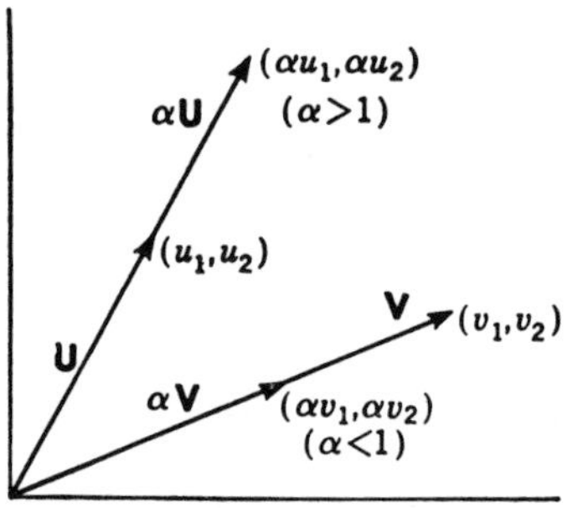

FIGURE 2.2

b. Addition of vectors (Fig. 2.3). To every pair $\mathbf{U} = (u_1,u_2)$ and $\mathbf{V} = (v_1,v_2)$ of vectors in $\mathbf{E}_2$ there corresponds a vector called the sum of $\mathbf{U}$ and $\mathbf{V}$, written $\mathbf{U} + \mathbf{V} = (u_1 + v_1,\ u_2 + v_2)$ and such that

(*a*) $\mathbf{U} + \mathbf{V} = \mathbf{V} + \mathbf{U}$ (Commutative law)

(*b*) $(\mathbf{U} + \mathbf{V}) + \mathbf{W} = \mathbf{U} + (\mathbf{V} + \mathbf{W})$ (Associative law)

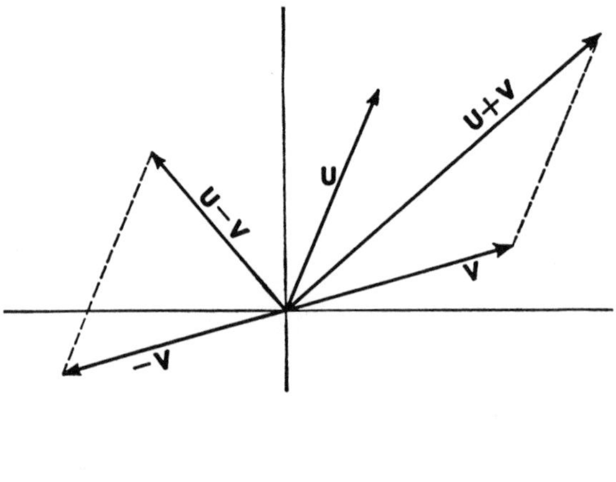

FIGURE 2.3

There exists in $\mathbf{E}_2$ a unique vector $\mathbf{O}$, called the origin, such that

(c) $\mathbf{U} + \mathbf{O} = \mathbf{U}$ for all $\mathbf{U}$ in $\mathbf{E}_2$

To each $\mathbf{U}$ in $\mathbf{E}_2$ there corresponds a unique vector, called the inverse of $\mathbf{U}$ and denoted by $-\mathbf{U}$, such that

(d) $\mathbf{U} + (-\mathbf{U}) = \mathbf{O}$

c. Inner product of vectors. To every pair of vectors $\mathbf{U}$ and $\mathbf{V}$ in $\mathbf{E}_2$ there corresponds a real number called the *inner product* of $\mathbf{U}$ and $\mathbf{V}$, written $\mathbf{U} \cdot \mathbf{V} = u_1v_1 + u_2v_2$ and such that

(a) $\mathbf{U} \cdot \mathbf{V} = \mathbf{V} \cdot \mathbf{U}$ (Commutative law)

(b) $(\alpha\mathbf{U} + \beta\mathbf{V}) \cdot \mathbf{W} = \alpha(\mathbf{U} \cdot \mathbf{W}) + \beta(\mathbf{V} \cdot \mathbf{W})$

for all scalars α and β and all vectors $\mathbf{U}$, $\mathbf{V}$, $\mathbf{W}$ in $\mathbf{E}_2$.

(c) $\mathbf{U} \cdot \mathbf{U} \geq 0$ and $\mathbf{U} \cdot \mathbf{U} = 0$ if and only if $\mathbf{U} = \mathbf{O}$

Two vectors $\mathbf{U}$ and $\mathbf{V}$ are said to be *orthogonal* to one another if $\mathbf{U} \cdot \mathbf{V} = 0$.

d. Length of a vector (Fig. 2.4). To every vector $\mathbf{U}$ in $\mathbf{E}_2$ there corresponds a real number, called the length of $\mathbf{U}$, written

$$\|\mathbf{U}\| = +\sqrt{u_1^2 + u_2^2}$$

and such that

(a) $\|\mathbf{U}\| \geq 0$ and $\|\mathbf{U}\| = 0$ if and only if $\mathbf{U} = \mathbf{O}$

(b) $\|\alpha\mathbf{U}\| = |\alpha| \cdot \|\mathbf{U}\|$

The length satisfies the triangle inequality

(c) $\|\mathbf{U} + \mathbf{V}\| \leq \|\mathbf{U}\| + \|\mathbf{V}\|$

for every $\mathbf{U}$ and $\mathbf{V}$ in $\mathbf{E}_2$.

(d) $\|\mathbf{U}\| = +\sqrt{\mathbf{U} \cdot \mathbf{U}}$

e. Distance between two vectors. To every pair of vectors $\mathbf{U}$ and $\mathbf{V}$ in $\mathbf{E}_2$ there corresponds a real number, called the distance between $\mathbf{U}$ and $\mathbf{V}$,

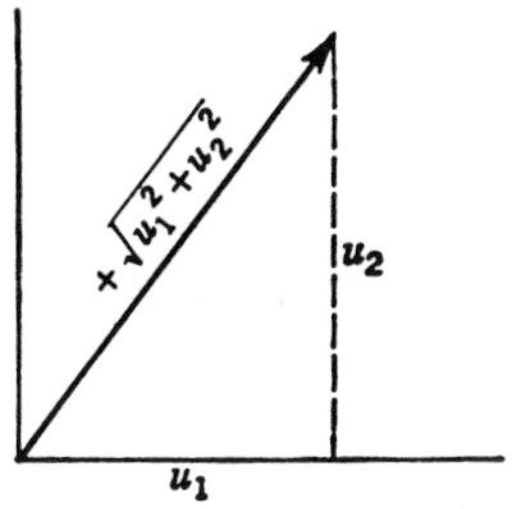

FIGURE 2.4

written $d(\mathbf{U},\mathbf{V})$, such that

$$d(\mathbf{U},\mathbf{V}) = \|\mathbf{U} - \mathbf{V}\| = +\sqrt{(u_1 - v_1)^2 + (u_2 - v_2)^2}$$

A set of vectors $\mathbf{U}_1, \mathbf{U}_2, \ldots, \mathbf{U}_n$ is called *linearly independent* if, for all numbers $\alpha_1, \alpha_2, \ldots, \alpha_n$,

$$\alpha_1\mathbf{U}_1 + \alpha_2\mathbf{U}_2 + \cdots + \alpha_n\mathbf{U}_n = \mathbf{O}$$

implies

$$\alpha_1 = \alpha_2 = \cdots = \alpha_n = 0$$

Otherwise the set is *linearly dependent.* For example, the set of two vectors

$$\mathbf{U}_1 = \begin{pmatrix}1\\0\end{pmatrix} \quad \text{and} \quad \mathbf{U}_2 = \begin{pmatrix}1\\1\end{pmatrix}$$

is linearly independent. To show this, we write

$$\alpha_1\mathbf{U}_1 + \alpha_2\mathbf{U}_2 = \alpha_1\begin{pmatrix}1\\0\end{pmatrix} + \alpha_2\begin{pmatrix}1\\1\end{pmatrix} = \begin{pmatrix}0\\0\end{pmatrix}$$

or

$$\begin{pmatrix}\alpha_1\\0\end{pmatrix} + \begin{pmatrix}\alpha_2\\\alpha_2\end{pmatrix} = \begin{pmatrix}0\\0\end{pmatrix}$$

$$\begin{pmatrix}\alpha_1 + \alpha_2\\\alpha_2\end{pmatrix} = \begin{pmatrix}0\\0\end{pmatrix}$$

Since in the last equation the corresponding elements in the two column vectors must be equal, we have $\alpha_2 = 0$ and hence $\alpha_1 = 0$. In a similar fashion we can demonstrate that the set of vectors

$$\mathbf{U}_1 = \begin{pmatrix}1\\0\end{pmatrix} \quad \mathbf{U}_2 = \begin{pmatrix}0\\1\end{pmatrix} \quad \text{and} \quad \mathbf{U}_3 = \begin{pmatrix}1\\1\end{pmatrix}$$

is linearly dependent. Here we have

$$\alpha_1\begin{pmatrix}1\\0\end{pmatrix} + \alpha_2\begin{pmatrix}0\\1\end{pmatrix} + \alpha_3\begin{pmatrix}1\\1\end{pmatrix} = \begin{pmatrix}0\\0\end{pmatrix} \tag{2.1}$$

or

$$\begin{pmatrix}\alpha_1 + \alpha_3\\\alpha_2 + \alpha_3\end{pmatrix} = \begin{pmatrix}0\\0\end{pmatrix}$$

We are then led to the conclusions that $\alpha_3 = -\alpha_1$ and $\alpha_3 = -\alpha_2$. Hence for $\alpha_1 = \alpha_2 = \alpha$ and $\alpha_3 = -\alpha$, where α can be *any* number, Eq. (2.1) holds.

This last example illustrates the property of $\mathbf{E}_2$ which makes it two-dimensional: that there exist two linearly independent vectors [e.g., the unit vectors (1,0) and (0,1)], whereas every set of three vectors is linearly dependent. Generalizing this notion of dimensionality, we have the following definition:

An *n-dimensional Euclidean space* $\mathbf{E}_n$ is a set of objects called vectors which satisfy properties described above under subheads *a* to *c*, with the property that there exist *n* linearly independent vectors while every set of $n + 1$ vectors is linearly dependent.

In $\mathbf{E}_n$ we have $\mathbf{U} = (u_1,u_2, \ldots, u_n)$ and

$$\alpha(u_1,u_2, \ldots ,u_n) = (\alpha u_1,\alpha u_2, \ldots ,\alpha u_n)$$

$$(u_1,u_2, \ldots ,u_n) + (v_1,v_2, \ldots ,v_n) = (u_1 + v_1, u_2 + v_2, \ldots , u_n + v_n)$$

$$(u_1,u_2, \ldots ,u_n) \cdot (v_1,v_2, \ldots ,v_n) = u_1v_1 + u_2v_2 + \cdots + u_nv_n$$

A *basis* for $\mathbf{E}_n$ is a set of *n* linearly independent vectors. Any vector in $\mathbf{E}_n$ can be expressed uniquely as a linear combination of the vectors of a given basis. The set of *n* unit vectors $(1,0, \ldots ,0)$, $(0,1, \ldots ,0)$, $\ldots$, $(0,0, \ldots ,1)$ is a basis for $\mathbf{E}_n$.

One linear condition in $\mathbf{E}_2$ (such as $a_1u_1 + a_2u_2 = b$, where a_1, a_2, and b are constants) defines a line; one linear condition in $\mathbf{E}_3$ defines a plane; the corresponding object defined by one linear condition in $\mathbf{E}_n$ is called a *hyperplane*. A hyperplane $\mathbf{H}(\mathbf{A},b)$ in $\mathbf{E}_n$ is the set of all vectors $\mathbf{U}$ such that $\mathbf{A} \cdot \mathbf{U} = b$ for a given vector $\mathbf{A} \neq \mathbf{O}$ and a real number b (Fig. 2.5).

A hyperplane divides $\mathbf{E}_n$ into two half spaces which we denote by

$$\mathbf{H}^+(\mathbf{A},b) = (\mathbf{U} \mid \mathbf{A} \cdot \mathbf{U} \geq b)$$

$$\mathbf{H}^-(\mathbf{A},b) = (\mathbf{U} \mid \mathbf{A} \cdot \mathbf{U} \leq b)$$

$\mathbf{H}^+(\mathbf{A},b)$ is the half space, i.e., that portion of $\mathbf{E}_n$, that contains the vectors $\mathbf{U}$ for which $\mathbf{A} \cdot \mathbf{U} \geq b$. $\mathbf{H}^-(\mathbf{A},b)$ is the half space that contains the vectors $\mathbf{U}$ for which $\mathbf{A} \cdot \mathbf{U} \leq b$.

For $\mathbf{A} = (2,-1)$ and $b = 1$, we have the inner product

$$\mathbf{A} \cdot \mathbf{U} = 2u_1 - u_2$$

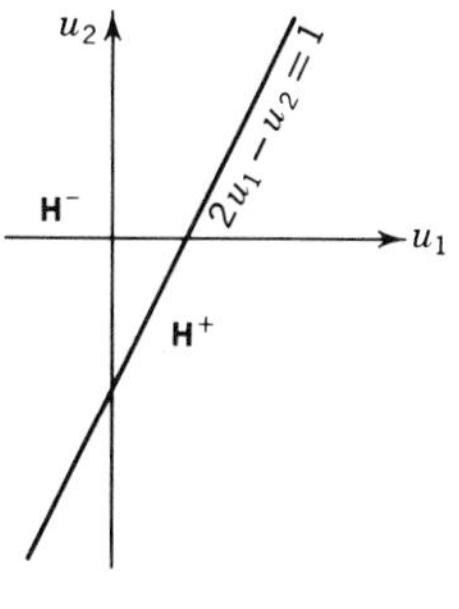

FIGURE 2.5

with the two half spaces

$$\mathbf{H}^{+}(\mathbf{A},b) = 2u_1 - u_2 \geq 1$$

and

$$\mathbf{H}^{-}(\mathbf{A},b) = 2u_1 - u_2 \leq 1$$

as depicted in Fig. 2.5. The points on the line $2u_1 - u_2 = 1$ belong to both half spaces.

3. CONVEX SETS

A *convex combination* of the points $\mathbf{U}_1, \mathbf{U}_2, \ldots, \mathbf{U}_n$ is a point

$$\mathbf{U} = \alpha_1\mathbf{U}_1 + \alpha_2\mathbf{U}_2 + \cdots + \alpha_n\mathbf{U}_n$$

where the α_i are scalars, $\alpha_i \geq 0$, and $\sum_i \alpha_i = 1$. A subset $\mathbf{C}$ of $\mathbf{E}_n$ is *convex* if and only if, for all pairs of points $\mathbf{U}_1$ and $\mathbf{U}_2$ in $\mathbf{C}$, any convex combination

$$\mathbf{U} = \alpha_1\mathbf{U}_1 + \alpha_2\mathbf{U}_2$$

is also in $\mathbf{C}$.

Examples of convex sets are the whole space $\mathbf{E}_n$, a circle, and a cube. The set of points that form the boundary of a circle is not a convex set.

We next show that, for a given convex set $\mathbf{C}$, any convex combination of any number of points in $\mathbf{C}$ is also in $\mathbf{C}$. Here we are given that the convex combination of any two points $\mathbf{U}_1$ and $\mathbf{U}_2$ in $\mathbf{C}$ is also in $\mathbf{C}$. First we prove that

$$\mathbf{U} = \alpha_1\mathbf{U}_1 + \alpha_2\mathbf{U}_2 + \alpha_3\mathbf{U}_3$$

for $\alpha_i \geq 0$, $\sum_i \alpha_i = 1$, is also in $\mathbf{C}$. Let

$$\alpha_i' = \frac{\alpha_i}{\alpha_1 + \alpha_2}$$

for $i = 1, 2$. We note that $\sum_i \alpha_i' = 1$ and $\alpha_i' \geq 0$. We have

$$\mathbf{U} = \alpha_1\mathbf{U}_1 + \alpha_2\mathbf{U}_2 + \alpha_3\mathbf{U}_3 = (\alpha_1 + \alpha_2)(\alpha_1'\mathbf{U}_1 + \alpha_2'\mathbf{U}_2) + \alpha_3\mathbf{U}_3$$

By hypothesis $\alpha_1'\mathbf{U}_1 + \alpha_2'\mathbf{U}_2$ is in $\mathbf{C}$. Let

$$\mathbf{U}_4 = \alpha_1'\mathbf{U}_1 + \alpha_2'\mathbf{U}_2$$

Then

$$\mathbf{U} = (\alpha_1 + \alpha_2)\mathbf{U}_4 + \alpha_3\mathbf{U}_3$$

where $\mathbf{U}$ is a convex combination of two points in $\mathbf{C}$, and hence $\mathbf{U}$ is in $\mathbf{C}$. This procedure can be generalized for any i.

Theorem 1. *Any point on the line segment joining two points in* $\mathbf{E}_n$ *can be expressed as a convex combination of the two points.*

Proof. Denote the two points by **U** and **V** and let **W** lie on the line segment joining **U** and **V**. This line segment is parallel to the line defined by vector $\mathbf{U} - \mathbf{V}$ (Fig. 2.6). By the rules of vector addition we have, for some $0 \leq \lambda \leq 1$,

$$\mathbf{V} + \lambda(\mathbf{U} - \mathbf{V}) = \mathbf{W}$$

or

$$(1 - \lambda)\mathbf{V} + \lambda\mathbf{U} = \mathbf{W}$$

which is the expression of the point **W** as a convex combination of **V** and **U**.

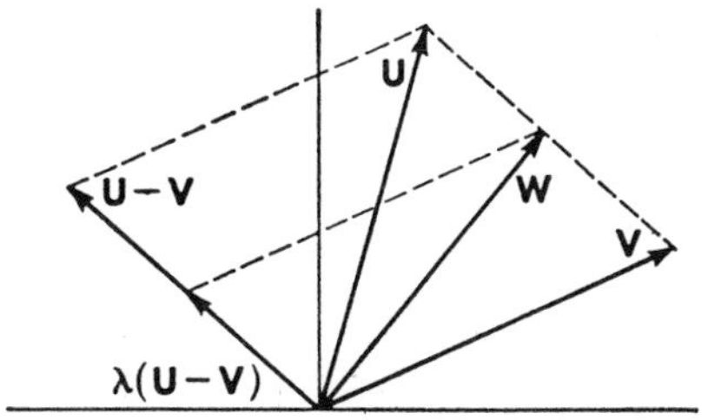

FIGURE 2.6

Theorem 2 (converse of theorem 1). *Any point that can be expressed as a convex combination of two points in* $\mathbf{E}_n$ *lies on the line segment joining the two points.*

Proof. Here we are given

$$\mathbf{W} = (1 - \lambda)\mathbf{V} + \lambda\mathbf{U}$$

or

$$\mathbf{W} - \mathbf{V} = \lambda(\mathbf{U} - \mathbf{V})$$

where $0 \leq \lambda \leq 1$. Hence the vector $\mathbf{W} - \mathbf{V}$ is some positive multiple of vector $\mathbf{U} - \mathbf{V}$, and these vectors *cannot* have the configuration as pictured in Fig. 2.7. Vector $\mathbf{W} - \mathbf{V}$ must coincide in direction

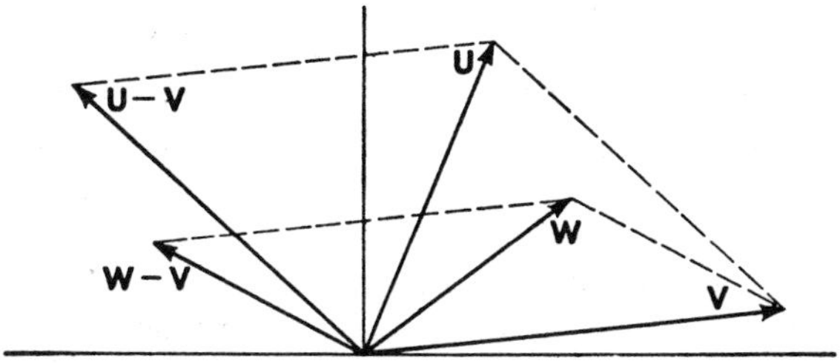

FIGURE 2.7

with **U** − **V**. Since the line segment joining **U** and **V** and the line segment joining **W** and **V** are parallel to the lines defined by **U** − **V** and **W** − **V**, respectively, point **W** must lie on the line segment joining **U** and **V**.

By Theorems 1 and 2 we see that geometrically a convex set is one which contains all line segments joining any two points in the set. The sets shown in Fig. 2.8*a* and *b* are convex sets, while those of Fig. 2.8*c* and *d* are not.

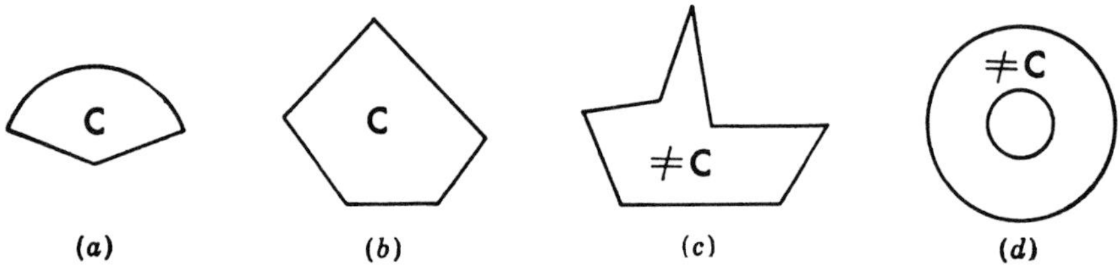

FIGURE 2.8

A point **U** in a convex set **C** is called an *extreme point* if **U** cannot be expressed as a convex combination of any other two distinct points in **C**. Every point on the boundary of a circle is an extreme point of the convex set which includes the boundary and the interior of the circle. If the convex set did not include the boundary, then this set would contain no extreme points. The extreme points of a triangle are its vertices.

The *convex hull* **C**(**S**) of any given set of points **S** is the set of all convex combinations of sets of points from **S**. **C**(**S**) is the smallest convex set containing **S**. If **S** is just the eight vertices of a cube, then **C**(**S**) is the whole cube, or, if **S** is the boundary of a circle, then **C**(**S**) is the whole circle.

If the set **S** consists of a finite number of points, the convex hull of **S** is termed a *convex polyhedron*. **C**(**S**) of the eight vertices of a cube is a convex polyhedron. If **C** is a closed and bounded set with a finite number of extreme points (e.g., a convex polyhedron), then any point in the set can be expressed as a convex combination of the extreme points. Thus **C** is the convex hull of its extreme points.

A set of vectors **S** is called a *cone* if, for every vector **U** in **S**, λ**U** is in **S**, where λ is a nonnegative number. Examples of a cone are the whole space, the origin, and the set **S** of Fig. 2.9. We note that a cone contains the origin, since λ can equal zero.

A *convex cone* is a cone which is convex. The cone of Fig. 2.9 is not convex. That part of **S** in the first quadrant is a convex cone.

A *simplex* is an n-dimensional convex polyhedron having exactly $n + 1$

vertices. The boundary of the simplex contains simplices of lower dimension which are called *simplicial faces*. The number of such faces of dimension i is $\binom{n+1}{i+1}$.† A simplex in zero dimension is a point; in one dimension it is a line; in two dimensions it is a triangle; and in three dimensions it is a tetrahedron. The equation of a simplex with unit intercepts is

$$x_i \geq 0$$
$$\sum_{i=1}^{m} x_i \leq 1$$

For $m = 3$ we have the tetrahedron with vertices (0,0,0), (1,0,0), (0,1,0), and (0,0,1).

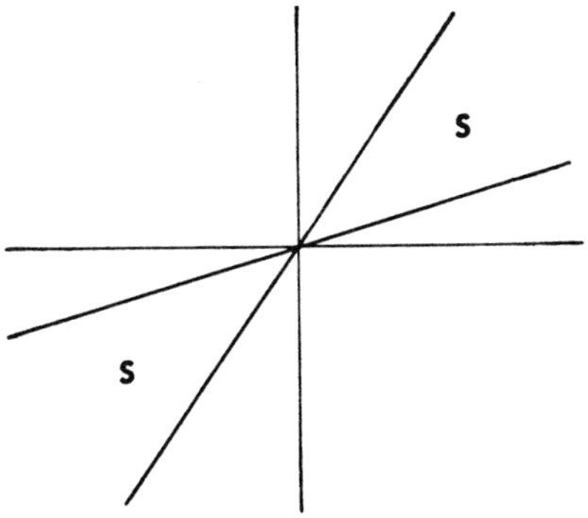

FIGURE 2.9

4. LINEAR INEQUALITIES

Let us consider the following set of five linear inequalities in two dimensions:

$$\begin{aligned}
&(a) \qquad & x_1 \qquad\quad &\geq 0 \\
&(b) \qquad & x_2 &\geq 0 \\
&(c) \qquad & x_1 + \ x_2 &\geq 1 \qquad\qquad (4.1)\\
&(d) \qquad & x_1 - \ x_2 &\geq 1 \\
&(e) \qquad & -x_1 + 2x_2 &\leq 0
\end{aligned}$$

We wish to determine which points (x_1,x_2) in two-dimensional Euclidean space satisfy the inequalities (4.1). We first note (Fig. 2.10) that the first two inequalities limit our search to the positive quadrant; i.e.,

† The notation $\binom{n}{m}$ represents the number $n!/m!(n - m)!$.

only those points having nonnegative coordinates need be considered. Inequality (4.1*a*) is satisfied only by points lying on or to the right of the line $x_1 = 0$; inequality (4.1*b*) is satisfied only by points lying on or above the line $x_2 = 0$. The only points which simultaneously satisfy both inequalities are those lying in the first, or positive, quadrant. Each inequality is satisfied by all the points lying in one of the half spaces that is generated when we treat the inequality as an equality. We note

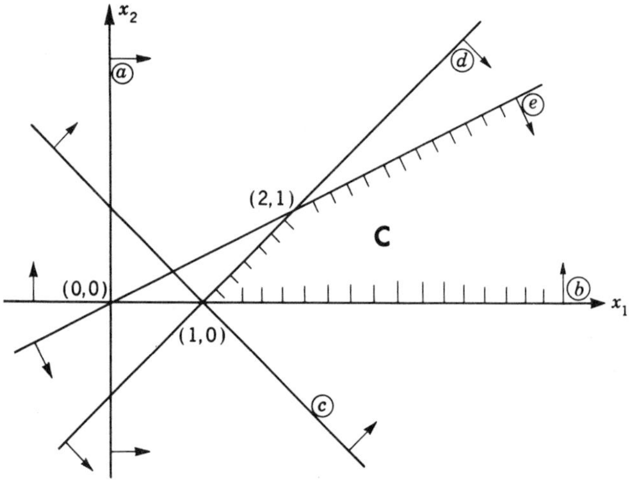

FIGURE 2.10

the solution half space by drawing arrows from the dividing hyperplane into the solution space. We next treat inequality (4.1*c*) as an equality and draw the line corresponding to $x_1 + x_2 = 1$. We determine that the half space which contains all the solutions to inequality (4.1*c*) lies upward to the right, by noting that the origin (0,0) does not satisfy the inequality; i.e., the solution space does not contain the origin. The solution space to inequalities (4.1*a*) to (4.1*c*) is that part of the positive quadrant whose points simultaneously satisfy the three inequalities. For inequality (4.1*d*), we draw the line $x_1 - x_2 = 1$ and note that the half space whose points satisfy the inequality does not contain the origin. The line which corresponds to inequality (4.1*e*), $-x_1 + 2x_2 = 0$, contains the origin. To determine which of the associated half spaces represents the solution space for inequality (4.1*e*), we take any point, say (1,0), and note whether or not it satisfies (4.1*e*). Since (1,0) does satisfy (4.1*e*), the half space which contains (1,0) is the solution space for (4.1*e*). Only those points in region **C** described by the hatched lines simultaneously satisfy the five inequalities. The solution space **C** is an unbounded, polygonal convex region whose boundaries are portions of the lines which correspond to inequalities (4.1*b*), (4.1*d*), and (4.1*e*). Inequalities (4.1*a*) and (4.1*c*) do not enter into the determination of the solution space. They are redundant to the system (4.1), but only some type of analysis, such as the one above, will reveal this fact.

We have interpreted each of the inequalities of (4.1) as defining a hyperplane in two-dimensional space. An alternate interpretation is obtained by writing the matrix representation of the system (4.1). We let matrix

$$\mathbf{A} = \begin{pmatrix} 1 & 0 \\ 0 & 1 \\ 1 & 1 \\ 1 & -1 \\ 1 & -2 \end{pmatrix}$$

be the matrix of coefficients, and write the matrix inequality

$$\begin{pmatrix} 1 & 0 \\ 0 & 1 \\ 1 & 1 \\ 1 & -1 \\ 1 & -2 \end{pmatrix} \begin{pmatrix} x_1 \\ x_2 \end{pmatrix} \geq \begin{pmatrix} 0 \\ 0 \\ 1 \\ 1 \\ 0 \end{pmatrix} \tag{4.2}$$

By multiplying the matrix $\mathbf{A}$ by the column vector whose elements are x_1 and x_2, we have the five inequalities of the original system. Here, inequality (4.1*e*) has been multiplied by -1 to reverse the direction of the inequality. We denote the first column of $\mathbf{A}$ by $\mathbf{P}_1$ and the second column by $\mathbf{P}_2$; that is,

$$\mathbf{P}_1 = \begin{pmatrix} 1 \\ 0 \\ 1 \\ 1 \\ 1 \end{pmatrix} \qquad \text{and} \qquad \mathbf{P}_2 = \begin{pmatrix} 0 \\ 1 \\ 1 \\ -1 \\ -2 \end{pmatrix}$$

and let

$$\mathbf{P}_0 = \begin{pmatrix} 0 \\ 0 \\ 1 \\ 1 \\ 0 \end{pmatrix}$$

the column vector of constants. We can then write the system (4.2) as

$$x_1\mathbf{P}_1 + x_2\mathbf{P}_2 \geq \mathbf{P}_0 \tag{4.3}$$

Here we interpret the column vectors $\mathbf{P}_1$, $\mathbf{P}_2$, and $\mathbf{P}_0$ as points in five-dimensional space, and the problem is to determine the weights x_1 and x_2 such that all the conditions defined by (4.3) are satisfied.

In general, the points which satisfy a linear inequality of the type

$$a_{11}x_1 + a_{12}x_2 + \cdots + a_{1n}x_n \geq b_1$$

define a half space in n-dimensional space. The joint solution space to the set of inequalities

$$\begin{aligned} a_{11}x_1 + a_{12}x_2 + \cdots + a_{1n}x_n &\geq b_1 \\ a_{21}x_1 + a_{22}x_2 + \cdots + a_{2n}x_n &\geq b_2 \\ \cdots\cdots\cdots\cdots\cdots\cdots \\ a_{m1}x_1 + a_{m2}x_2 + \cdots + a_{mn}x_n &\geq b_m \end{aligned} \tag{4.4}$$

is a convex region in n-dimensional space. (We shall prove the convexity of the solution space in Chap. 3.) Since the direction of the inequality can be reversed by multiplication by -1, any set of inequalities can be written in the form of (4.4). The set of inequalities (4.4) can be transformed into a set of equalities by subtracting from each inequality an unknown nonnegative number. These numbers are termed *slack variables*. For (4.4), this transformation yields the set of equalities

$$\begin{aligned} a_{11}x_1 + a_{12}x_2 + \cdots + a_{1n}x_n - x_{n+1} \qquad\qquad &= b_1 \\ a_{21}x_1 + a_{22}x_2 + \cdots + a_{2n}x_n \qquad - x_{n+2} \qquad &= b_2 \\ \cdots\cdots\cdots\cdots\cdots\cdots\cdots\cdots \\ a_{m1}x_1 + a_{m2}x_2 + \cdots + a_{mn}x_n \qquad\qquad - x_{n+m} &= b_m \end{aligned}$$

where $x_{n+i} \geq 0$ $(i = 1, \ldots, m)$. Since we can represent each variable x_i as the difference of two nonnegative variables,[1] the system (4.4) has the following equivalent system in terms of nonnegative variables and equalities:

$$\begin{aligned} a_{11}(x_1' - x_1'') + a_{12}(x_2' - x_2'') + \cdots + a_{1n}(x_n' - x_n'') - x_{n+1} \qquad\qquad &= b_1 \\ a_{21}(x_1' - x_1'') + a_{22}(x_2' - x_2'') + \cdots + a_{2n}(x_n' - x_n'') \qquad - x_{n+2} \qquad &= b_2 \\ \cdots\cdots\cdots\cdots\cdots\cdots\cdots\cdots\cdots\cdots \\ a_{m1}(x_1' - x_1'') + a_{m2}(x_2' - x_2'') + \cdots + a_{mn}(x_n' - x_n'') \qquad\qquad - x_{n+m} &= b_m \end{aligned}$$

where

$$\begin{aligned} x_j &= x_j' - x_j'' \\ x_j' &\geq 0 \qquad j = 1, \ldots n \\ x_j'' &\geq 0 \\ x_{n+i} &\geq 0 \qquad i = 1, \ldots m \end{aligned}$$

[1] The reader should note that any number can be written as the difference of two nonnegative numbers.

An inequality of the form

$$a_{11}x_1 + a_{12}x_2 + \cdots + a_{1n}x_n \leq b_1$$

can be transformed into an equality by adding to it an unknown non-negative number (slack variable), e.g.,

$$a_{11}x_1 + a_{12}x_2 + \cdots + a_{1n}x_n + x_{n+1} = b_1$$

where $x_{n+1} \geq 0$.

It should be noted that the solution space for a set of inequalities might be void. There are no solutions to the set of inequalities

$$x_1 + x_2 \leq 1$$
$$2x_1 + 2x_2 \geq 3$$

As a mathematical problem, the general linear-programming problem can be described as follows: We are given a convex set which is defined by a set of linear constraints in n-dimensional Euclidean space. From all the points that belong to the convex set, we wish to determine a subset of points (which will contain either one or many points) for which a linear objective function is optimized. To illustrate how this can be done for two-dimensional constraints, we have represented in Fig. 2.11 the convex set **C**, that is, the solution space, for Eqs. (4.1). Let us determine the subset of points for which the linear objective function

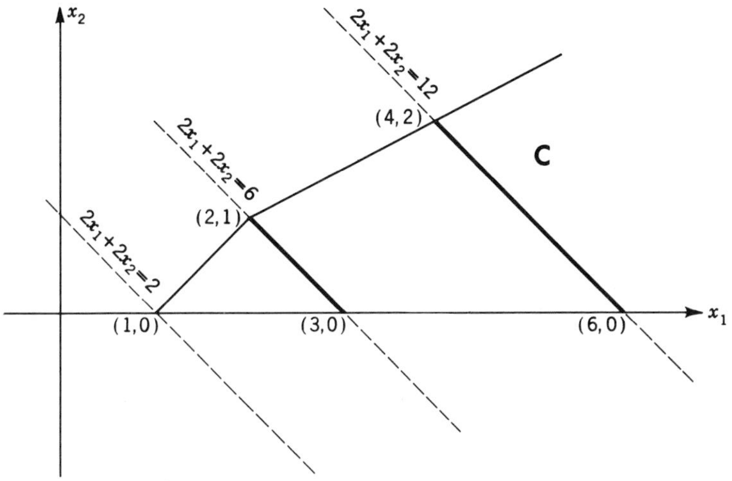

FIGURE 2.11

$2x_1 + 2x_2$ is a minimum. In Fig. 2.11 we have also drawn the line

$$2x_1 + 2x_2 = b$$

for $b = 12$, 6, and 2. The intersections of these lines with **C** have been represented by heavy lines, with the last intersection being a point. Every point (x_1,x_2) of the first intersection will yield a value of the objective function equal to 12; all points of the second intersection yield a value of 6; and finally we see that the last intersection, which consists of only the point (1,0), gives the minimum value of 2. Hence, out of the infinite number of points contained in **C** only one point minimizes the objective function. This point is also an extreme point of **C**, and the full consequences of this will be discussed in Chap. 3. For any other objective function we can similarly determine the subset of points for which it is optimized. We need only to plot the objective function for some particular value and then move this line parallel to itself in the appropriate direction for minimizing or maximizing, as was done in Fig. 2.11. We move the line until the intersection of the line and convex set is reduced either to a point, a portion of the boundary of **C**, or until it is apparent that the maximum or minimum value is infinite. These last two situations are illustrated by minimizing $2x_1 - 2x_2$ and maximizing $2x_1 + 2x_2$ for **C** of Fig. 2.11. If our convex set were a convex polyhedron, then all objective functions would have a finite minimum or maximum.

5. SOLUTION OF A SET OF SIMULTANEOUS LINEAR EQUATIONS

Any set of simultaneous linear equations has a convenient representation using matrix notation. The system

$$\begin{aligned} a_{11}x_1 + a_{12}x_2 + \cdots + a_{1n}x_n &= b_1 \\ a_{21}x_1 + a_{22}x_2 + \cdots + a_{2n}x_n &= b_2 \\ \cdots\cdots\cdots\cdots\cdots\cdots\cdots\cdots & \\ a_{m1}x_1 + a_{m2}x_2 + \cdots + a_{mn}x_n &= b_m \end{aligned} \tag{5.1}$$

can be written as

$$\mathbf{AX} = \mathbf{b}$$

where

$$\mathbf{A} = (a_{ij}) \qquad \mathbf{X} = \begin{pmatrix} x_1 \\ \cdot \\ \cdot \\ \cdot \\ x_n \end{pmatrix} \qquad \text{and} \qquad \mathbf{b} = \begin{pmatrix} b_1 \\ \cdot \\ \cdot \\ \cdot \\ b_m \end{pmatrix}^{\dagger}$$

† The reader is referred to the discussion of matrix multiplication in Sec. 1 for the validity of this representation.

If $\mathbf{A}$ is square (that is, if $m = n$) and nonsingular, the solution vector is given by

$$\mathbf{X} = \mathbf{A}^{-1}\mathbf{b}$$

There is a simple computational scheme that can be applied to any nonsingular system to obtain the solution vector and/or the inverse. This procedure, the complete elimination method of Jordan and Gauss, has a finite number of steps, or iterations. In just $m(=n)$ iterations, the procedure multiplies the system (5.1) by $\mathbf{A}^{-1}$ to obtain $\mathbf{X} = \mathbf{A}^{-1}\mathbf{b}$. If it is so desired, $\mathbf{A}^{-1}$ may be explicitly developed at the same time. We shall illustrate the method by an example.

Consider the set of three equations in three variables:

$$\begin{aligned} x_1 + x_2 - x_3 &= 2 \\ -2x_1 + x_2 + x_3 &= 3 \\ x_1 + x_2 + x_3 &= 6 \end{aligned} \tag{5.2}$$

Let

$$\mathbf{A} = \begin{pmatrix} 1 & 1 & -1 \\ -2 & 1 & 1 \\ 1 & 1 & 1 \end{pmatrix}$$

be the matrix of coefficients. It is easy to verify that $\mathbf{A}$ is nonsingular. We can rewrite (5.2) as the matrix equation

$$\begin{pmatrix} 1 & 1 & -1 \\ -2 & 1 & 1 \\ 1 & 1 & 1 \end{pmatrix} \begin{pmatrix} x_1 \\ x_2 \\ x_3 \end{pmatrix} = \begin{pmatrix} 2 \\ 3 \\ 6 \end{pmatrix}$$

or, by letting

$$\mathbf{P}_1 = \begin{pmatrix} 1 \\ -2 \\ 1 \end{pmatrix} \quad \mathbf{P}_2 = \begin{pmatrix} 1 \\ 1 \\ 1 \end{pmatrix} \quad \mathbf{P}_3 = \begin{pmatrix} -1 \\ 1 \\ 1 \end{pmatrix} \quad \mathbf{P}_0 = \begin{pmatrix} 2 \\ 3 \\ 6 \end{pmatrix}$$

we can rewrite (5.2) as

$$x_1\mathbf{P}_1 + x_2\mathbf{P}_2 + x_3\mathbf{P}_3 = \mathbf{P}_0$$

Since $\mathbf{A}$ is nonsingular, the set of vectors $\mathbf{P}_1$, $\mathbf{P}_2$, and $\mathbf{P}_3$ is linearly independent and hence forms a basis in three-dimensional space.[1] To solve

[1] To prove that the vectors $\mathbf{P}_1, \mathbf{P}_2, \ldots, \mathbf{P}_n$ of a nonsingular matrix $\mathbf{A}$ form a linearly independent set, we first assume that the set is linearly dependent. Hence, for some set of α_j we must have

$$\alpha_1\mathbf{P}_1 + \alpha_2\mathbf{P}_2 + \cdots + \alpha_n\mathbf{P}_n = \mathbf{0}$$

with at least one $\alpha_j \neq 0$. We can then solve for some vector $\mathbf{P}_j$, for example, $\mathbf{P}_1$, in

(5.2), we need to find the unique linear combination (i.e., the numbers x_1, x_2, and x_3) of $\mathbf{P}_1$, $\mathbf{P}_2$, and $\mathbf{P}_3$ which equals vector $\mathbf{P}_0$. We shall first obtain the solution vector to (5.2). The complete elimination procedure successively eliminates the first, the second, the third, etc., variable from all equations except the first, the second, the third, etc., respectively. For (5.2), the first step is to eliminate the first variable from all the equations except the first. This elimination is accomplished by adding and subtracting suitable multiples of the first equation to the second and third equations.

Step 1

$$\begin{aligned} x_1 + x_2 - x_3 &= 2 \\ -2x_1 + x_2 + x_3 &= 3 \\ x_1 + x_2 + x_3 &= 6 \end{aligned}$$

Since matrix **A** is nonsingular, Eqs. (5.2) can be listed in such an order that the coefficient of the first variable in the first equation does not equal zero. Here $a_{11} = 1$. To eliminate x_1 from the second equation, we multiply the first equation by 2 and add it to the second. To eliminate the first variable from the third equation, we multiply the first equation by -1 and add it to the third. These two substeps transform the original equations into the equations

$$\begin{aligned} x_1 + x_2 - x_3 &= 2 \\ 3x_2 - x_3 &= 7 \\ 2x_3 &= 4 \end{aligned}$$

Step 2. Here we eliminate the second variable from all the equations obtained as a result of Step 1 except the second. Again the equations can be arranged to make the coefficient of x_2 in the second equation not equal to zero (this would have been done if the third equation had originally been the second). Since at the start of an iteration it is convenient to make the coefficient of the variable being eliminated equal to 1, we

---

terms of the other vectors to obtain

$$\mathbf{P}_1 = \beta_2\mathbf{P}_2 + \cdots + \beta_n\mathbf{P}_n$$

where $\beta_j = -\alpha_j/\alpha_1$. We substitute this expression for vector $\mathbf{P}_1$ in matrix **A**. Then by successively adding to the first column $-\beta_j\mathbf{P}_j$ for $j = 2, 3, \ldots, n$, we reduce the first column to the zero vector. From the properties of determinants given in Sec. 1, the addition of these vectors does not change the value of the determinant, and since a column is now equal to zero, the value of the determinant is zero. However, this contradicts the given information that **A** is a nonsingular matrix. Hence, our assumption of linear dependence has led to a contradiction, and therefore the set of vectors of **A** must be linearly independent.

divide the second equation by 3. The coefficient 3 is referred to as the *pivot element.* (In Step 1 the pivot element was $a_{11} = 1$.) We then have to multiply this transformed second equation by -1 and add it to the first to obtain the equations

$$\begin{aligned} x_1 \qquad - \tfrac{2}{3}x_3 &= -\tfrac{1}{3} \\ x_2 - \tfrac{1}{3}x_3 &= \tfrac{7}{3} \\ 2x_3 &= 4 \end{aligned}$$

Step 3. Here the pivot element is 2. We first divide the third equation obtained in Step 2 by 2, and then multiply it by $\frac{2}{3}$ and $\frac{1}{3}$ and add the corresponding results to the first and second equations, respectively. The system (5.2) has been reduced to the solution

$$\begin{aligned} x_1 \qquad &= 1 \\ x_2 \qquad &= 3 \\ x_3 &= 2 \end{aligned} \tag{5.3}$$

The operations of the complete elimination procedure have transformed the matrix of coefficients

$$\mathbf{A} = \begin{pmatrix} 1 & 1 & -1 \\ -2 & 1 & 1 \\ 1 & 1 & 1 \end{pmatrix}$$

into the identity matrix [the matrix of coefficients for the transformed Eqs. (5.3)]

$$\mathbf{I}_3 = \begin{pmatrix} 1 & 0 & 0 \\ 0 & 1 & 0 \\ 0 & 0 & 1 \end{pmatrix}$$

We then see that the total transformation accomplished by the complete elimination procedure is equivalent to multiplying the system (5.1) by $\mathbf{A}^{-1}$.

It is very often necessary not only to solve a system of simultaneous linear equations, but also to obtain the inverse of the matrix of coefficients. This is accomplished by attaching an $m \times m$ identity matrix to the right of the original matrix of coefficients and applying the elimination transformations to the extended matrix. The inverse is generated in place of the identity matrix. If we write the partitioned matrix

$$(\mathbf{A} \mid \mathbf{I} \mid \mathbf{P}_0)$$

(i.e., the matrix $\mathbf{A}$ with an $m \times m$ identity matrix written beside it and the $\mathbf{P}_0$ vector attached to the end) and apply the complete elimination transformation, we have

$$(\mathbf{A}^{-1}\mathbf{A} \mid \mathbf{A}^{-1}\mathbf{I} \mid \mathbf{A}^{-1}\mathbf{P}_0) = (\mathbf{I} \mid \mathbf{A}^{-1} \mid \mathbf{X})$$

We shall illustrate this procedure for the previous example. The partitioned matrix is

$$\left(\begin{array}{rrr|rrr|r} 1 & 1 & -1 & 1 & 0 & 0 & 2 \\ -2 & 1 & 1 & 0 & 1 & 0 & 3 \\ 1 & 1 & 1 & 0 & 0 & 1 & 6 \end{array}\right)$$

The successive steps, as before, yield

Step 1

$$\left(\begin{array}{rrr|rrr|r} 1 & 1 & -1 & 1 & 0 & 0 & 2 \\ 0 & 3 & -1 & 2 & 1 & 0 & 7 \\ 0 & 0 & 2 & -1 & 0 & 1 & 4 \end{array}\right)$$

Step 2

$$\left(\begin{array}{rrr|rrr|r} 1 & 0 & -2/3 & 1/3 & -1/3 & 0 & -1/3 \\ 0 & 1 & -1/3 & 2/3 & 1/3 & 0 & 7/3 \\ 0 & 0 & 2 & -1 & 0 & 1 & 4 \end{array}\right)$$

Step 3

$$\left(\begin{array}{rrr|rrr|r} 1 & 0 & 0 & 0 & -1/3 & 1/3 & 1 \\ 0 & 1 & 0 & 1/2 & 1/3 & 1/6 & 3 \\ 0 & 0 & 1 & -1/2 & 0 & 1/2 & 2 \end{array}\right)$$

We then have

$$\mathbf{A}^{-1} = \begin{pmatrix} 0 & -1/3 & 1/3 \\ 1/2 & 1/3 & 1/6 \\ -1/2 & 0 & 1/2 \end{pmatrix}$$

The unique linear combination of the vectors $\mathbf{P}_1$, $\mathbf{P}_2$, and $\mathbf{P}_3$ which equals $\mathbf{P}_0$ has been determined to be

$$1\mathbf{P}_1 + 3\mathbf{P}_2 + 2\mathbf{P}_3 = \mathbf{P}_0$$

Since the vectors $\mathbf{P}_1$, $\mathbf{P}_2$, $\mathbf{P}_3$ form a basis in three-dimensional space, any other vector in three-space can also be expressed as a unique linear combination of these three vectors. For any such vector, say $\mathbf{P}_4$, we wish to determine y_1, y_2, and y_3 such that

$$y_1\mathbf{P}_1 + y_2\mathbf{P}_2 + y_3\mathbf{P}_3 = \mathbf{P}_4$$

or

$$\mathbf{AY} = \mathbf{P}_4 \tag{5.4}$$

To solve for $\mathbf{Y}$, we must multiply (5.4) by $\mathbf{A}^{-1}$ to obtain

$$\mathbf{A}^{-1}\mathbf{A}\mathbf{Y} = \mathbf{A}^{-1}\mathbf{P}_4$$
$$\mathbf{Y} = \mathbf{A}^{-1}\mathbf{P}_4$$

For example, let

$$\mathbf{P}_4 = \begin{pmatrix} 4 \\ -2 \\ 4 \end{pmatrix}$$

$\mathbf{Y}$ is given by

$$\mathbf{Y} = \begin{pmatrix} 0 & -\frac{1}{3} & \frac{1}{3} \\ \frac{1}{2} & \frac{1}{3} & \frac{1}{6} \\ -\frac{1}{2} & 0 & \frac{1}{2} \end{pmatrix} \begin{pmatrix} 4 \\ -2 \\ 4 \end{pmatrix} = \begin{pmatrix} 2 \\ 2 \\ 0 \end{pmatrix}$$

$\mathbf{P}_4$ expressed as a linear combination of $\mathbf{P}_1$, $\mathbf{P}_2$, and $\mathbf{P}_3$ is then

$$2\mathbf{P}_1 + 2\mathbf{P}_2 + 0\mathbf{P}_3 = \mathbf{P}_4$$

Here $\mathbf{P}_4$ is equal to a linear combination of only two of the three basis vectors. The three-dimensional vectors $\mathbf{P}_1$, $\mathbf{P}_2$, $\mathbf{P}_4$ are linearly dependent since

$$2\mathbf{P}_1 + 2\mathbf{P}_2 - \mathbf{P}_4 = \mathbf{O}$$

with the coefficients of vectors $\mathbf{P}_1$, $\mathbf{P}_2$, and $\mathbf{P}_4$ not all zero. The situation where a vector can be expressed as a linear combination of less than m vectors from a given basis is termed *degenerate.*

A system of linear equations $\mathbf{AX} = \mathbf{b}$ is said to be *homogeneous* if $\mathbf{b} = \mathbf{O}$. Such a system always has the trivial solution $\mathbf{X} = \mathbf{O}$.

REMARKS

Much of the material in Sec. 1 is from Hildebrand [55*a*], and much of that in Sec. 2 is from Kuhn [67]. For additional reading concerning the theory of convex sets, the reader is referred to Part Three of Koopmans [65], Kuhn and Tucker [68], Hadley [52*b*], and Nef [77*d*].

EXERCISES

1. Given the matrices

$$\mathbf{A} = \begin{pmatrix} 2 & -6 & 0 \\ 4 & 3 & 2 \end{pmatrix} \qquad \mathbf{B} = \begin{pmatrix} -4 & 2 & 1 \\ 5 & 0 & 3 \end{pmatrix}$$

$$\mathbf{C} = \begin{pmatrix} 2 & 4 & 4 \\ -1 & 3 & 0 \\ 9 & 5 & 2 \end{pmatrix} \qquad \mathbf{D} = \begin{pmatrix} -2 & 2 & 0 \\ 0 & 4 & 2 \\ 2 & 0 & 1 \end{pmatrix}$$

compute the following:

$(\mathbf{A} + \mathbf{B})$, $\mathbf{CD}$, $\mathbf{DC}$, $|\mathbf{C}|$, $|\mathbf{D}|$, $\mathbf{C}^{-1}$, $\mathbf{D}^{-1}$, $\mathbf{C}'$, $\mathbf{D}'$, $(\mathbf{CD})'$, $\mathbf{D}'\mathbf{C}'$, $(\mathbf{CD})^{-1}$

2. Graph the following vectors:

$$\mathbf{U} = \begin{pmatrix} -1 \\ 2 \end{pmatrix} \quad \mathbf{V} = \begin{pmatrix} 4 \\ 6 \end{pmatrix} \quad \mathbf{W} = \begin{pmatrix} -5 \\ -4 \end{pmatrix}$$

$\mathbf{U} + \mathbf{V}$, $\mathbf{U} - \mathbf{V}$, $2\mathbf{U} - \frac{1}{2}\mathbf{V}$, $\mathbf{U} - \mathbf{V} + \mathbf{W}$, $\mathbf{U} + \mathbf{V} + \mathbf{W}$

3. Graph the convex hull of the points

$(0,0)$, $(1,1)$, $(-1,-1)$, $(-2,2)$, $(1,4)$, $(0,3)$, $(-1,1)$, $(\frac{1}{2},4)$, $(-1,2)$, $(2,5)$

Find convex combinations of the extreme points which express those given points that are interior points of the convex hull.

4. Graph the following linear constraints, mark the area that satisfies the constraints, and determine the extreme points of this convex set:

$$\begin{aligned} -2x_1 + 5x_2 - 10 &\leq 0 \\ 2x_1 + x_2 - 6 &\leq 0 \\ x_1 + 2x_2 - 2 &\geq 0 \\ -x_1 + 3x_2 - 3 &\leq 0 \end{aligned}$$

5. Write the system of inequalities of Exercise 4 as a system of equalities in nonnegative variables.

6. Assume a given basis of the vectors

$$\mathbf{P}_1 = \begin{pmatrix} 1 \\ 0 \\ 1 \end{pmatrix} \quad \mathbf{P}_2 = \begin{pmatrix} 1 \\ 1 \\ 1 \end{pmatrix} \quad \mathbf{P}_3 = \begin{pmatrix} 2 \\ 1 \\ 1 \end{pmatrix}$$

Compute the inverse of the matrix associated with this basis and determine the linear combination of the basis vectors that equals vector $\mathbf{P}_4 = \begin{pmatrix} 1 \\ 3 \\ 4 \end{pmatrix}$.

7. The convex region shown in the figure below is the set of solutions determined by a set of linear inequalities. At what point or points are the following linear functions optimized?

a. $x_1 + x_2 - 1$, to be a maximum
b. $3x_1 - x_2 + 6$, to be a minimum
c. $-2x_1 - 2x_2 + 2$, to be a maximum

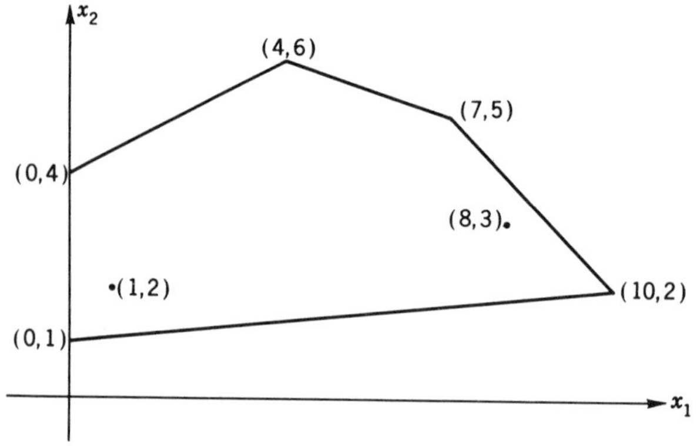

FIGURE E7

8. Given the set of equations

$$a_{11}x_1 + a_{12}x_2 + a_{13}x_3 = a_{10}$$
$$a_{21}x_1 + a_{22}x_2 + a_{23}x_3 = a_{20}$$
$$a_{31}x_1 + a_{32}x_2 + a_{33}x_3 = a_{30}$$

Assume that the given matrix of coefficients (a_{ij}) is nonsingular. Reduce the system to the unknown vector (x_1,x_2,x_3) by the complete elimination procedure, and develop general formulas for the transformations that are performed in each iteration.

9. Solve the following set of equations, using the complete elimination formulas developed in Exercise 8. Also compute $\mathbf{A}^{-1}$ by means of the adjoint matrix and check the result by computing $\mathbf{AA}^{-1}$.

$$2x_1 + 2x_2 - x_3 = 4$$
$$3x_1 - x_2 - 3x_3 = 7$$
$$x_1 + x_2 + 2x_3 = 3$$

10. Solve the following linear-programming problems by graphical methods:

a. Constraints:

$$2x_1 - x_2 \geq -2$$
$$x_1 + 2x_2 \leq 8$$
$$x_1 \geq 0$$
$$x_2 \geq 0$$

Objective functions: maximize x_2, maximize $3x_1 + 2x_2$, minimize $2x_1 - 2x_2$, maximize $2x_1 + 4x_2$, minimize $-3x_1 - 2x_2$.

b. Constraints:

$$3x_1 + 2x_2 \leq 6$$
$$x_1 - x_2 \geq -1$$
$$-x_1 - 2x_2 \geq 1$$
$$x_1 \geq 0$$
$$x_2 \geq 0$$

Objective function: maximize $2x_1 - 6x_2$.

c. Constraints:

$$x_1 - 3x_2 \leq 6$$
$$2x_1 + 4x_2 \geq 8$$
$$x_1 - 3x_2 \geq -6$$
$$x_1 \geq 0$$
$$x_2 \geq 0$$

Objective functions: maximize $2x_1 + 3x_2$, minimize $x_1 + 2x_2$, maximize $x_1 - 2x_2$, maximize $x_1 - 3x_2$, maximize $x_1 - 6x_2$.

d. Constraints:

$$\begin{aligned} x_1 + x_2 + x_3 &\leq 2 \\ x_1 + x_2 - x_3 &\leq 1 \\ x_1 \qquad\qquad &\geq 0 \\ x_2 \qquad &\geq 0 \\ x_3 &\geq 0 \end{aligned}$$

Objective functions: maximize $x_1 + x_2 + x_3$, minimize $x_1 + x_3$, maximize $x_1 + x_2 - 3x_3$.

11. For a given set of inequalities

$$\begin{aligned} a_{i1}x_1 + \cdots + a_{in}x_n &\leq 1 \qquad i = 1, \ldots, m \\ x_j &\geq 0 \qquad j = 1, \ldots, n \end{aligned}$$

discuss conditions that indicate that an inequality does not form part of the boundary of the solution space.

12. Define the intersection of two sets $\mathbf{K}_1$ and $\mathbf{K}_2$ to be the set of points belonging to both $\mathbf{K}_1$ and $\mathbf{K}_2$. If $\mathbf{K}_1$ and $\mathbf{K}_2$ are both convex, prove that their intersection is also convex.

13. For a convex cone, show that the positive sum of any two vectors in the cone is also in the cone.

14. Solve Exercise 5 of Chap. 1.

PART TWO / **METHODS: THEORETICAL AND COMPUTATIONAL**

CHAPTER 3 / THE GENERAL LINEAR-PROGRAMMING PROBLEM

1. THE LINEAR-PROGRAMMING PROBLEM

The general linear-programming problem is to find a vector $(x_1,x_2, \ldots, x_j, \ldots, x_n)$ which minimizes the linear form (i.e., the objective function)

$$c_1x_1 + c_2x_2 + \cdots + c_jx_j + \cdots + c_nx_n \tag{1.1}$$

subject to the linear constraints

$$x_j \geq 0 \qquad j = 1, 2, \ldots, n \tag{1.2}$$

and

$$\begin{aligned} a_{11}x_1 + a_{12}x_2 + \cdots + a_{1j}x_j + \cdots + a_{1n}x_n &= b_1 \\ a_{21}x_1 + a_{22}x_2 + \cdots + a_{2j}x_j + \cdots + a_{2n}x_n &= b_2 \\ \cdots\cdots\cdots\cdots\cdots\cdots\cdots\cdots & \cdots\cdots \\ a_{i1}x_1 + a_{i2}x_2 + \cdots + a_{ij}x_j + \cdots + a_{in}x_n &= b_i \\ \cdots\cdots\cdots\cdots\cdots\cdots\cdots\cdots & \cdots\cdots \\ a_{m1}x_1 + a_{m2}x_2 + \cdots + a_{mj}x_j + \cdots + a_{mn}x_n &= b_m \end{aligned} \tag{1.3}$$

where the a_{ij}, b_i, and c_j are given constants and $m < n$. We shall always assume that the Eqs. (1.3) have been multiplied by -1 where necessary to make all $b_i \geq 0$. Because of the variety of notations in common use, one will find the general linear-programming problem stated in many forms. The more common are the following:

a. Minimize

$$\sum_{1}^{n} c_jx_j$$

subject to

$$x_j \geq 0 \qquad j = 1, 2, \ldots, n$$

and

$$\sum_{j=1}^{n} a_{ij}x_j = b_i \qquad i = 1, 2, \ldots, m$$

b. Minimize

$$\mathbf{cX}$$

subject to

$$\mathbf{X} \geq \mathbf{O}$$

and

$$\mathbf{AX} = \mathbf{b}$$

where $\mathbf{c} = (c_1,c_2, \ldots ,c_n)$ is a row vector, $\mathbf{X} = (x_1,x_2, \ldots ,x_n)$ is a column vector, $\mathbf{A} = (a_{ij})$, $\mathbf{b} = (b_1,b_2, \ldots ,b_m)$ is a column vector, and $\mathbf{O}$ is an n-dimensional null column vector.

c. Minimize

$$\mathbf{cX}$$

subject to

$$\mathbf{X} \geq \mathbf{O}$$

and

$$x_1\mathbf{P}_1 + x_2\mathbf{P}_2 + \cdots + x_n\mathbf{P}_n = \mathbf{P}_0$$

where $\mathbf{P}_j$ for $j = 1, 2, \ldots, n$ is the jth column of the matrix $\mathbf{A}$ and $\mathbf{P}_0 = \mathbf{b}$.

2. PROPERTIES OF A SOLUTION TO THE LINEAR-PROGRAMMING PROBLEM

In this section we shall state a number of standard definitions and describe the more important characteristics of a solution to the general linear-programming problem. Much of this material and that in Chap. 4 is contained in Dantzig [17] and Charnes, Cooper, and Henderson [12].

Definition 1. A *feasible solution* to the linear-programming problem is a vector $\mathbf{X} = (x_1,x_2, \ldots ,x_n)$ which satisfies conditions (1.2) and (1.3).

Definition 2*a*. A *basic solution* to (1.3) is a solution obtained by setting $n - m$ variables equal to zero and solving for the remaining m variables, provided that the determinant of the coefficients of these m variables is nonzero. The m variables are called *basic variables.*

Definition 2*b*. A *basic feasible solution* is a basic solution which also satisfies (1.2); that is, all basic variables are nonnegative.

Definition 3. A *nondegenerate basic feasible solution* is a basic feasible solution with *exactly* m positive x_i; that is, all basic variables are positive.

Definition 4. A *minimum feasible solution* is a feasible solution which also minimizes (1.1).

Unless otherwise stated, when we refer to a solution, we shall mean any feasible solution.

Definition 5. A *linear functional* $f(\mathbf{X})$ is a real-valued function defined on an n-dimensional vector space such that, for every vector $\mathbf{X} = \alpha\mathbf{U} + \beta\mathbf{V}$, $f(\mathbf{X}) = f(\alpha\mathbf{U} + \beta\mathbf{V}) = \alpha f(\mathbf{U}) + \beta f(\mathbf{V})$ for all n-dimensional vectors $\mathbf{U}$ and $\mathbf{V}$ and all scalars α and β. For example, let $\mathbf{U} = (9,3)$, $\mathbf{V} = (6,6)$, $\alpha = 1/3$, $\beta = -2/3$, and $f(\mathbf{X}) = 2x_1 + x_2$. We have $\alpha\mathbf{U} + \beta\mathbf{V} = (-1,-3)$ and $f(\alpha\mathbf{U} + \beta\mathbf{V}) = -5$; $\alpha f(\mathbf{U}) = 7$ and $\beta f(\mathbf{V}) = -12$.

We note that the objective function (1.1) is a linear functional for those $\mathbf{X}$ satisfying (1.2) and (1.3).

Theorem 1. *The set of all feasible solutions to the linear-programming problem is a convex set.*

Proof. We need to show that every convex combination of any two feasible solutions is also a feasible solution. (The theorem is true, of course, if the set of solutions has only one element.) Assume there are at least two solutions $\mathbf{X}_1$ and $\mathbf{X}_2$. We have

$$\mathbf{AX}_1 = \mathbf{b} \qquad \text{for } \mathbf{X}_1 \geq \mathbf{O}$$

and

$$\mathbf{AX}_2 = \mathbf{b} \qquad \text{for } \mathbf{X}_2 \geq \mathbf{O}$$

For $0 \leq \alpha \leq 1$, let $\mathbf{X} = \alpha\mathbf{X}_1 + (1 - \alpha)\mathbf{X}_2$ be any convex combination of $\mathbf{X}_1$ and $\mathbf{X}_2$. We note that all the elements of the vector $\mathbf{X}$ are nonnegative; that is, $\mathbf{X} \geq \mathbf{O}$. We then see that $\mathbf{X}$ is a feasible solution, for we have

$$\begin{aligned}\mathbf{AX} &= \mathbf{A}[\alpha\mathbf{X}_1 + (1 - \alpha)\mathbf{X}_2] = \alpha\mathbf{AX}_1 + (1 - \alpha)\mathbf{AX}_2 \\ &= \alpha\mathbf{b} + \mathbf{b} - \alpha\mathbf{b} = \mathbf{b}\end{aligned}$$

In a similar manner, one can prove that the sets of solutions to the inequalities (4.4) and the equalities (5.1) of Chap. 2 are convex sets.

We shall denote the convex set of solutions to the linear-programming problem by $\mathbf{K}$. Since $\mathbf{K}$ is determined by the intersection of the finite set of linear constraints (1.2) and (1.3), the boundary of **K** (if **K** is not void) will consist of sections of some of the corresponding hyperplanes. **K** will be a region of $\mathbf{E}_n$ and can either be void, a convex polyhedron, or a convex region which is unbounded in some direction. If **K** is void, then our problem does not have any solutions; if it is a convex polyhedron, then our problem has a solution with a finite minimum value for the objective function; and if **K** is unbounded, the problem has a solution, but the minimum *might* be unbounded. Valid linear-programming models should yield **K**'s of the second or possibly the third type, as they are models of situations that have a number of possible solutions. By Theorem 1, if a problem has more than one solution, it has, in reality, an infinite number of solutions. Out of all these solutions, it is our task to determine the one which minimizes the corresponding objective function. This work is somewhat simplified by the results of Theorem 2 below. Before proceeding with this theorem, we should note the following: If **K** is a convex polyhedron, then **K** is the convex hull of the extreme points of **K**. That is, every feasible solution in **K** can be represented as a convex combination of the extreme feasible solutions in **K**. (By definition, a convex polyhedron has a finite number of extreme points.) An unbounded **K** also has a finite number of extreme points, but not all points of **K** can be represented as convex combinations of these extreme points. For ease in discussion, we can assume that all our problems have a **K** that is a convex polyhedron. As will be shown in later chapters, computational devices exist that determine whether **K** is void or whether a problem has an unbounded minimum. (For computational purposes, we always assume that **K** is a convex polyhedron.)

With the assumption that **K** is a convex polyhedron, we can surmise from the above discussion that we need only to look at the extreme points of the convex polyhedron in order to determine the minimum feasible solution. We prove this with the following theorem:

Theorem 2. *The objective function* (1.1) *assumes its minimum at an extreme point of the convex set* **K** *generated by the set of feasible solutions to the linear-programming problem. If it assumes its minimum at more than one extreme point, then it takes on the same value for every convex combination of those particular points.*

Proof. Since we have assumed **K** to be a convex polyhedron, **K** has a finite number of extreme points. In two dimensions, **K** might look like Fig. 3.1. Let us denote the objective function by $f(\mathbf{X})$, the extreme points by $\bar{\mathbf{X}}_1, \bar{\mathbf{X}}_2, \ldots, \bar{\mathbf{X}}_p$, and the minimum feasible solution by $\mathbf{X}_0$. This means that $f(\mathbf{X}_0) \leq f(\mathbf{X})$ for all $\mathbf{X}$ in **K**. If $\mathbf{X}_0$ is an extreme point, the first part of the theorem is true. Suppose

$\mathbf{X}_0$ is not an extreme point (as indicated in Fig. 3.1). We can then write $\mathbf{X}_0$ as a convex combination of the extreme points of $\mathbf{K}$, that is,

$$\mathbf{X}_0 = \sum_{i=1}^{p} \alpha_i \bar{\mathbf{X}}_i$$

for $\alpha_i \geq 0$ and $\sum_i \alpha_i = 1$. Then, since $f(\mathbf{X})$ is a linear functional, we have

$$f(\mathbf{X}_0) = f\left(\sum_{i=1}^{p} \alpha_i \bar{\mathbf{X}}_i\right) = f(\alpha_1 \bar{\mathbf{X}}_1 + \alpha_2 \bar{\mathbf{X}}_2 + \cdots + \alpha_p \bar{\mathbf{X}}_p)$$

$$= \alpha_1 f(\bar{\mathbf{X}}_1) + \alpha_2 f(\bar{\mathbf{X}}_2) + \cdots + \alpha_p f(\bar{\mathbf{X}}_p) = m \tag{2.1}$$

where m is the minimum of $f(\mathbf{X})$ for all $\mathbf{X}$ in $\mathbf{K}$.

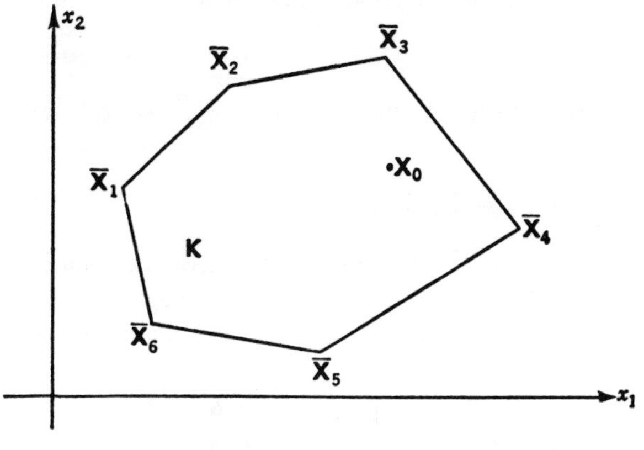

FIGURE 3.1

Since all $\alpha_i \geq 0$, we do not increase the sum (2.1) if we substitute for each $f(\bar{\mathbf{X}}_i)$ the minimum of the values $f(\bar{\mathbf{X}}_i)$. Let $f(\bar{\mathbf{X}}_m) = \min_i f(\bar{\mathbf{X}}_i)$. Substituting in (2.1) we have, since $\sum_i \alpha_i = 1$,

$$f(\mathbf{X}_0) \geq \alpha_1 f(\bar{\mathbf{X}}_m) + \alpha_2 f(\bar{\mathbf{X}}_m) + \cdots + \alpha_p f(\bar{\mathbf{X}}_m) = f(\bar{\mathbf{X}}_m)$$

Since we assumed $f(\mathbf{X}_0) \leq f(\mathbf{X})$ for all $\mathbf{X}$ in $\mathbf{K}$, we must have

$$f(\mathbf{X}_0) = f(\bar{\mathbf{X}}_m) = m$$

Therefore, there is an extreme point, $\bar{\mathbf{X}}_m$, at which the objective function assumes its minimum value.

To prove the second part of the theorem, let $f(\mathbf{X})$ assume its minimum at more than one extreme point, say at $\bar{\mathbf{X}}_1, \bar{\mathbf{X}}_2, \ldots, \bar{\mathbf{X}}_q$. Here we have $f(\bar{\mathbf{X}}_1) = f(\bar{\mathbf{X}}_2) = \cdots = f(\bar{\mathbf{X}}_q) = m$. If $\mathbf{X}$ is any convex combination of the above $\bar{\mathbf{X}}_i$, say

$$\mathbf{X} = \sum_{i=1}^{q} \alpha_i \bar{\mathbf{X}}_i$$

for $\alpha_i \geq 0$ and $\sum_i \alpha_i = 1$, then

$$\begin{aligned} f(\mathbf{X}) &= f(\alpha_1 \bar{\mathbf{X}}_1 + \alpha_2 \bar{\mathbf{X}}_2 + \cdots + \alpha_q \bar{\mathbf{X}}_q) \\ &= \alpha_1 f(\bar{\mathbf{X}}_1) + \alpha_2 f(\bar{\mathbf{X}}_2) + \cdots + \alpha_q f(\bar{\mathbf{X}}_q) = \sum_i \alpha_i m \\ &= m \end{aligned}$$

The proof is now completed. By making the obvious changes, the theorem can be proved for the case where (1.1) is to be maximized. By Theorem 2, we need only to consider the extreme points of **K** in our search for a minimum feasible solution to the linear-programming problem.

Recall that a feasible solution is a vector $\mathbf{X} = (x_1, x_2, \ldots, x_n)$, with all $x_i \geq 0$, such that

$$x_1 \mathbf{P}_1 + x_2 \mathbf{P}_2 + \cdots + x_n \mathbf{P}_n = \mathbf{P}_0$$

Assume we have found a set of k vectors that is linearly independent and that there exists a nonnegative combination of these vectors that is equal to $\mathbf{P}_0$. Let this set of vectors be $\mathbf{P}_1, \mathbf{P}_2, \ldots, \mathbf{P}_k$. We then have the following theorem:

Theorem 3. *If a set of $k \leq m$ vectors $\mathbf{P}_1, \mathbf{P}_2, \ldots, \mathbf{P}_k$ can be found that is linearly independent and such that*

$$x_1 \mathbf{P}_1 + x_2 \mathbf{P}_2 + \cdots + x_k \mathbf{P}_k = \mathbf{P}_0$$

and all $x_i \geq 0$, then the point $\mathbf{X} = (x_1, x_2, \ldots, x_k, 0, \ldots, 0)$ is an extreme point of the convex set of feasible solutions. Here $\mathbf{X}$ is an n-dimensional vector whose last $n - k$ elements are zero.

Proof. Suppose **X** is not an extreme point. Then, since **X** is a feasible solution, it can be written as a convex combination of two other points $\mathbf{X}_1$ and $\mathbf{X}_2$ in **K**. We have $\mathbf{X} = \alpha \mathbf{X}_1 + (1 - \alpha)\mathbf{X}_2$ for $0 < \alpha < 1$. Since all the elements x_i of **X** are nonnegative and since $0 < \alpha < 1$, the last $n - k$ elements of $\mathbf{X}_1$ and $\mathbf{X}_2$ must also equal zero; that is,

$$\mathbf{X}_1 = (x_1^{(1)}, x_2^{(1)}, \ldots, x_k^{(1)}, 0, \ldots, 0)$$
$$\mathbf{X}_2 = (x_1^{(2)}, x_2^{(2)}, \ldots, x_k^{(2)}, 0, \ldots, 0)$$

Since $\mathbf{X}_1$ and $\mathbf{X}_2$ are feasible solutions, we have

$$\mathbf{AX}_1 = \mathbf{b}$$

and

$$\mathbf{AX}_2 = \mathbf{b}$$

Rewriting these equations in vector notation, we have

$$x_1^{(1)}\mathbf{P}_1 + x_2^{(1)}\mathbf{P}_2 + \cdots + x_k^{(1)}\mathbf{P}_k = \mathbf{P}_0$$

and

$$x_1^{(2)}\mathbf{P}_1 + x_2^{(2)}\mathbf{P}_2 + \cdots + x_k^{(2)}\mathbf{P}_k = \mathbf{P}_0$$

But $\mathbf{P}_1, \mathbf{P}_2, \ldots, \mathbf{P}_k$ are linearly independent, and hence $\mathbf{P}_0$ can be expressed as a *unique* linear combination in terms of $\mathbf{P}_1, \mathbf{P}_2, \ldots, \mathbf{P}_k$.† This implies that $x_i = x_i^{(1)} = x_i^{(2)}$. Therefore, $\mathbf{X}$ cannot be expressed as a convex combination of two distinct points in $\mathbf{K}$ and must be an extreme point of $\mathbf{K}$.

Theorem 4. *If $\mathbf{X} = (x_1, x_2, \ldots, x_n)$ is an extreme point of $\mathbf{K}$, then the vectors associated with positive x_i form a linearly independent set. From this it follows that, at most, m of the x_i are positive.*

Proof. Let the nonzero coefficients be the first k coefficients, so that $\sum_{i=1}^{k} x_i\mathbf{P}_i = \mathbf{P}_0$. We prove the main part of the theorem by contradiction. Assume that $\mathbf{P}_1, \mathbf{P}_2, \ldots, \mathbf{P}_k$ are linearly dependent. Then there exists a linear combination of these vectors which equals the zero vector,

$$d_1\mathbf{P}_1 + d_2\mathbf{P}_2 + \cdots + d_k\mathbf{P}_k = \mathbf{O} \tag{2.2}$$

with at least one $d_i \neq 0$. From the hypothesis of the theorem, we have

$$x_1\mathbf{P}_1 + x_2\mathbf{P}_2 + \cdots + x_k\mathbf{P}_k = \mathbf{P}_0 \tag{2.3}$$

† This can be shown as follows: Let $\mathbf{P}_1, \mathbf{P}_2, \ldots, \mathbf{P}_k$ be a set of linearly independent vectors and assume that we can represent the vector $\mathbf{P}_0$ in terms of these vectors by two different linear combinations, for example,

$$e_1\mathbf{P}_1 + e_2\mathbf{P}_2 + \cdots + e_k\mathbf{P}_k = \mathbf{P}_0$$

and

$$f_1\mathbf{P}_1 + f_2\mathbf{P}_2 + \cdots + f_k\mathbf{P}_k = \mathbf{P}_0$$

To show that these combinations have to be identical, we subtract the second from the first to obtain

$$(e_1 - f_1)\mathbf{P}_1 + (e_2 - f_2)\mathbf{P}_2 + \cdots + (e_k - f_k)\mathbf{P}_k = \mathbf{O}$$

By the definition of linear independence we must have each $e_i - f_i = 0$, which implies that $e_i = f_i$.

For some $d > 0$, we multiply (2.2) by d and add and subtract the result from (2.3) to obtain the two equations

$$\sum_{i=1}^{k} x_i\mathbf{P}_i + d \sum_{i=1}^{k} d_i\mathbf{P}_i = \mathbf{P}_0$$

$$\sum_{i=1}^{k} x_i\mathbf{P}_i - d \sum_{i=1}^{k} d_i\mathbf{P}_i = \mathbf{P}_0$$

We then have the two solutions to (1.3) (note that they might not be feasible solutions):

$$\mathbf{X}_1 = (x_1 + dd_1, x_2 + dd_2, \ldots, x_k + dd_k, 0, \ldots, 0)$$

and

$$\mathbf{X}_2 = (x_1 - dd_1, x_2 - dd_2, \ldots, x_k - dd_k, 0, \ldots, 0)$$

Since all $x_i > 0$, we can let d be as small as necessary, but still positive, to make the first k components of both $\mathbf{X}_1$ and $\mathbf{X}_2$ positive. Then $\mathbf{X}_1$ and $\mathbf{X}_2$ are feasible solutions. But $\mathbf{X} = \frac{1}{2}\mathbf{X}_1 + \frac{1}{2}\mathbf{X}_2$, which contradicts the hypothesis that $\mathbf{X}$ is an extreme point. The assumption of linear dependence for the vectors $\mathbf{P}_1, \mathbf{P}_2, \ldots, \mathbf{P}_k$ has thus led to a contradiction and hence must be false; i.e., the set of vectors $\mathbf{P}_1, \mathbf{P}_2, \ldots, \mathbf{P}_k$ is linearly independent.

Since every set of $m + 1$ vectors in m-dimensional space is necessarily linearly dependent, we cannot have more than m positive x_i. For assume that we did. Then the above proof of the main part of the theorem would imply that there exist vectors $\mathbf{P}_1, \ldots, \mathbf{P}_m, \mathbf{P}_{m+1}$ that are linearly independent.

We can, without any loss of generality, assume that the set of vectors $\mathbf{P}_1, \mathbf{P}_2, \ldots, \mathbf{P}_n$ of the linear-programming problem always contains a set of m linearly independent vectors. If this property is not evident when a particular problem is being solved, the original set of vectors is augmented by a set of m linearly independent vectors, and we then seek a solution to the extended problem. This procedure will be explained in detail in the succeeding chapters.

Corollary 1. *Associated with every extreme point of* $\mathbf{K}$ *is a set of* m *linearly independent vectors from the given set* $\mathbf{P}_1, \mathbf{P}_2, \ldots, \mathbf{P}_n$.

Proof. Theorem 4 has shown that there are $k \leq m$ such vectors. For $k = m$, the corollary is proved. Assume that $k < m$ and that we can find only additional vectors $\mathbf{P}_{k+1}, \ldots, \mathbf{P}_r$ such that the set

$$\mathbf{P}_1, \ldots, \mathbf{P}_k, \mathbf{P}_{k+1}, \ldots, \mathbf{P}_r$$

for $r < m$ is linearly independent. Then this implies that the remaining $n - r$ vectors are dependent on $\mathbf{P}_1, \ldots, \mathbf{P}_r$. But this contradicts the assumption that we always have a set of m linearly independent vectors in the given set of $\mathbf{P}_1, \ldots, \mathbf{P}_n$. Therefore, there must be m linearly independent vectors $\mathbf{P}_1, \ldots, \mathbf{P}_m$ associated with every extreme point, such that

$$\sum_{i=1}^{k} x_i\mathbf{P}_i + \sum_{i=k+1}^{m} 0\mathbf{P}_i = \mathbf{P}_0$$

We can sum up the preceding theorems by the following:

Theorem 5. $\mathbf{X} = (x_1,x_2, \ldots ,x_n)$ *is an extreme point of* $\mathbf{K}$ *if and only if the positive* x_j *are coefficients of linearly independent vectors* $\mathbf{P}_j$ *in*

$$\sum_{j=1}^{n} x_j\mathbf{P}_j = \mathbf{P}_0$$

As a result of the assumptions and theorems of this section, we have:

1. There is an extreme point of $\mathbf{K}$ at which the objective function takes on its minimum.
2. Every basic feasible solution corresponds to an extreme point of $\mathbf{K}$.
3. Every extreme point of $\mathbf{K}$ has m linearly independent vectors of the given set of n associated with it.

From the above we can conclude that we need only investigate extreme-point solutions and hence only those feasible solutions generated by m linearly independent vectors. Since there are at most $\binom{n}{m}$ sets of m linearly independent vectors from the given set of n, the value $\binom{n}{m}$ is the upper bound to the number of possible solutions to the problem.[1] For large n and m it would be an impossible task to evaluate all the possible solutions and select one that minimizes the objective function. What is required is a computational scheme that selects, in an orderly fashion, a small subset of the possible solutions which converges to a minimum solution. The *simplex procedure,* devised by G. B. Dantzig, is such a scheme.[2] This procedure finds an extreme point and determines whether it is the minimum. If it is not, the procedure finds a neighboring extreme point[3] whose corresponding value of the objective

[1] See Saaty [88] and Quandt and Kuhn [85*c*] for discussions of the upper bound to the number of possible solutions.

[2] The name simplex method is due to the use of the equation $\sum_j x_j = 1$ as a constraint in a geometric interpretation of this procedure, as described in Dantzig [17].

[3] Two extreme points are said to be neighbors if they are joined by a boundary of the convex polyhedron.

function is less than or equal to the preceding value. In a finite number of such steps (usually between m and $2m$), a minimum feasible solution is found. The simplex method makes it possible to discover whether the problem has no finite minimum solutions or no feasible solutions. It is a powerful scheme for solving *any* linear-programming problem.

Before going into the validity and full computational aspects of the simplex procedure, we wish to introduce the following computational element of the procedure.

3. GENERATING EXTREME-POINT SOLUTIONS

Here we assume that an extreme-point solution in terms of m vectors $\mathbf{P}_j$ of the original set of n vectors is known. We can let this set of m linearly independent vectors be the first m, and let

$$\mathbf{X} = (x_1, x_2, \ldots, x_m, 0, \ldots, 0)$$

be the solution vector. We then have

$$x_1\mathbf{P}_1 + x_2\mathbf{P}_2 + \cdots + x_m\mathbf{P}_m = \mathbf{P}_0 \tag{3.1}$$

where all $x_i \geq 0$. With this information, the problem is to determine, in a computationally efficient manner, a new extreme-point solution. (We shall, of course, assume that a different extreme solution exists.) Since the vectors $\mathbf{P}_1, \mathbf{P}_2, \ldots, \mathbf{P}_m$ are linearly independent, they form a basis in m-dimensional vector space. We can then express every vector of the given n as a linear combination of these basis vectors. We can write

$$\sum_{i=1}^{m} x_{ij}\mathbf{P}_i = \mathbf{P}_j \qquad j = 1, \ldots, n$$

Assume that some vector not in the given basis, say $\mathbf{P}_{m+1}$, has at least one element $x_{i,\,m+1} > 0$ in the expression

$$x_{1,\,m+1}\mathbf{P}_1 + x_{2,\,m+1}\mathbf{P}_2 + \cdots + x_{m,\,m+1}\mathbf{P}_m = \mathbf{P}_{m+1} \tag{3.2}$$

Let θ be any number, and multiply (3.2) by θ and subtract the result from (3.1) to obtain

$$(x_1 - \theta x_{1,\,m+1})\mathbf{P}_1 + (x_2 - \theta x_{2,\,m+1})\mathbf{P}_2 + \cdots + (x_m - \theta x_{m,\,m+1})\mathbf{P}_m + \theta\mathbf{P}_{m+1} = \mathbf{P}_0 \tag{3.3}$$

The vector $\mathbf{X}' = (x_1 - \theta x_{1,\,m+1},\ x_2 - \theta x_{2,\,m+1},\ \ldots,\ x_m - \theta x_{m,\,m+1},\ \theta)$ is a solution to the problem, and if all the elements of $\mathbf{X}'$ are nonnegative, $\mathbf{X}'$ is a feasible solution. Since we want $\mathbf{X}'$ to be a feasible solution different from $\mathbf{X}$, we restrict θ to be greater than zero.[1] With this restriction, all the elements of $\mathbf{X}'$ that have a negative or zero $x_{i,\,m+1}$ will also have a nonnegative $x_i - \theta x_{i,\,m+1}$. We need only concern ourselves with those elements having a positive $x_{i,\,m+1}$. We wish to find a $\theta > 0$ such that

$$x_i - \theta x_{i,\,m+1} \geq 0 \tag{3.4}$$

for all $x_{i,\,m+1} > 0$.

From (3.4) we have

$$\frac{x_i}{x_{i,\,m+1}} \geq \theta$$

and hence any θ for which

$$0 < \theta \leq \min_i \frac{x_i}{x_{i,\,m+1}}$$

will give a feasible solution for (3.3). However, as we are looking for an extreme-point solution, we know by our theorems in Sec. 2 that we cannot have all the $m + 1$ elements of $\mathbf{X}'$ positive. We then must force at least one of the elements of $\mathbf{X}'$ to be exactly equal to zero. We see that, if we let

$$\theta = \theta_0 = \min_i \frac{x_i}{x_{i,\,m+1}}$$

for $x_{i,\,m+1} > 0$, then the element in $\mathbf{X}'$ for which this minimum is attained will reduce to zero. Let this element be the first, that is,

$$\theta_0 = \min_i \frac{x_i}{x_{i,\,m+1}} = \frac{x_1}{x_{1,\,m+1}}$$

We have now obtained a new feasible solution

$$x_2'\mathbf{P}_2 + x_3'\mathbf{P}_3 + \cdots + x_m'\mathbf{P}_m + x_{m+1}'\mathbf{P}_{m+1} = \mathbf{P}_0$$

where

$$x_i' = x_i - \theta_0 x_{i,\,m+1} \qquad i = 2, \ldots, m$$

$$x_{m+1}' = \theta_0$$

[1] The reader should note that this restriction causes us to assume (for illustrative purposes) that associated with an $x_{i,\,m+1} > 0$ is an $x_i > 0$. This, in general, is not the case and will be true only if nondegeneracy is assumed for all basic feasible solutions. As discussed in Chap. 4, a value of $\theta = 0$ is an acceptable one; i.e., some $x_i = 0$ for an $x_{i,\,m+1} > 0$. With $\theta = 0$, the transformation retains the old extreme point but selects a new feasible basis. In sum, the simplex procedure allows for all values of $\theta \geq 0$. The reader should review the applicability of the above discussion to the case $\theta = \min_i (x_i/x_{i,\,m+1}) = 0$ for $x_{i,\,m+1} > 0$.

If all the $x_{i,\,m+1}$ had been equal to or less than zero, then we would not have been able to select a positive θ that would have eliminated at least one of the vectors $\mathbf{P}_1, \ldots, \mathbf{P}_m$ from the basis. For this situation we obtain for *any* $\theta > 0$ a non-extreme-point feasible solution associated with the $m + 1$ vectors $\mathbf{P}_1, \ldots, \mathbf{P}_m, \mathbf{P}_{m+1}$. As will be shown in Chap. 4, this situation indicates that the problem does not have a finite minimum solution.

To show that $\mathbf{X}' = (x'_2, \ldots, x'_m, x'_{m+1})$ is an extreme point, we have to prove $\mathbf{P}_2, \ldots, \mathbf{P}_m, \mathbf{P}_{m+1}$ are linearly independent. Assume they are linearly dependent. We can then find (from the definition of linear dependence)

$$d_2\mathbf{P}_2 + d_3\mathbf{P}_3 + \cdots + d_m\mathbf{P}_m + d_{m+1}\mathbf{P}_{m+1} = \mathbf{O} \tag{3.5}$$

where not all $d_i = 0$. Since any subset of a set of linearly independent vectors is also a set of linearly independent vectors, $\mathbf{P}_2, \ldots, \mathbf{P}_m$ are linearly independent. This implies that $d_{m+1} \neq 0$. From (3.5) we have

$$e_2\mathbf{P}_2 + e_3\mathbf{P}_3 + \cdots + e_m\mathbf{P}_m = \mathbf{P}_{m+1} \tag{3.6}$$

where

$$e_i = \frac{-d_i}{d_{m+1}}$$

Subtracting (3.6) from (3.2), we obtain

$$x_{1,\,m+1}\mathbf{P}_1 + (x_{2,\,m+1} - e_2)\mathbf{P}_2 + (x_{3,\,m+1} - e_3)\mathbf{P}_3 + \cdots + (x_{m,\,m+1} - e_m)\mathbf{P}_m = \mathbf{O} \tag{3.7}$$

Since $\mathbf{P}_1, \mathbf{P}_2, \ldots, \mathbf{P}_m$ are linearly independent, all the coefficients in (3.7) must equal zero. But $x_{1,\,m+1}$ was assumed to be positive. Hence the assumption of linear dependence for $\mathbf{P}_2, \ldots, \mathbf{P}_{m+1}$ has led to a contradiction, and these vectors must be linearly independent.

In order to continue this process of obtaining new extreme feasible solutions, we need the representation of any vector not in the new basis $\mathbf{P}_2, \mathbf{P}_3, \ldots, \mathbf{P}_{m+1}$ in terms of this basis. From (3.2) we have

$$\mathbf{P}_1 = \frac{1}{x_{1,\,m+1}}(\mathbf{P}_{m+1} - x_{2,\,m+1}\mathbf{P}_2 - \cdots - x_{m,\,m+1}\mathbf{P}_m) \tag{3.8}$$

Let

$$\mathbf{P}_j = x_{1j}\mathbf{P}_1 + x_{2j}\mathbf{P}_2 + \cdots + x_{mj}\mathbf{P}_m \tag{3.9}$$

be any vector not in the new basis. Substitute the expression (3.8) for $\mathbf{P}_1$ in (3.9) to obtain

$$\mathbf{P}_j = \left(x_{2j} - \frac{x_{1j}}{x_{1,\,m+1}}\,x_{2,\,m+1}\right)\mathbf{P}_2 + \left(x_{3j} - \frac{x_{1j}}{x_{1,\,m+1}}\,x_{3,\,m+1}\right)\mathbf{P}_3 + \cdots + \left(x_{mj} - \frac{x_{1j}}{x_{1,\,m+1}}\,x_{m,\,m+1}\right)\mathbf{P}_m + \frac{x_{1j}}{x_{1,\,m+1}}\,\mathbf{P}_{m+1}$$

The reader will note that the formulas for complete elimination required in Exercise 8, Chap. 2, are equivalent to the transformation that describes $\mathbf{P}_0$ and $\mathbf{P}_j$ in terms of the new basis. The procedure for obtaining new extreme-point solutions is that of selecting a new variable to be introduced into the system, determining which variable has to be removed from the solution in order to preserve feasibility, and applying the complete elimination formulas to obtain the new solution and the new representations of the vectors not in the basis. The criterion used to determine which variable is to be introduced into the solution is a feature of the simplex procedure and will be considered in Chap. 4.

Example. We are given the following set of equations:

$\mathbf{P}_1$	$\mathbf{P}_2$	$\mathbf{P}_3$	$\mathbf{P}_4$	$\mathbf{P}_5$	$\mathbf{P}_6$	$\mathbf{P}_0$
$3x_1$	$-\ x_2$	$+\ 2x_3$	$+\ x_4$			$=\ 7$
$2x_1$	$-\ 4x_2$			$+\ x_5$		$=\ 12$
$-4x_1$	$-\ 3x_2$	$+\ 8x_3$			$+\ x_6$	$=\ 10$

We have as an initial extreme-point solution $x_1 = 0$, $x_2 = 0$, $x_3 = 0$, $x_4 = 7$, $x_5 = 12$, $x_6 = 10$, which in vector notation is given by

$$7\mathbf{P}_4 + 12\mathbf{P}_5 + 10\mathbf{P}_6 = \mathbf{P}_0 \tag{3.10}$$

Here the basis vectors $\mathbf{P}_4$, $\mathbf{P}_5$, $\mathbf{P}_6$ are unit vectors. We wish to introduce vector $\mathbf{P}_1$ to obtain another extreme-point solution. The representation of $\mathbf{P}_1$ in terms of the basis vectors is simply

$$3\mathbf{P}_4 + 2\mathbf{P}_5 - 4\mathbf{P}_6 = \mathbf{P}_1 \tag{3.11}$$

that is,

$$x_{41} = 3 \qquad x_{51} = 2 \qquad x_{61} = -4$$

If we multiply (3.11) by θ and subtract the result from (3.10), we have

$$(7 - 3\theta)\mathbf{P}_4 + (12 - 2\theta)\mathbf{P}_5 + (10 + 4\theta)\mathbf{P}_6 + \theta\mathbf{P}_1 = \mathbf{P}_0 \tag{3.12}$$

Since $x_{41} = 3$ and $x_{51} = 2$ are both positive, we determine θ_0 by evaluating, for these positive x_{i1},

$$\theta = \theta_0 = \min \frac{x_i}{x_{i1}} = 7/3\dagger$$

Substituting this value in (3.12), we eliminate $\mathbf{P}_4$ from the basis to obtain

$$22/3\,\mathbf{P}_5 + 58/3\,\mathbf{P}_6 + 7/3\,\mathbf{P}_1 = \mathbf{P}_0$$

† We form the ratios for those i in the current solution. Here $i = 4, 5, 6$.

or the extreme-point solution $x_1 = 7/3, x_2 = 0, x_3 = 0, x_4 = 0, x_5 = 22/3, x_6 = 58/3$.

If, instead of $\mathbf{P}_1$, we tried in a similar manner to obtain an extreme solution with $\mathbf{P}_2$, where

$$-\mathbf{P}_4 - 4\mathbf{P}_5 - 3\mathbf{P}_6 = \mathbf{P}_2$$

we would have developed the following expression for $\mathbf{P}_0$ in terms of $\mathbf{P}_4, \mathbf{P}_5, \mathbf{P}_6, \mathbf{P}_2$:

$$(7 + \theta)\mathbf{P}_4 + (12 + 4\theta)\mathbf{P}_5 + (10 + 3\theta)\mathbf{P}_6 + \theta\mathbf{P}_2 = \mathbf{P}_0 \qquad (3.13)$$

From (3.13) we see that any $\theta > 0$ yields a feasible solution $x_1 = 0$, $x_2 = \theta$, $x_3 = 0$, $x_4 = 7 + \theta$, $x_5 = 12 + 4\theta$, $x_6 = 10 + 3\theta$. Here, since all $x_{i2} < 0$, we do not obtain a new extreme-point solution.

A more efficient way of interpreting the problem is as a transformation accomplished by the elimination procedure. Here we detach the coefficients of the equations and set up the following tableau:

$\mathbf{P}_1$	$\mathbf{P}_2$	$\mathbf{P}_3$	$\mathbf{P}_4$	$\mathbf{P}_5$	$\mathbf{P}_6$	$\mathbf{P}_0$	θ
③	−1	2	1	0	0	7	$7/3 = \theta_0$
2	−4	0	0	1	0	12	6
−4	−3	8	0	0	1	10	

As we want to introduce $\mathbf{P}_1$ into the basis, we again form the ratios x_i/x_{i1} for $x_{i1} > 0$. Since $\theta_0 = 7/3$ is the minimum of these ratios, we let the element 3 of $\mathbf{P}_1$ be the pivot element of the elimination procedure, as denoted by the circle. That is, we shall eliminate x_1 from all the equations except the first. If we carry out the elimination transformation, we obtain a new tableau. Here $x_1 = 7/3$, $x_5 = 22/3$, $x_6 = 58/3$, and $x_2 = x_3 = x_4 = 0$.

$\mathbf{P}'_1$	$\mathbf{P}'_2$	$\mathbf{P}'_3$	$\mathbf{P}'_4$	$\mathbf{P}'_5$	$\mathbf{P}'_6$	$\mathbf{P}'_0$	θ
1	$-1/3$	$2/3$	$1/3$	0	0	$7/3$	
0	$-10/3$	$-4/3$	$-2/3$	1	0	$22/3$	
0	$-13/3$	$32/3$	$4/3$	0	1	$58/3$	

We now have a basis of $\mathbf{P}_1$, $\mathbf{P}_5$, $\mathbf{P}_6$, with $\mathbf{P}_2$, $\mathbf{P}_3$, $\mathbf{P}_4$ explicitly given in terms of these basis vectors, that is,

$$-\tfrac{1}{3}\mathbf{P}_1 - \tfrac{10}{3}\mathbf{P}_5 - \tfrac{13}{3}\mathbf{P}_6 = \mathbf{P}_2$$

$$\tfrac{2}{3}\mathbf{P}_1 - \tfrac{4}{3}\mathbf{P}_5 + \tfrac{32}{3}\mathbf{P}_6 = \mathbf{P}_3$$

$$\tfrac{1}{3}\mathbf{P}_1 - \tfrac{2}{3}\mathbf{P}_5 + \tfrac{4}{3}\mathbf{P}_6 = \mathbf{P}_4$$

Hence if we wanted to obtain an extreme-point solution with $\mathbf{P}_3$ in the basis, we could start with the second tableau and determine θ_0 as before and transform this tableau by the elimination formulas. The resulting tableau will yield the representation of the vectors not in the basis in terms of the new basis vectors.

EXERCISES

1. Find all basic feasible solutions for the equations

$$2x_1 + 6x_2 + 2x_3 + x_4 = 3$$
$$6x_1 + 4x_2 + 4x_3 + 6x_4 = 2$$

and determine the associated general convex combination of extreme-point solutions.

2. Construct and graph linear-programming problems in three variables with a unique extreme-point optimum solution, three extreme-point optimum solutions, and four extreme-point optimum solutions.

3. In the discussion in Sec. 3, the extreme-point solution (3.1) can be degenerate. Discuss the computation of θ and the transformation to a new basic feasible solution if some $x_i = 0$. Can there be any assurance that $\theta_0 > 0$? Also discuss and interpret the situation where the selection of $\min_i (x_i/x_{ij})$ for $x_{ij} > 0$ is not unique.

4. The following set of equations has a given extreme-point solution $\mathbf{X} = (x_1,x_2,x_3) = (4,3,6)$. By the algebraic procedure of Sec. 3, obtain two basic solutions for the bases $\mathbf{P}_1$, $\mathbf{P}_2$, $\mathbf{P}_4$ and $\mathbf{P}_2$, $\mathbf{P}_3$, $\mathbf{P}_4$. In each case, determine the expression for $\mathbf{P}_5$ and the vector eliminated in terms of the new basis.

$\mathbf{P}_1$	$\mathbf{P}_2$	$\mathbf{P}_3$	$\mathbf{P}_4$	$\mathbf{P}_5$	$\mathbf{P}_0$
x_1			$+2x_4$	$-\ x_5$	$= 4$
	x_2		$-\ x_4$	$+\ x_5$	$= 3$
		x_3	$+3x_4$	$-2x_5$	$= 6$

5. Do Exercise 4 by setting up the tableau and applying the general formulas developed for the complete elimination procedure in Exercise 8 of Chap. 2. For both bases, the variable x_4 is eliminated from all the equations except one.

6. For the 3×3 matrix formed by vectors $\mathbf{P}_1$, $\mathbf{P}_2$, $\mathbf{P}_4$ of Exercise 4 compute its inverse by means of the adjoint matrix as described in Chap. 2. Do the same for vectors $\mathbf{P}_2$, $\mathbf{P}_3$, $\mathbf{P}_4$.

7. Given the following set of equations:

$$x_1 + 4x_2 - x_3 = 3$$
$$5x_1 + 2x_2 + 3x_3 = 4$$

Determine the basic feasible solution involving x_1 and x_2. Do basic feasible solutions exist for x_1 and x_3 and for x_2 and x_3?

Discuss the graphical solution to this problem if each column of the matrix is assumed to be a vector in two-dimensional space.

CHAPTER 4 / THE SIMPLEX COMPUTATIONAL PROCEDURE

We shall next discuss and prove the validity of the basic elements of the simplex procedure and related computational algorithms. By the simplex procedure, we can, once any basic (extreme-point) feasible solution has been determined, obtain a minimum feasible solution in a finite number of steps. These steps, or iterations, consist in finding a new feasible solution whose corresponding value of the objective function is less than the value of the objective function for the preceding solution. This process is continued until a minimum solution has been reached. From the discussion of Chap. 3, we have that all extreme-point solutions, and especially the minimum solution, have m linearly independent vectors associated with them. We then limit our search to those solutions that are generated by m linearly independent vectors. We note that there are a finite number of such solutions.

1. DEVELOPMENT OF A MINIMUM FEASIBLE SOLUTION

We assume that the linear-programming problem is feasible, that every basic feasible solution is nondegenerate, and that we are given a basic feasible solution.[1] These assumptions, as will be discussed later, are made without any loss in generality. Let the given solution be $\mathbf{X}_0 = (x_{10},x_{20}, \ldots ,x_{m0})$† and the associated set of linearly independent vectors be $\mathbf{P}_1, \mathbf{P}_2, \ldots \mathbf{P}_m$. We then have

$$x_{10}\mathbf{P}_1 + x_{20}\mathbf{P}_2 + \cdots + x_{m0}\mathbf{P}_m = \mathbf{P}_0 \tag{1.1}$$

$$x_{10}c_1 + x_{20}c_2 + \cdots + x_{m0}c_m = z_0 \tag{1.2}$$

[1] The definitions of these terms are given in Sec. 2 of Chap. 3.

† In order to generalize the simplex transformations we now denote the solution vector $\mathbf{X} = (x_1,x_2, \ldots ,x_m)$ by the vector $\mathbf{X}_0 = (x_{10},x_{20}, \ldots ,x_{m0})$. We should note that the remaining $n - m$ values of the solution vector have arbitrarily been set equal to zero.

where all $x_{i0} > 0$, the c_i are the cost coefficients of the objective function, and z_0 is the corresponding value of the objective function for the given solution. Since $\mathbf{P}_1, \mathbf{P}_2, \ldots, \mathbf{P}_m$ are linearly independent, we can express any vector from the set $\mathbf{P}_1, \mathbf{P}_2, \ldots, \mathbf{P}_n$ in terms of $\mathbf{P}_1, \mathbf{P}_2, \ldots, \mathbf{P}_m$. Let $\mathbf{P}_j$ be given by

$$x_{1j}\mathbf{P}_1 + x_{2j}\mathbf{P}_2 + \cdots + x_{mj}\mathbf{P}_m = \mathbf{P}_j \qquad j = 1, \ldots, n \tag{1.3}$$

and *define*

$$x_{1j}c_1 + x_{2j}c_2 + \cdots + x_{mj}c_m = z_j \qquad j = 1, \ldots, n \tag{1.4}$$

where the c_i are the cost coefficients corresponding to the $\mathbf{P}_i$.

Theorem 1. *If, for any fixed j, the condition $z_j - c_j > 0$ holds, then a set of feasible solutions can be constructed such that $z < z_0$ for any member of the set, where the lower bound of z is either finite or infinite. (z is the value of the objective function for a particular member of the set of feasible solutions.)*

Case I. If the lower bound is finite, a new feasible solution consisting of exactly m positive variables can be constructed whose value of the objective function is less than the value for the preceding solution.

Case II. If the lower bound is infinite, a new feasible solution consisting of exactly $m + 1$ positive variables can be constructed whose value of the objective function can be made arbitrarily small.

The following analysis applies to the proof of both cases:

Multiplying (1.3) by some number θ and subtracting from (1.1), and similarly multiplying (1.4) by the same θ and subtracting from (1.2), for $j = 1, 2, \ldots, n$ we get

$$(x_{10} - \theta x_{1j})\mathbf{P}_1 + (x_{20} - \theta x_{2j})\mathbf{P}_2 + \cdots + (x_{m0} - \theta x_{mj})\mathbf{P}_m + \theta\mathbf{P}_j = \mathbf{P}_0 \tag{1.5}$$

$$(x_{10} - \theta x_{1j})c_1 + (x_{20} - \theta x_{2j})c_2 + \cdots + (x_{m0} - \theta x_{mj})c_m + \theta c_j = z_0 - \theta(z_j - c_j) \tag{1.6}$$

where θc_j has been added to both sides of (1.6). If all the coefficients of the vectors $\mathbf{P}_1, \mathbf{P}_2, \ldots, \mathbf{P}_m, \mathbf{P}_j$ in (1.5) are nonnegative, then we have determined a new feasible solution whose value of the objective function is, by (1.6), $z = z_0 - \theta(z_j - c_j)$. Since the variables $x_{10}, x_{20}, \ldots, x_{m0}$ in (1.5) are all positive, it is clear, from our discussion in Sec. 3 of Chap. 3, that there is a value of $\theta > 0$ (either finite or infinite) for which the coefficients of the vectors in (1.5) remain positive. From the assumption that, for a fixed j, $z_j - c_j > 0$, we have

$$z = z_0 - \theta(z_j - c_j) < z_0$$

for $\theta > 0$. We see that in either event a new feasible solution can be obtained whose corresponding value of the objective function is less than the value for the preceding solution.

The proof of Case I follows:

If, for the fixed j, at least one $x_{ij} > 0$ in (1.3) for $i = 1, 2, \ldots, m$, the largest value of θ for which all coefficients of (1.5) remain nonnegative is given by

$$\theta_0 = \min_i \frac{x_{i0}}{x_{ij}} > 0 \tag{1.7}$$

for $x_{ij} > 0$ (see Sec. 3 of Chap. 3). Since we assumed that the problem is nondegenerate, i.e., that all basic feasible solutions have m positive elements, the minimum in (1.7) will be obtained for a unique i. If θ_0 is substituted for θ in (1.5) and (1.6), the coefficient corresponding to this unique i will vanish. We have then constructed a new basic feasible solution consisting of $\mathbf{P}_j$ and $m - 1$ vectors of the original basis. This new basis can be used as the previous one. If a new $z_j - c_j > 0$ and a corresponding $x_{ij} > 0$, another solution can be obtained which has a smaller value of the objective function. This process will continue either until all $z_j - c_j \leq 0$, or until, for some $z_j - c_j > 0$, all $x_{ij} \leq 0$. If all $z_j - c_j \leq 0$, the process terminates.

For Case II we have:

If at any stage we have, for some j, $z_j - c_j > 0$ and all $x_{ij} \leq 0$, then there is no upper bound to θ and the objective function has a lower bound of $-\infty$. We see for this case that, for any $\theta > 0$, all the coefficients of (1.5) are positive. We then have a feasible solution consisting of $m + 1$ positive elements. Hence, by taking θ large enough, the corresponding value of the objective function given by the right-hand side of (1.6) can be made arbitrarily small.

Theorem 2. *If for any basic feasible solution* $\mathbf{X} = (x_{10}, x_{20}, \ldots, x_{m0})$ *the conditions* $z_j - c_j \leq 0$ *hold for all* $j = 1, 2, \ldots, n$, *then* (1.1) *and* (1.2) *constitute a minimum feasible solution.*[1]

[1] This optimality criterion is sometimes varied. For a minimization problem we could have computed, instead of the $z_j - c_j$, the numbers $c_j - z_j$ and selected as the vector to be introduced into the basis the one corresponding to the min $(c_j - z_j)$. An optimum has been reached when all $c_j - z_j \geq 0$. If the problem was originally to be maximized, we could use the following criterion instead of changing to a minimization problem: Compute the $z_j - c_j$ and select a new vector corresponding to min $(z_j - c_j)$; an optimum solution has been found when all $z_j - c_j \geq 0$. Or compute the $c_j - z_j$ elements with the new vector corresponding to max $(c_j - z_j)$, the procedure stopping when all $c_j - z_j \leq 0$. It is much more efficient, however, especially in developing a computational procedure for an electronic computer, to select one criterion to be used in solving all problems. See page 71 for the discussion on the criterion used to select a vector to be introduced into the basis.

Proof. Let

$$y_{10}\mathbf{P}_1 + y_{20}\mathbf{P}_2 + \cdots + y_{n0}\mathbf{P}_n = \mathbf{P}_0 \tag{1.8}$$

and

$$y_{10}c_1 + y_{20}c_2 + \cdots + y_{n0}c_n = z^* \tag{1.9}$$

be any other feasible solution with z^* the corresponding value of the objective function. We shall show that $z_0 \leq z^*$. (Note that the nondegeneracy assumption is not required for this theorem.)

By hypothesis, $z_j - c_j \leq 0$ for all j, so that replacing c_j by z_j in (1.9) yields

$$y_{10}z_1 + y_{20}z_2 + \cdots + y_{n0}z_n \leq z^* \tag{1.10}$$

For each j we substitute the expression for $\mathbf{P}_j$ given by (1.3) into (1.8), to obtain

$$y_{10}\left(\sum_{i=1}^{m} x_{i1}\mathbf{P}_i\right) + y_{20}\left(\sum_{i=1}^{m} x_{i2}\mathbf{P}_i\right) + \cdots + y_{n0}\left(\sum_{i=1}^{m} x_{in}\mathbf{P}_i\right) = \mathbf{P}_0$$

or, by regrouping terms,

$$\left(\sum_{j=1}^{n} y_{j0}x_{1j}\right)\mathbf{P}_1 + \left(\sum_{j=1}^{n} y_{j0}x_{2j}\right)\mathbf{P}_2 + \cdots + \left(\sum_{j=1}^{n} y_{j0}x_{mj}\right)\mathbf{P}_m = \mathbf{P}_0 \tag{1.11}$$

Similarly, for each j we substitute the expression for z_j given by (1.4) into (1.10) to obtain

$$\left(\sum_{j=1}^{n} y_{j0}x_{1j}\right)c_1 + \left(\sum_{j=1}^{n} y_{j0}x_{2j}\right)c_2 + \cdots + \left(\sum_{j=1}^{n} y_{j0}x_{mj}\right)c_m \leq z^* \tag{1.12}$$

Since the set of vectors $\mathbf{P}_1, \mathbf{P}_2, \ldots, \mathbf{P}_m$ is linearly independent, the coefficients of the corresponding vectors in (1.1) and (1.11) must be equal,[1] and hence (1.12) becomes

$$x_{10}c_1 + x_{20}c_2 + \cdots + x_{m0}c_m \leq z^*$$

or, by (1.2), $z_0 \leq z^*$.

The results of Theorems 1 and 2 enable us to start with a basic feasible solution and generate a set of new basic feasible solutions that converge to the minimum solution or determine that a finite solution does not exist.

The nondegeneracy assumption was invoked to ensure the convergence to the minimum solution. If we did not make this assumption, it

[1] See footnote of Theorem 3, Chap. 3, for a proof of the equality of the coefficients of (1.1) and (1.11).

would be possible to have at least one of the m x_{i0} of the given solution equal to zero. If this were the case, then θ_0 could equal zero and the value of the objective function for the new solution would then equal the value of the old solution. This lack of improvement in the solution could continue for a number of successive steps. For this situation, the procedure could conceivably repeat a basis and hence keep returning to this basis. The simplex procedure is then said to have *cycled*, and the computational routine for determining the minimum solution breaks down. During actual computation, the phenomenon of degeneracy is reflected by basic solutions with less than m positive x_{i0} and/or by more than one i yielding $\theta_0 = \min_i (x_{i0}/x_{ij})$ for $x_{ij} > 0$. If the i is not unique, some of the x_{i0} in the new solution will equal zero.

Dantzig, Orden, and Wolfe [32] and Charnes, Cooper, and Henderson [12] have resolved degeneracy from both the theoretical and computational points of view. Computational experience, however, does not warrant incorporating their "degeneracy techniques" into the standard simplex procedure. Out of the many linear-programming problems considered by investigators in the field, only three have been known to cycle. These were artificially constructed by Hoffman [58] and Beale [4] to demonstrate that cycling could occur. What is normally done in a computation is to treat degenerate solutions as nothing unusual and to compute with $\theta_0 = 0$ whenever the procedure yields such a value for θ. If ties occur when determining θ_0, the usual rule is to select $\theta_0 = \min_i (x_{i0}/x_{ij})$ for the smallest index i. Degeneracy techniques and a cycling problem of Beale are discussed in Chap. 7.

2. COMPUTATIONAL PROCEDURE

In this section we assume either that (1) we have selected m linearly independent vectors that yield a feasible solution and have expressed all other vectors in terms of this basis or that (2) our problem matrix contains m vectors that can be explicitly arranged to form a unit matrix of order m.

For the first case let the m linearly independent vectors be $\mathbf{P}_1$, $\mathbf{P}_2$, . . . , $\mathbf{P}_m$ and denote the $m \times m$ matrix $(\mathbf{P}_1\mathbf{P}_2 \cdots \mathbf{P}_m)$ by $\mathbf{B}$. The matrix $\mathbf{B}$ is termed an *admissible basis*. To compute the corresponding solution vector $\mathbf{X}$ and the representation of the other vectors in terms of the basis, we must first compute $\mathbf{B}^{-1}$. Since

$$\mathbf{B}\mathbf{X}_0 = \mathbf{P}_0$$

we have

$$\mathbf{X}_0 = \mathbf{B}^{-1}\mathbf{P}_0$$

and

$$\mathbf{X}_j = \mathbf{B}^{-1}\mathbf{P}_j$$

where

$$\mathbf{X}_0 = (x_{10},x_{20}, \ . \ . \ . \ ,x_{m0}) \qquad x_{i0} \geq 0$$

and

$$\mathbf{X}_j = (x_{1j},x_{2j}, \ . \ . \ . \ ,x_{mj})$$

are both column vectors.

To start the simplex process, we group the vectors of the problem matrix as follows:

$$(\mathbf{P}_0 \mid \mathbf{P}_1\mathbf{P}_2 \cdots \mathbf{P}_m \mid \mathbf{P}_{m+1} \cdots \mathbf{P}_n)$$

or

$$(\mathbf{P}_0 \mid \mathbf{B} \mid \mathbf{P}_{m+1} \cdots \mathbf{P}_n) \tag{2.1}$$

By multiplying the elements in the paritioned matrix (2.1) by $\mathbf{B}^{-1}$, we obtain

$$(\mathbf{X}_0 \mid \mathbf{I}_m \mid \mathbf{X}_{m+1} \cdots \mathbf{X}_n)$$

Since we know the c_j, we next compute the $z_j - c_j$ and determine whether, for any j, the corresponding $z_j - c_j > 0$. If so, we carry out the computational procedure described in Theorem 1. If not, we have found the minimum feasible solution. In general, since we have no assurance that an arbitrary set of m vectors from the given set will be linearly independent, let alone yield a feasible solution, our assumption in the first case is not the usual situation encountered in practice. (There have been computational procedures proposed that do start with an arbitrary, but educated, selection of m vectors. Some of these procedures are described in Sec. 1 of Chap. 9.) The condition of the second case is quite common, and we shall describe it in great detail.

In the second case we assume that the given set of n vectors $\mathbf{P}_1, \ . \ . \ . \ ,$ $\mathbf{P}_n$ contains m unit vectors that can be grouped together to form an $m \times m$ unit matrix. It will be shown in Sec. 3 below that this assumption is not restrictive. We let these vectors be $\mathbf{P}_1, \mathbf{P}_2, \ . \ . \ . \ , \mathbf{P}_m$ and take as our admissible basis

$$\mathbf{B} = (\mathbf{P}_1\mathbf{P}_2 \cdots \mathbf{P}_m) = \mathbf{I}_m$$

Since $\mathbf{B}^{-1} = \mathbf{I}_m$ and since all the elements of $\mathbf{P}_0$ were originally assumed to be nonnegative, we have the initial extreme-point solution

$$\mathbf{X}_0 = \mathbf{P}_0$$

and

$$\mathbf{X}_j = \mathbf{P}_j$$

where

$$\mathbf{X}_0 = (x_{10},x_{20}, \ . \ . \ . \ ,x_{m0}) \qquad x_{i0} \geq 0$$

and

$$\mathbf{X}_j = (x_{1j},x_{2j}, \ . \ . \ . \ ,x_{mj})$$

To start the simplex procedure, we arrange the problem matrix as shown in Tableau 4.1. (In practice, one does not have to group the unit

Tableau 4.1. Initial step of computational procedure

i	Basis	$\mathbf{c}$		c_1	c_2	·	c_l	·	c_m	c_{m+1}	·	c_j	·	c_k	·	c_n
			$\mathbf{P}_0$	$\mathbf{P}_1$	$\mathbf{P}_2$	·	$\mathbf{P}_l$	·	$\mathbf{P}_m$	$\mathbf{P}_{m+1}$	·	$\mathbf{P}_j$	·	$\mathbf{P}_k$	·	$\mathbf{P}_n$
1	$\mathbf{P}_1$	c_1	x_{10}	1	0	·	0	·	0	$x_{1,\ m+1}$	·	x_{1j}	·	x_{1k}	·	x_{1n}
2	$\mathbf{P}_2$	c_2	x_{20}	0	1	·	0	·	0	$x_{2,\ m+1}$	·	x_{2j}	·	x_{2k}	·	x_{2n}
·	·	·	·	·	·	·	·	·	·	·	·	·	·	·	·	·
·	·	·	·	·	·	·	·	·	·	·	·	·	·	·	·	·
l	$\mathbf{P}_l$	c_l	x_{l0}	0	0	·	1	·	0	$x_{l,\ m+1}$	·	x_{lj}	·	x_{lk}	·	x_{ln}
·	·	·	·	·	·	·	·	·	·	·	·	·	·	·	·	·
·	·	·	·	·	·	·	·	·	·	·	·	·	·	·	·	·
m	$\mathbf{P}_m$	c_m	x_{m0}	0	0	·	0	·	1	$x_{m,\ m+1}$	·	x_{mj}	·	x_{mk}	·	x_{mn}
$m+1$			z_0	0	0	·	0	·	0	$z_{m+1} - c_{m+1}$	·	$z_j - c_j$	·	$z_k - c_k$	·	$z_n - c_n$

vectors together, but we shall do this for illustrative purposes.) From the original equations of the problem given by $\mathbf{AX} = \mathbf{b}$, we have let $x_{i0} = b_i$ and $x_{ij} = a_{ij}$. z_j for $j = 0, 1, \ . \ . \ . \ , n$ is obtained by taking the inner product of the jth vector with the column vector labeled $\mathbf{c}$, that is,

$$z_0 = \sum_{i=1}^{m} c_i x_{i0}$$

$$z_j = \sum_{i=1}^{m} c_i x_{ij} \qquad j = 1, 2, \ . \ . \ . \ , n$$

The elements z_0 and $z_j - c_j$ are entered in the $(m + 1)$st row of their respective columns. The $z_j - c_j$ for those vectors in the basis will always equal zero. If all the numbers $z_j - c_j \leq 0$ for $j = 1, 2, \ . \ . \ . \ , n$, then the solution $\mathbf{X}_0 = (x_{10},x_{20}, \ . \ . \ . \ ,x_{m0}) = (b_1,b_2, \ . \ . \ . \ ,b_m)$ is a minimum feasible solution, and the corresponding value of the objective function is z_0. We shall assume at least one $z_j - c_j > 0$ and compute a new feasible solution whose basis contains $m - 1$ vectors of the original basis $\mathbf{P}_1,\mathbf{P}_2, \ . \ . \ . \ , \mathbf{P}_m$. In searching for a new vector to enter the basis, we can theoretically select any vector whose corresponding $z_j - c_j > 0$. As Dantzig [17] points out, the number of iterations necessary to obtain a minimum solution can, in general, be greatly reduced by not selecting at random any vector $\mathbf{P}_j$ with its $z_j - c_j > 0$, but by selecting the one which gives the greatest immediate decrease in the value of the objective function. The vector $\mathbf{P}_j$ should then be the one which corresponds to the

$$\max_j \theta_0(z_j - c_j) \tag{2.1a}$$

where, for each j, θ_0 is given by (1.7). If there are a number of j for which $z_j - c_j > 0$, the above rule is rather complicated to apply. A much simpler criterion for selecting the vector to be introduced is to select the one which corresponds to the

$$\max_j (z_j - c_j) \tag{2.1b}$$

If there are ties, the rule is to select the vector with the lowest (or the highest) index j. This criterion is the standard one employed in most computation centers and has proved to be an excellent one. When using this rule, approximately m changes of basis are required to go from the first feasible solution to the minimum solution. We shall employ the second criterion, (2.1b).[1]

In our example, let

$$\max_j (z_j - c_j) = z_k - c_k > 0$$

The vector $\mathbf{P}_k$ is to be introduced into the basis. We next compute

$$\theta_0 = \min_i \frac{x_{i0}}{x_{ik}}$$

for $x_{ik} > 0$.† If all $x_{ik} \leq 0$, we can then find a feasible solution whose value of the objective function can be made arbitrarily small (Theorem 1, Case II). Our computation is then complete. Assume, however, some $x_{ik} > 0$ and

$$\theta_0 = \min_i \frac{x_{i0}}{x_{ik}} = \frac{x_{l0}}{x_{lk}}$$

Vector $\mathbf{P}_l$ will be the one eliminated from the basis. Our new feasible solution will have a new basis consisting of $\mathbf{P}_1, \ldots, \mathbf{P}_{l-1}, \mathbf{P}_{l+1}, \ldots, \mathbf{P}_m, \mathbf{P}_k$. We next wish to compute the new solution explicitly and to express each vector not in the basis in terms of the new basis.

Since our initial basis is $(\mathbf{P}_1\mathbf{P}_2 \cdots \mathbf{P}_m) = \mathbf{I}_m$, we can readily express all the vectors $\mathbf{P}_j$ in terms of this basis. We then have

[1] See the work of Wolfe and Cutler [105g] and Quandt and Kuhn [85c] which describe and evaluate alternate selection rules.

† For the general situation the index i ranges over those i whose corresponding variables are in the basic solution.

$$\mathbf{P}_0 = x_{10}\mathbf{P}_1 + \cdots + x_{l0}\mathbf{P}_l + \cdots + x_{m0}\mathbf{P}_m \tag{2.2}$$

$$\mathbf{P}_k = x_{1k}\mathbf{P}_1 + \cdots + x_{lk}\mathbf{P}_l + \cdots + x_{mk}\mathbf{P}_m \tag{2.3}$$

and

$$\mathbf{P}_j = x_{1j}\mathbf{P}_1 + \cdots + x_{lj}\mathbf{P}_l + \cdots + x_{mj}\mathbf{P}_m \tag{2.4}$$

From (2.3)

$$\mathbf{P}_l = \frac{1}{x_{lk}}(\mathbf{P}_k - x_{1k}\mathbf{P}_1 - \cdots - x_{mk}\mathbf{P}_m) \tag{2.5}$$

Substituting the above expression for $\mathbf{P}_l$ into (2.2), we obtain

$$\mathbf{P}_0 = x_{10}\mathbf{P}_1 + \cdots + x_{l0}\left[\frac{1}{x_{lk}}(\mathbf{P}_k - x_{1k}\mathbf{P}_1 - \cdots - x_{mk}\mathbf{P}_m)\right] + \cdots + x_{m0}\mathbf{P}_m$$

or

$$\mathbf{P}_0 = \left(x_{10} - \frac{x_{l0}}{x_{lk}}x_{1k}\right)\mathbf{P}_1 + \cdots + \frac{x_{l0}}{x_{lk}}\mathbf{P}_k + \cdots + \left(x_{m0} - \frac{x_{l0}}{x_{lk}}x_{mk}\right)\mathbf{P}_m\dagger$$

The new feasible solution $\mathbf{X}_0' = (x_{10}', \ldots, x_{k0}', \ldots, x_{m0}')$, $x_{i0}' \geq 0$, is given by

$$\mathbf{P}_0 = x_{10}'\mathbf{P}_1 + \cdots + x_{k0}'\mathbf{P}_k + \cdots + x_{m0}'\mathbf{P}_m$$

where

$$\begin{aligned} x_{i0}' &= x_{i0} - \frac{x_{l0}}{x_{lk}}x_{ik} \qquad \text{for } i = 1, 2, \ldots, l-1, l+1, \ldots, m \\ x_{k0}' &= \frac{x_{l0}}{x_{lk}} \end{aligned} \tag{2.6}$$

Similarly, by substituting (2.5) into (2.4), we can obtain the expression for each $\mathbf{P}_j$ not in the new basis in terms of this basis. This yields

$$\mathbf{P}_j = x_{1j}'\mathbf{P}_1 + \cdots + x_{kj}'\mathbf{P}_k + \cdots + x_{mj}'\mathbf{P}_m$$

where

$$\begin{aligned} x_{ij}' &= x_{ij} - \frac{x_{lj}}{x_{lk}}x_{ik} \qquad \text{for } i \neq l \\ x_{kj}' &= \frac{x_{lj}}{x_{lk}} \end{aligned} \tag{2.7}$$

Since

$$z_j' - c_j = x_{1j}'c_1 + \cdots + x_{kj}'c_k + \cdots + x_{mj}'c_m - c_j$$

† This expression is equivalent to Eq. (1.5) with $j = k$ and $\theta = x_{l0}/x_{lk} = \theta_0$.

one can readily verify by substituting the values (2.7) for x'_{ij} that

$$z'_j - c_j = z_j - c_j - \frac{x_{lj}}{x_{lk}}(z_k - c_k) \tag{2.7a}$$

and by substituting the values (2.6) for x'_{i0} into

$$z'_0 = c_1 x'_{10} + \cdots + c_k x'_{k0} + \cdots + c_m x'_{m0}$$

that

$$z'_0 = z_0 - \frac{x_{l0}}{x_{lk}}(z_k - c_k) \tag{2.7b}$$

We then note that, in order to obtain the new solution $\mathbf{X}'_0$, the new vectors $\mathbf{X}'_j$, and the corresponding $z'_j - c_j$, every element in Tableau 4.1 for rows $i = 1, \ldots, m + 1$ and columns $j = 0, 1, \ldots, n$ is transformed by the formulas

$$x'_{ij} = x_{ij} - \frac{x_{lj}}{x_{lk}} x_{ik} \qquad \text{for } i \neq l$$

$$x'_{lj} = \frac{x_{lj}}{x_{lk}} \tag{2.8}$$

where

$$z'_0 = x'_{m+1,\,0} \qquad z'_j - c_j = x'_{m+1,\,j}$$

Here we are letting the general formulas (2.8) apply to all elements of the computational tableau including the $\mathbf{P}_0$ column and the $(m + 1)$st row. The transformation defined by (2.8) is equivalent to the complete elimination formulas when the pivot element is x_{lk}.†

Once an initial computational tableau has been constructed, the simplex procedure calls for the successive application (i.e., an iteration) of:

1. The testing of the $z_j - c_j$ elements to determine whether a minimum solution has been found, i.e., whether $z_j - c_j \leq 0$ for all j.

† The reader should note that we are now letting the indices i and l refer to the corresponding rows in the computational tableau. We keep track of which variables are in the current basic solution by means of the column labeled Basis. Hence, the general elimination formulas given by Eqs. (2.8) are applied to each element x_{ij}, where i refers to the row and j to the column of the tableau. Element x_{lk} is the pivot element of the elimination transformation, with row l being the pivot row and column k being the pivot column.

In applying Eqs. (2.8) to the tableau we can use two schemes. For manual computation it is best to compute first the new elements of the pivot row and add suitable multiples of this row to the other rows of the tableau in order to carry out the elimination of x_{k0} from all rows except the lth. For automatic computation by an electronic computer, it is best to compute first the ratios x_{ik}/x_{lk} corresponding to the pivot column. The elimination transformation is then performed by applying Eqs. (2.8) to each column. Here it is probably best to carry the number $z_j - c_j$ as the first element of each column.

2. The selection of the vector to be introduced into the basis if some $z_j - c_j > 0$, i.e., selection of the vector with maximum $z_j - c_j$.
3. The selection of the vector to be eliminated from the basis to ensure feasibility of the new solution, i.e., the vector with $\min_i (x_{i0}/x_{ik})$ for those $x_{ik} > 0$, where k corresponds to the vector selected in Step 2. If all $x_{ik} \leq 0$, then the solution is unbounded.
4. The transformation of the tableau by the complete elimination procedure to obtain the new solution and associated elements.

Each such iteration produces a new basic feasible solution, and by the discussions of Theorems 1 and 2 we shall eventually obtain a minimum solution or determine an unbounded solution.

An application of the simplex procedure to the initial tableau yields the transformed values of Tableau 4.2.

Tableau 4.2. Second step of computational procedure

i	Basis	c		c_1	c_2	·	c_l	·	c_m	c_{m+1}	·	c_j	·	c_k	·	c_n
			$\mathbf{P}_0$	$\mathbf{P}_1$	$\mathbf{P}_2$	·	$\mathbf{P}_l$	·	$\mathbf{P}_m$	$\mathbf{P}_{m+1}$	·	$\mathbf{P}_j$	·	$\mathbf{P}_k$	·	$\mathbf{P}_n$
1	$\mathbf{P}_1$	c_1	x'_{10}	1	0	·	x'_{1l}	·	0	$x'_{1,\,m+1}$	·	x'_{1j}	·	0	·	x'_{1n}
2	$\mathbf{P}_2$	c_2	x'_{20}	0	1	·	x'_{2l}	·	0	$x'_{2,\,m+1}$	·	x'_{2j}	·	0	·	x'_{2n}
·	·	·	·	·	·	·	·	·	·	·	·	·	·	·	·	·
·	·	·	·	·	·	·	·	·	·	·	·	·	·	·	·	·
l	$\mathbf{P}_k$	c_k	x'_{k0}	0	0	·	x'_{ll}	·	0	$x'_{l,\,m+1}$	·	x'_{lj}	·	1	·	x'_{ln}
·	·	·	· ·	·	·	·	·	·	·	·	·	·	·	·	·	·
·	·	·	·	·	·	·	·	·	·	·	·	·	·	·	·	·
m	$\mathbf{P}_m$	c_m	x'_{m0}	0	0	·	x'_{ml}	·	1	$x'_{m,\,m+1}$	·	x'_{mj}	·	0	·	x'_{mn}
$m+1$			z'_0	0	0	·	$z'_l - c_l$	·	0	$z'_{m+1} - c_{m+1}$	·	$z_j' - c_j$	·	0	·	$z'_n - c_n$

Example. As an example, let us solve the following linear-programming problem by means of the simplex procedure: Minimize

$$x_2 - 3x_3 \qquad + 2x_5$$

subject to

$$x_j \geq 0$$

and

$$\begin{aligned} x_1 + 3x_2 - x_3 + 2x_5 &= 7 \\ -2x_2 + 4x_3 + x_4 &= 12 \\ -4x_2 + 3x_3 + 8x_5 + x_6 &= 10 \end{aligned}$$

Tableau 4.3

Initial Step

i	Basis	c		0	1	-3	0	2	0
			$\mathbf{P}_0$	$\mathbf{P}_1$	$\mathbf{P}_2$	$\mathbf{P}_3$	$\mathbf{P}_4$	$\mathbf{P}_5$	$\mathbf{P}_6$
1	$\mathbf{P}_1$	0	7	1	3	-1	0	2	0
2	$\mathbf{P}_4$	0	12	0	-2	$\textcircled{4}$	1	0	0
3	$\mathbf{P}_6$	0	10	0	-4	3	0	8	1
4			0	0	-1	3	0	-2	0

Second Step

			$\mathbf{P}_0$	$\mathbf{P}_1$	$\mathbf{P}_2$	$\mathbf{P}_3$	$\mathbf{P}_4$	$\mathbf{P}_5$	$\mathbf{P}_6$
1	$\mathbf{P}_1$	0	10	1	$\textcircled{5/2}$	0	$1/4$	2	0
2	$\mathbf{P}_3$	-3	3	0	$-1/2$	1	$1/4$	0	0
3	$\mathbf{P}_6$	0	1	0	$-5/2$	0	$-3/4$	8	1
4			-9	0	$1/2$	0	$-3/4$	-2	0

Third Step

			$\mathbf{P}_0$	$\mathbf{P}_1$	$\mathbf{P}_2$	$\mathbf{P}_3$	$\mathbf{P}_4$	$\mathbf{P}_5$	$\mathbf{P}_6$
1	$\mathbf{P}_2$	1	4	$2/5$	1	0	$1/10$	$4/5$	0
2	$\mathbf{P}_3$	-3	5	$1/5$	0	1	$3/10$	$2/5$	0
3	$\mathbf{P}_6$	0	11	1	0	0	$-1/2$	10	1
4			-11	$-1/5$	0	0	$-4/5$	$-12/5$	0

Our initial basis (see Tableau 4.3) consists of $\mathbf{P}_1$, $\mathbf{P}_4$, $\mathbf{P}_6$, and the corresponding solution is $\mathbf{X}_0 = (x_1,x_4,x_6) = (7,12,10)$. Since

$$c_1 = c_4 = c_6 = 0$$

the corresponding value of the objective function, z_0, equals zero. $\mathbf{P}_3$ is selected to go into the basis, since

$$\max_j (z_j - c_j) = z_3 - c_3 = 3 > 0$$

θ_0 is the minimum of x_{i0}/x_{i3} for $x_{i3} > 0$, that is,

$$\min (12/4, 10/3) = 12/4 = \theta_0$$

and hence $\mathbf{P}_4$ is eliminated. We transform the tableau (see Second Step, Tableau 4.3) and obtain a new solution

$$\mathbf{X}_0' = (x_1,x_3,x_6) = (10,3,1)$$

and the value of the objective function is -9. In the second step, since

$$\max_j \,(z_j' - c_j) = z_2' - c_2 = \tfrac{1}{2} > 0$$

and

$$\theta_0 = \frac{10}{5/2}$$

$\mathbf{P}_2$ is introduced into the basis and $\mathbf{P}_1$ is eliminated. We transform the second-step values of Tableau 4.3 and obtain the third solution

$$\mathbf{X}_0'' = (x_2,x_3,x_6) = (4,5,11)$$

with a value of the objective function equal to -11. Since

$$\max \,(z_j'' - c_j) = 0$$

this solution is a minimum feasible solution. As a check on each complete elimination transformation, one should explicitly compute the individual z_0 and $z_j - c_j$ and compare them with the transformed values of z_0 and $z_j - c_j$. If for the minimum feasible solution some $z_j - c_j' = 0$ for a vector $\mathbf{P}_j$ not in the final basis, then this vector can be introduced into the basis without changing the final value of the objective function. The resulting solution will also be a minimum feasible solution, and hence we have determined multiple minimum solutions. Any convex combination of these minimum solutions will also be a minimum solution.

The determinants of the bases used in the simplex computation can be readily obtained as a by-product of the computation. Let the basis of the pth iteration be

$$\mathbf{B}_p = (\mathbf{P}_1\mathbf{P}_2 \cdots \mathbf{P}_l \cdots \mathbf{P}_m)$$

and the $(p + 1)$st basis be

$$\mathbf{B}_{p+1} = (\mathbf{P}_1\mathbf{P}_2 \cdots \mathbf{P}_k \cdots \mathbf{P}_m)$$

where

$$\mathbf{P}_k = \sum_{i=1}^{m} x_{ik}\mathbf{P}_i$$

We see that

$$\mathbf{B}_{p+1} = \mathbf{B}_p\mathbf{C}_p = \mathbf{B}_p \begin{pmatrix} 1 & \cdots & x_{1k} & \cdots & 0 \\ \cdot & \cdot & \cdot & \cdot & \cdot \\ 0 & \cdots & x_{lk} & \cdots & 0 \\ \cdot & \cdot & \cdot & \cdot & \cdot \\ 0 & \cdots & x_{mk} & \cdots & 1 \end{pmatrix}$$

where $\mathbf{C}_p$ differs from the identity matrix in the lth column. We then have

$$|\mathbf{B}_{p+1}| = |\mathbf{B}_p| \cdot |\mathbf{C}_p| = x_{lk}|\mathbf{B}_p|$$

In the simplex process for $p = 0$, $\mathbf{B}_0 = \mathbf{I}$; hence the determinant of the pth basis of the simplex method is given by the product

$$|\mathbf{B}_p| = \prod_{q=0}^{p} x_{lk}{}^{(q)}$$

where $x_{lk}{}^{(0)} = 1$, and $x_{lk}{}^{(q)}$ are the pivot elements of the successive bases. We note that $|\mathbf{B}_p| > 0$.

3. THE ARTIFICIAL-BASIS TECHNIQUE

Up to this point, we have always assumed that the given linear-programming problem was feasible and contained a unit matrix that could be used for the initial basis. Although a correct formulation of a problem will usually guarantee that the problem will be feasible, many problems do not contain a unit matrix. For such problems, the method of the *artificial basis* (see Orden [84]) is a satisfactory way to start the simplex process. This procedure also determines whether or not the problem has any feasible solutions.

The general linear-programming problem is to minimize

$$c_1x_1 + \cdots + c_nx_n$$

subject to

$$\begin{aligned} a_{11}x_1 + \cdots + a_{1n}x_n &= b_1 \\ a_{21}x_1 + \cdots + a_{2n}x_n &= b_2 \\ \cdots\cdots\cdots&\cdots\cdots \\ a_{m1}x_1 + \cdots + a_{mn}x_n &= b_m \end{aligned}$$

and

$$x_j \geq 0$$

For the method of the artificial basis we augment the above system as follows: Minimize

$$c_1x_1 + \cdots + c_nx_n + wx_{n+1} + wx_{n+2} + \cdots + wx_{n+m}$$

subject to

$$\begin{aligned} a_{11}x_1 + \cdots + a_{1n}x_n + x_{n+1} \qquad\qquad\qquad\qquad &= b_1 \\ a_{21}x_1 + \cdots + a_{2n}x_n \qquad\quad + x_{n+2} \qquad\qquad &= b_2 \\ \cdots\cdots\cdots\cdots\cdots\cdots\cdots\cdots&\cdots\cdots \\ a_{m1}x_1 + \cdots + a_{mn}x_n \qquad\qquad\qquad + x_{n+m} &= b_m \end{aligned}$$

and to $x_j \geq 0$ for $j = 1, \ldots, n, n+1, \ldots, n+m$. The quantity w is taken to be an unspecified large positive number. The vectors $\mathbf{P}_{n+1}, \mathbf{P}_{n+2}, \ldots, \mathbf{P}_{n+m}$ form a basis—an artificial basis—for the augmented system. If there is at least one feasible solution to the original problem, then this solution is also a feasible one for the augmented system. The simplex procedure will then ensure our obtaining the minimum solution, in which it is impossible for one of the artificial variables, x_{n+i}, to appear with a positive value. If the original problem is not feasible, then the minimum feasible solution to the augmented problem will contain at least one $x_{n+i} > 0$. As we shall show below, it is not necessary to assign a specific value to w.

For the augmented problem, the first feasible solution is

$$\mathbf{X}_0 = (x_{n+1,0}, x_{n+2,0}, \ldots, x_{n+m,0}) = (b_1, b_2, \ldots, b_m)$$

with a corresponding value of the objective function $z_0 = w \sum_{i=1}^{m} b_i$. Since the basis is a unit matrix, $\mathbf{X}_j = (x_{1j}, x_{2j}, \ldots, x_{mj}) = (a_{1j}, a_{2j}, \ldots, a_{mj})$, and $z_j = w \sum_{i=1}^{m} x_{ij}$. As long as there are artificial vectors in the basis, each $z_j - c_j$ will be a linear function of w. For the first solution,

$$z_j - c_j = w \sum_{i=1}^{m} x_{ij} - c_j$$

Each $z_j - c_j$ will then have a w coefficient and a non-w coefficient which are independent of each other. We next set up the associated computational procedure as Tableau 4.4. For each j, the non-w component and the w component of $z_j - c_j$ have been placed in the $(m+1)$st and $(m+2)$nd rows, respectively, of that column.

Tableau 4.4. First step of artificial-basis computational procedure

i	Basis	c		c_1	c_2	·	c_k	·	c_n	w	·	w	·	w
			$\mathbf{P}_0$	$\mathbf{P}_1$	$\mathbf{P}_2$	·	$\mathbf{P}_k$	·	$\mathbf{P}_n$	$\mathbf{P}_{n+1}$	·	$\mathbf{P}_{n+l}$	·	$\mathbf{P}_{n+m}$
1	$\mathbf{P}_{n+1}$	w	$x_{n+1,0}$	x_{11}	x_{12}	·	x_{1k}	·	x_{1n}	1	·	0	·	0
2	$\mathbf{P}_{n+2}$	w	$x_{n+2,0}$	x_{21}	x_{22}	·	x_{2k}	·	x_{2n}	0	·	0	·	0
·	·	·	·	·	·	·	·	·	·	·	·	·	·	·
·	·	·	·	·	·	·	·	·	·	·	·	·	·	·
l	$\mathbf{P}_{n+l}$	w	$x_{n+l,0}$	x_{l1}	x_{l2}	·	x_{lk}	·	x_{ln}	0	·	1	·	0
·	·	·	·	·	·	·	·	·	·	·	·	·	·	·
·	·	·	·	·	·	·	·	·	·	·	·	·	·	·
m	$\mathbf{P}_{n+m}$	w	$x_{n+m,0}$	x_{m1}	x_{m2}	·	x_{mk}	·	x_{mn}	0	·	0	·	1
$m+1$			0	$-c_1$	$-c_2$	·	$-c_k$	·	$-c_n$	0	·	0	·	0
$m+2$			Σx_{n+i}	Σx_{i1}	Σx_{i2}	·	Σx_{ik}	·	Σx_{in}	0	·	0	·	0

We treat this tableau exactly like the original simplex tableau (Tableau 4.1), except that the vector introduced into the basis is associated with the largest positive element in the $(m + 2)$nd row. For the first iteration, the vector corresponding to $\max_j \sum_{i=1}^{m} x_{ij}$ is introduced into the basis. The elements in the $(m + 2)$nd row are also transformed by the usual elimination procedure. Once an artificial vector is eliminated from the basis, it is never selected to reenter the basis. Hence we do not have to transform the last m columns of the tableau. If we are interested in the inverse of the final basis, then these last m vectors should be transformed as usual.[1] It should be noted that, even if there are artificial vectors in the basis, the iteration may not eliminate one of them. When using the full artificial basis, approximately $2m$ iterations are required to reach the minimum feasible solution.

We continue to select a vector to be introduced into the basis, using the element in the $(m + 2)$nd row as criterion, until either (1) all the artificial vectors are eliminated from the basis or (2) no positive $(m + 2)$nd element exists. The first alternative implies that all the elements in the $(m + 2)$nd row equal zero and that the corresponding basis is a feasible basis for the original problem. We than apply the regular simplex algorithm to determine the minimum feasible solution. In the second alternative, if the $(m + 2, 0)$ element, i.e., the artificial part of the corresponding value of the objective function, is greater than zero, then the original problem is not feasible. Theorem 2 tells us that there are no other feasible solutions whose value of the objective function is smaller than this final solution. If the $(m + 2, 0)$ element is equal to zero, then we have a degenerate feasible solution to the original problem which contains at least one artificial vector. The artificial variables have values of zero. We have not, however, reached the minimum feasible solution. We continue the iterations by introducing a vector that corresponds to the maximum positive element in the $(m + 1)$st row which is above a zero element in the $(m + 2)$nd row.[2] This criterion is used until the minimum solution has been reached, i.e., until there are no more positive $(m + 1)$st elements over a zero in the $(m + 2)$nd row. The final solution may or may not contain artificial variables with values equal to zero. When we use this criterion, the elements in the $(m + 2)$nd row are never transformed, since the $z_j - c_j$ of the vector being introduced is equal to $0 \cdot w + (z_j - c_j)$.

For both alternatives (1) and (2), all the $(m + 2, j)$ elements are less than or equal to zero, with the possible exception of the $(m + 2, 0)$

[1] See Orden [84] for a discussion of the above and the application of the simplex procedure to related matrix problems.

[2] This procedure forces the artificial variables still in the solution always to maintain a value of zero and also eliminates the selection of certain vectors that would not reduce the value of the objective function.

element. The latter element is always nonnegative, and its value is nonincreasing. As Orden [84] points out, if the original problem is feasible but has an unbounded minimum, the method of the artificial basis will determine the feasibility before the unboundedness.

Whenever the original problem contains some unit vectors, these vectors along with the necessary artificial ones should be used for the initial basis. Doing this will tend to decrease the total number of iterations. The following will illustrate the artificial-basis technique:

Example.[1] Maximize

$$x_1 + 2x_2 + 3x_3 - x_4$$

subject to

$$\begin{aligned} x_1 + 2x_2 + 3x_3 \qquad &= 15 \\ 2x_1 + x_2 + 5x_3 \qquad &= 20 \\ x_1 + 2x_2 + x_3 + x_4 &= 10 \end{aligned}$$

and $x_j \geq 0$. Since we are always minimizing, the corresponding objective is to minimize $-x_1 - 2x_2 - 3x_3 + x_4$.

Since the given system contains a unit vector $\mathbf{P}_4$, we need only two artificial vectors $\mathbf{P}_5$ and $\mathbf{P}_6$. The first solution (see Tableau 4.5) is $\mathbf{X}_0 = (x_5,x_6,x_4) = (15,20,10)$, with a value of the objective function equal to $10 + 35w$. Each z_j is computed by taking the inner product of $\mathbf{P}_j$ with the column vector **c**. For example,

$$z_1 - c_1 = w + 2w + 1 - (-1) = 2 + 3w$$

Vector $\mathbf{P}_3$ is introduced into the basis because the maximum $(m + 2, j)$ element is $(m + 2, 3) = 8$. The corresponding $\theta_0 = 20/5$, and the artificial vector $\mathbf{P}_6$ is eliminated from the basis. We transform all the elements of the initial step of Tableau 4.5 by the elimination formulas. The new solution is $\mathbf{X}_0' = (x_5,x_3,x_4) = (3,4,6)$, with a value of the objective function of $-6 + 3w$. $\mathbf{P}_2$ goes into the basis, and $\mathbf{P}_5$ is eliminated. In the third step, since all the $(m + 2, j) \leq 0$ and $(m + 2, 0) = 0$, we have a feasible solution to the original problem; this solution is

$$\mathbf{X}_0'' = (x_2,x_3,x_4) = (15/7, 25/7, 15/7)$$

[1] This simple example used to illustrate the computational technique also points out the advantages of applying a bit of analysis to the equations of a linear-programming model before solving it. Here, since $x_1 + 2x_2 + 3x_3 = 15$, the problem is really to maximize the objective function $15 - x_4$. Hence, for $x_4 \geq 0$, the maximum basic feasible solution involves x_1, x_2, and x_3 with a maximum value of 15.

Tableau 4.5

Initial Step

i	Basis	c	$\mathbf{P}_0$	-1	-2	-3	1	w	w
				$\mathbf{P}_1$	$\mathbf{P}_2$	$\mathbf{P}_3$	$\mathbf{P}_4$	$\mathbf{P}_5$	$\mathbf{P}_6$
1	$\mathbf{P}_5$	w	15	1	2	3	0	1	0
2	$\mathbf{P}_6$	w	20	2	1	(5)	0	0	1
3	$\mathbf{P}_4$	1	10	1	2	1	1	0	0
4			10	2	4	4	0	0	0
5			35	3	3	8	0	0	0

Second Step

$\mathbf{P}_5$	w	3	$-1/5$	($7/5$)	0	0	1
$\mathbf{P}_3$	-3	4	$2/5$	$1/5$	1	0	0
$\mathbf{P}_4$	1	6	$3/5$	$9/5$	0	1	0
		-6	$2/5$	$16/5$	0	0	0
		3	$-1/5$	$7/5$	0	0	0

Third Step

$\mathbf{P}_2$	-2	$15/7$	$-1/7$	1	0	0
$\mathbf{P}_3$	-3	$25/7$	$3/7$	0	1	0
$\mathbf{P}_4$	1	$15/7$	($6/7$)	0	0	1
		$-90/7$	$6/7$	0	0	0
		0	0	0	0	0

Fourth Step

$\mathbf{P}_2$	-2	$5/2$	0	1	0	$1/6$
$\mathbf{P}_3$	-3	$5/2$	0	0	1	$-3/6$
$\mathbf{P}_1$	-1	$5/2$	1	0	0	$7/6$
		-15	0	0	0	-1

with the objective function equal to $-90/7$. The fourth step yields a minimum feasible solution

$$\mathbf{X}_0''' = (x_2,x_3,x_1) = (5/2,5/2,5/2)$$

with $x_4 = 0$ and the objective function equal to -15. The correct value of the objective function is $+15$, since we were originally dealing with a maximization problem.

4. A FIRST FEASIBLE SOLUTION USING SLACK VARIABLES[1]

If the linear-programming problem is originally of the form $\mathbf{AX} \leq \mathbf{b}$, its corresponding initial computational tableau will contain an $m \times m$ unit matrix. This matrix appears when we write each inequality as an equality by adding a nonnegative variable called a *slack variable.* These slack variables have associated cost coefficients that are usually set equal to zero.

For example, given the linear-programming problem: Maximize

$$2x_1 + 3x_2$$

subject to

$$\begin{aligned} -x_1 + 2x_2 &\leq 4 \\ x_1 + x_2 &\leq 6 \\ x_1 + 3x_2 &\leq 9 \\ x_j &\geq 0 \end{aligned}$$

Transforming to equations by adding a new nonnegative slack variable to all inequalities except the nonnegativity inequalities, we obtain the equivalent linear-programming problem: Maximize

$$2x_1 + 3x_2$$

subject to

$$\begin{aligned} -x_1 + 2x_2 + x_3 \qquad\qquad\qquad &= 4 \\ x_1 + x_2 \qquad + x_4 \qquad\quad &= 6 \\ x_1 + 3x_2 \qquad\qquad + x_5 &= 9 \\ x_j &\geq 0 \end{aligned}$$

We see that these equations contain a starting basis of the *slack vectors* $(\mathbf{P}_3\mathbf{P}_4\mathbf{P}_5)$ with the associated first feasible solution of $x_1 = 0$, $x_2 = 0$, $x_3 = 4$, $x_4 = 6$, $x_5 = 9$. The corresponding value of the objective function is zero, as all cost coefficients for the slack variables are here assumed to be zero. Some applications might call for these cost coefficients to be other than zero, e.g., penalty costs for not using certain raw materials, storage costs.

If the problem was originally $\mathbf{AX} \geq \mathbf{b}$, the equivalent set of equations is obtained by the subtraction of a new nonnegative slack variable from each inequality. Here the equations contain a negative unit basis which, since we assume $\mathbf{b} \geq \mathbf{O}$, cannot be used to initiate the computational procedure. For this case the problem, in general, must employ at least one

[1] See Sec. 4 of Chap. 2 for previous discussion of slack variables.

artificial vector. This is further described in Sec. 1 of Chap. 9. A problem can have a mixture of $\leq$, $\geq$, and $=$ constraints and is put into equation form by suitable additions or subtractions of slack variables.

5. GEOMETRIC INTERPRETATION OF THE SIMPLEX PROCEDURE

The preceding algebraic description of the simplex procedure has the following geometric interpretation, described in Hoffman, Mannos, Sokolowsky, and Wiegmann [61*a*]. It was first given in Dantzig [17].

Let the $(m + 1)$-dimensional vectors $\mathbf{P}'_1, \mathbf{P}'_2, \ldots, \mathbf{P}'_n$ be obtained by appending the cost coefficients $c_1, c_2, \ldots, c_n$, respectively, to the m-dimensional column vectors $\mathbf{P}_1, \mathbf{P}_2, \ldots, \mathbf{P}_n$; that is, for $\mathbf{P}'_j$ the corresponding c_j becomes the $(m + 1)$st coordinate. Let $\mathbf{C}$ be the convex cone in $(m + 1)$-space determined by $\mathbf{P}'_1, \mathbf{P}'_2, \ldots, \mathbf{P}'_n$. Let $\mathbf{B}$ be the line in $(m + 1)$-space consisting of all points whose first m coordinates are $b_1, b_2, \ldots, b_m$. This line can be thought of as being generated by an $(m + 1)$-dimensional vector $\mathbf{P}'_0$ whose first m coordinates are $b_1, b_2, \ldots, b_m$ and whose $(m + 1)$st coordinate takes on all values. The object of the computation is to find the lowest point of $\mathbf{B}$ which is also in $\mathbf{C}$, that is, the point of $\mathbf{B}$ in $\mathbf{C}$ whose $(m + 1)$st coordinate is a minimum.

The computational technique is as follows: Assume that m of the vectors $\mathbf{P}'_1, \mathbf{P}'_2, \ldots, \mathbf{P}'_n$, say the first m, are linearly independent and have the property that the m-dimensional cone $\mathbf{D}$ which they form contains a point of $\mathbf{B}$. This property is equivalent to saying that the m vectors are an admissible basis, i.e., correspond to a feasible solution. These vectors can either be given, as in Sec. 2, or be artificial, as in Sec. 3. The hyperplane containing $\mathbf{D}$ divides the remaining vectors $\mathbf{P}'_j$ into two groups. One group is on the side of the hyperplane which contains the positive $(m + 1)$st coordinate axis. The second group contains all the vectors which are below the hyperplane and are eligible to be selected as new basis vectors. Each vector of this second group can be joined to the hyperplane containing $\mathbf{D}$ by a line segment parallel to the $(m + 1)$st coordinate axis. Let $\mathbf{P}'_k$ be the vector with the property that its line segment is the longest. This corresponds to selecting the vector with $\max_j (z_j - c_j) > 0$. Then $\mathbf{P}'_k$ and a certain set of $m - 1$ of the vectors $\mathbf{P}'_1, \ldots, \mathbf{P}'_m$ have the property that the m-dimensional cone they form contains a point of $\mathbf{B}$, and this point will be lower than the intersection of $\mathbf{D}$ with $\mathbf{B}$. We replace the eliminated vector by $\mathbf{P}'_k$ and continue the process until a minimum solution has been found.

We next illustrate the above with an example. Consider the problem of minimizing

$$5x_1 + 3x_2 + 4x_3 + 6x_4$$

subject to the conditions

$$x_1 + 3x_2 + 5x_3 + 6x_4 = 3$$
$$6x_1 + 4x_2 + 2x_3 + x_4 = 2$$

and

$$x_j \geq 0$$

We have

$$\mathbf{P}_1' = \begin{pmatrix} 1 \\ 6 \\ 5 \end{pmatrix} \qquad \mathbf{P}_2' = \begin{pmatrix} 3 \\ 4 \\ 3 \end{pmatrix} \qquad \mathbf{P}_3' = \begin{pmatrix} 5 \\ 2 \\ 4 \end{pmatrix} \qquad \mathbf{P}_4' = \begin{pmatrix} 6 \\ 1 \\ 6 \end{pmatrix}$$

We plot these points as shown in Fig. 4.1. Since the line **B** intersects the cone **C**, feasible solutions to the problems exist. We can readily show that $\mathbf{P}_1'$ and $\mathbf{P}_4'$ are linearly independent and that $\mathbf{P}_0$ can be expressed as a

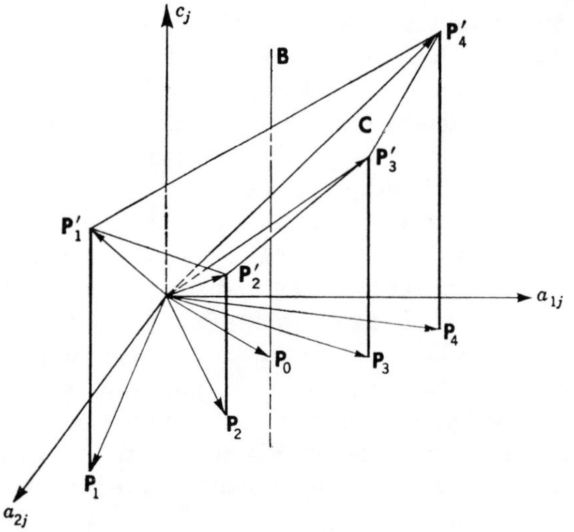

FIGURE 4.1

positive combination of $\mathbf{P}_1$ and $\mathbf{P}_4$. We then take as our first basis vectors $\mathbf{P}_1$ and $\mathbf{P}_4$, and we determine the two-dimensional cone formed by vectors $\mathbf{P}_1'$ and $\mathbf{P}_4'$ (Fig. 4.2). The point $\mathbf{P}_2$ is below the hyperplane containing **D** by a distance of z_2, where z_2 is defined in Eq. (1.4). $\mathbf{P}_3$ is a distance z_3 below this hyperplane. Point $\mathbf{P}_2'$ is below the hyperplane containing **D** by a distance of $z_2 - c_2$. The corresponding value of the objective function is given by z. Since $z_2 - c_2 > z_3 - c_3$, we introduce $\mathbf{P}_2$ into the basis. From Fig. 4.2 we see that $\mathbf{P}_0$ can be expressed as a positive combination of vectors $\mathbf{P}_2$ and $\mathbf{P}_4$ and cannot be expressed as a positive combination of $\mathbf{P}_2$ and $\mathbf{P}_1$; that is, $\mathbf{P}_0$ is not contained in the cone generated by vectors $\mathbf{P}_1$ and $\mathbf{P}_2$. Hence $\mathbf{P}_2$ is introduced into the basis and elimi-

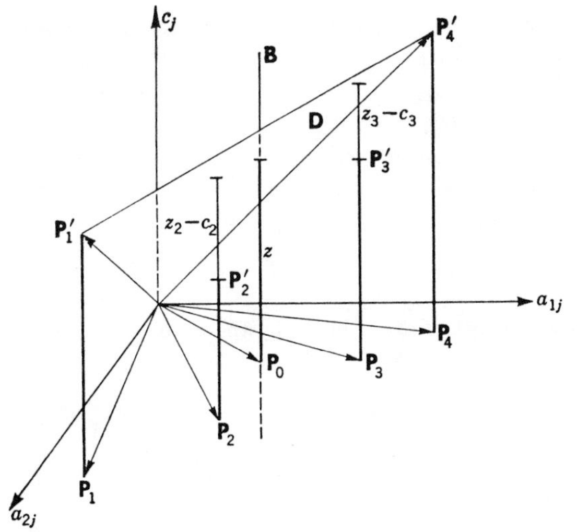

FIGURE 4.2

nates $\mathbf{P}_1$, giving a new basis of $\mathbf{P}_2$ and $\mathbf{P}_4$. For this basis we can perform a similar analysis, which will introduce $\mathbf{P}_3$ into the basis and eliminate $\mathbf{P}_4$. This basis of vectors $\mathbf{P}_2$ and $\mathbf{P}_3$ corresponds to the minimum solution.

We may also interpret the simplex procedure in terms of moving from one extreme point to an adjacent extreme point (see Sec. 4 of Chap. 2 and Saaty [88]). The procedure starts with the linear form passing through an extreme point of **K** (Fig. 4.3). Here we have pictured an arbitrary two-dimensional set of linear inequalities and its corresponding region of solution.

The next iteration would advance the linear form parallel to itself until it passes through $\bar{\mathbf{X}}_3$. In two additional iterations the linear form would pass through $\bar{\mathbf{X}}_1$, and we would have the minimum solution. The

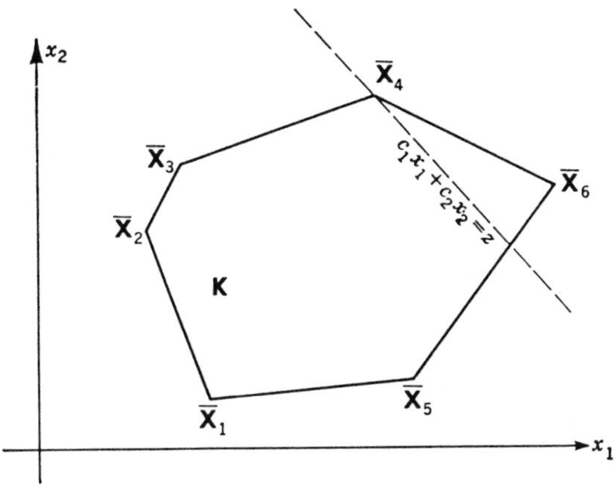

FIGURE 4.3

total number of iterations necessary to reach the minimum depends on which feasible solution is used to initiate the computation. If, as above, we started with the solution that corresponds to the extreme point $\bar{\mathbf{X}}_4$, then it would take us three iterations. If the solution associated with $\bar{\mathbf{X}}_5$ were taken as the first, then it would require only one iteration. If the coefficients of the linear form and the equations are normalized, then z is the distance of the linear form from the origin.

EXERCISES

1. Solve the following linear-programming problems by the simplex method:

a. Minimize

$$x_1 + x_2 + x_3$$

subject to

$$\begin{aligned} x_1 \qquad\qquad - x_4 \qquad\quad - 2x_6 &= 5 \\ x_2 \qquad + 2x_4 - 3x_5 + x_6 &= 3 \\ x_3 + 2x_4 - 5x_5 + 6x_6 &= 5 \end{aligned}$$

and $x_j \geq 0$.

b. Minimize

$$x_1 - x_2 + x_3 + x_4 + x_5 - x_6$$

subject to

$$\begin{aligned} x_1 \qquad\qquad + x_4 \qquad + 6x_6 &= 9 \\ 3x_1 + x_2 - 4x_3 \qquad\qquad + 2x_6 &= 2 \\ x_1 \qquad + 2x_3 \qquad + x_5 + 2x_6 &= 6 \end{aligned}$$

and $x_j \geq 0$.

c. Minimize

$$2x_1 + x_2 - x_3 - x_4$$

subject to

$$\begin{aligned} x_1 - x_2 + 2x_3 - x_4 &= 2 \\ 2x_1 + x_2 - 3x_3 + x_4 &= 6 \\ x_1 + x_2 + x_3 + x_4 &= 7 \end{aligned}$$

and $x_j \geq 0$.

d. Minimize

$$-3x_1 + x_2 + 3x_3 - x_4$$

subject to

$$\begin{aligned} x_1 + 2x_2 - x_3 + x_4 &= 0 \\ 2x_1 - 2x_2 + 3x_3 + 3x_4 &= 9 \\ x_1 - x_2 + 2x_3 - x_4 &= 6 \end{aligned}$$

and $x_j \geq 0$.

e. Maximize

$$x_4 - x_5$$

subject to

$$\begin{aligned} 2x_2 - x_3 - x_4 + x_5 &\geq 0 \\ -2x_1 + 2x_3 - x_4 + x_5 &\geq 0 \\ x_1 - 2x_2 - x_4 + x_5 &\geq 0 \\ x_1 + x_2 + x_3 &= 1 \end{aligned}$$

and $x_j \geq 0$.

f. Maximize

$$x_1 - x_2 + x_3 - 3x_4 + x_5 - x_6 - 3x_7$$

subject to

$$\begin{aligned} 3x_3 + x_5 + x_6 &= 6 \\ x_2 + 2x_3 - x_4 &= 10 \\ -x_1 + x_6 &= 0 \\ x_3 + x_6 + x_7 &= 6 \end{aligned}$$

and $x_j \geq 0$.

g. Minimize

$$x_1 - 2x_2 + 3x_3$$

subject to

$$\begin{aligned} -2x_1 + x_2 + 3x_3 &= 2 \\ 2x_1 + 3x_2 + 4x_3 &= 1 \end{aligned}$$

and $x_j \geq 0$.

h. Maximize

$$3x_1 - x_2$$

subject to

$$\begin{aligned} 2x_1 + x_2 &\geq 2 \\ x_1 + 3x_2 &\leq 3 \\ x_2 &\leq 4 \end{aligned}$$

and $x_j \geq 0$.

i. Minimize

$$2x_1 - 3x_2 + 6x_3$$

subject to

$$\begin{aligned} 3x_1 - 4x_2 - 6x_3 &\leq 2 \\ 2x_1 + x_2 + 2x_3 &\geq 11 \\ x_1 + 3x_2 - 2x_3 &\leq 5 \end{aligned}$$

and $x_j \geq 0$.

j. Maximize

$$x_1 + x_2$$

subject to

$$x_1 + x_2 \geq 1$$
$$x_1 - x_2 \leq 1$$
$$-x_1 + x_2 \leq 1$$

and $x_j \geq 0$.

2. Given a linear-programming problem that has a finite minimum feasible solution. Let the minimum value of the objective function be z_0. The problem is being solved by the simplex procedure, and in the kth iteration a degenerate basic feasible solution with exactly one $x_i = 0$ appears. The corresponding value of the objective function is $\bar{z}$ with $\bar{z} > z_0$. Prove that this kth basis cannot reappear in the succeeding iterations.

3. Solve Exercise 1*c* by selecting $\mathbf{P}_1$, $\mathbf{P}_2$, and $\mathbf{P}_3$ as an initial basis; similarly, solve Exercise 1*e* with $\mathbf{P}_1$, $\mathbf{P}_2$, $\mathbf{P}_3$, and $\mathbf{P}_4$ as an initial basis. Also compute the determinant of the final basis.

4. (A test for a near-optimum solution in the simplex procedure due to L. Goldstein.) Given a linear-programming problem with a known upper bound S to the sum of the variables, e.g., the transportation problem. Let $\bar{z}$ be the value of the objective function for a given basic feasible solution and z_0 be the unknown minimum value of the objective function. Let $m = z_k - c_k = \max (z_j - c_j)$ for the given basic solution. Show that $\bar{z} - z_0 \leq \epsilon$, is a predetermined tolerance limit, if $m \leq \epsilon/S$.

5. Solve Exercise 1*b*, 1*d*, and 1*i* using the criterion (2.1*a*) for selecting the variable to be introduced into the solution.

6. Prove the validity of the transformation formulas (2.7*a*) and (2.7*b*).

7. Prove the following statement: If the original problem is feasible but has an unbounded minimum, the method of the artificial basis will determine the feasibility before the unboundedness.

8. Prove that if a set of m linear equations in n unknowns, where $m \leq n$, has a feasible solution, there exists a basic feasible solution. (Assume that the rank of the system is m.)

9. For the standard linear-programming problem, let

$$x_0 = c_1x_1 + c_2x_2 + \cdots + c_nx_n$$

Write an equivalent linear-programming problem which uses this equation as a constraint and discuss the application of the simplex algorithm to the new problem.

10. For Fig. 4.2 show that for some $\mathbf{P}_j$ not in the basis, e.g., vector $\mathbf{P}_2$, the corresponding z_j does measure the vertical distance of the point $\mathbf{P}_j$ from the plane $\mathbf{D}$.

CHAPTER 5 / THE DUALITY PROBLEMS OF LINEAR PROGRAMMING

Associated with every linear-programming problem as defined in Sec. 1 of Chap. 4 is a corresponding optimization problem called the *dual problem*. The original problem is termed the *primal*. The optimum solution of either problem reveals information concerning the optimum solution of the other. In fact, if the initial simplex tableau for the primal problem contains an $m \times m$ unit matrix, then the solution of either problem by the simplex procedure yields an explicit solution to the other. We shall state and prove the theorems associated with the dual problems as given by Dantzig and Orden [31]. For additional reading, the reader is referred to Gale, Kuhn, and Tucker [45], Vajda [97], and Kuhn and Tucker [68].

1. THE UNSYMMETRIC PRIMAL-DUAL PROBLEMS

The primal problem.[1] Find a column vector $\mathbf{X} = (x_1, x_2, \ldots, x_n)$ which *minimizes* the linear functional

$$f(\mathbf{X}) = \mathbf{cX} \tag{1.1}$$

subject to the conditions

$$\mathbf{AX} = \mathbf{b} \tag{1.2}$$

and

$$\mathbf{X} \geq \mathbf{O} \tag{1.3}$$

This statement of the general linear-programming problem assumes that the number of rows of **A** is less than the number of columns of **A**.

[1] For discussion purposes we have designated the minimization problem as the primal. In general, either problem can be considered the primal, with the remaining problem the dual.

The dual Problem. Find a row vector $\mathbf{W} = (w_1, w_2, \ldots, w_m)$ which *maximizes* the linear functional

$$g(\mathbf{W}) = \mathbf{Wb} \tag{1.4}$$

subject to the conditions

$$\mathbf{WA} \leq \mathbf{c} \tag{1.5}$$

In the dual problem, the variables w_i are not restricted to be nonnegative. For both problems we have that $\mathbf{c} = (c_1, c_2, \ldots, c_n)$ is a row vector, $\mathbf{b} = (b_1, b_2, \ldots, b_m)$ is a column vector, and $\mathbf{A} = (a_{ij})$ is the matrix of coefficients. Multiplying the $1 \times m$ row vector $\mathbf{W}$ by the $m \times n$ matrix $\mathbf{A}$, we have the following explicit representation of (1.5):

$$\begin{aligned} a_{11}w_1 + a_{21}w_2 + \cdots + a_{m1}w_m &\leq c_1 \\ a_{12}w_1 + a_{22}w_2 + \cdots + a_{m2}w_m &\leq c_2 \\ \cdots\cdots\cdots\cdots\cdots\cdots & \\ a_{1n}w_1 + a_{2n}w_2 + \cdots + a_{mn}w_m &\leq c_n \end{aligned}$$

The matrix of coefficients for the above inequalities is given by $\mathbf{A}'$.

Associated with these problems is the following important theorem:

The duality theorem. *If either the primal or the dual problem has a finite optimum solution, then the other problem has a finite optimum solution and the extremes of the linear functions are equal, that is,* $\min f(\mathbf{X}) = \max g(\mathbf{W})$.

If either problem has an unbounded optimum solution, then the other problem has no feasible solutions.[1]

Proof. We first assume that the primal is feasible and that a minimum feasible solution has been obtained by the simplex procedure. For discussion purposes, let the first m vectors $\mathbf{P}_1, \mathbf{P}_2, \ldots, \mathbf{P}_m$ be the final basis vectors. Let $\mathbf{B}$ equal the $m \times m$ matrix $(\mathbf{P}_1\mathbf{P}_2 \cdots \mathbf{P}_m)$. The final computational tableau contains the vectors of the original system, $\mathbf{P}_1, \mathbf{P}_2, \ldots, \mathbf{P}_m, \mathbf{P}_{m+1}, \ldots, \mathbf{P}_n$, in terms of the final basis vectors; i.e., for each vector $\mathbf{P}_j$ the final tableau contains the vector $\mathbf{X}_j$, where $\mathbf{P}_j = \mathbf{BX}_j$. Let the $m \times n$ matrix

$$\bar{\mathbf{X}} = (\mathbf{X}_1\mathbf{X}_2 \cdots \mathbf{X}_m\mathbf{X}_{m+1} \cdots \mathbf{X}_n)$$

be the matrix of coefficients contained in the final simplex tableau. Since we assumed that the final basis contained the first m vectors,

[1] For the dual problem, a solution is feasible if it satisfies (1.5). The variables are not required to be nonnegative.

we have

$$\bar{\mathbf{X}} = \begin{pmatrix} 1 & 0 & \cdots & 0 & x_{1,\,m+1} & \cdots & x_{1n} \\ 0 & 1 & \cdots & 0 & x_{2,\,m+1} & \cdots & x_{2n} \\ \cdot & \cdot & \cdot & \cdot & \cdot & \cdot & \cdot \\ 0 & 0 & \cdots & 1 & x_{m,\,m+1} & \cdots & x_{mn} \end{pmatrix}$$

The minimum-solution vector is given by $\mathbf{X}^0 = \mathbf{B}^{-1}\mathbf{b}$. For this final solution we then have the following relationships:

$$\mathbf{A} = \mathbf{B}\bar{\mathbf{X}} \qquad \mathbf{B}^{-1}\mathbf{A} = \bar{\mathbf{X}} \tag{1.6}$$

$$\mathbf{b} = \mathbf{B}\mathbf{X}^0 \qquad \mathbf{B}^{-1}\mathbf{b} = \mathbf{X}^0 \tag{1.7}$$

$$\min f(\mathbf{X}) = \mathbf{c}^0\mathbf{X}^0 \tag{1.8}$$

$$Z = \mathbf{c}^0\bar{\mathbf{X}} - \mathbf{c} \leq \mathbf{O} \tag{1.9}$$

where $\mathbf{c}^0 = (c_1,c_2 \ldots ,c_m)$ is a row vector and, by (1.4) of Chap. 4,

$$\begin{aligned} \mathbf{Z} &= (\mathbf{c}^0\mathbf{X}_1 - c_1,\, \mathbf{c}^0\mathbf{X}_2, - c_2,\, \ldots,\, \mathbf{c}^0\mathbf{X}_n - c_n) \\ &= (z_1 - c_1,\, z_2 - c_2,\, \ldots,\, z_n - c_n) \end{aligned}$$

is a row vector whose elements are nonpositive, as they are the $z_j - c_j$ elements corresponding to an optimal solution. Here the vector $\mathbf{O}$ is a null n-dimensional row vector.

Let $\mathbf{W}^0 = (w_1^0,w_2^0, \ldots ,w_m^0)$ be defined by

$$\mathbf{W}^0 = \mathbf{c}^0\mathbf{B}^{-1}$$

Then, by (1.6) and (1.9), we have

$$\mathbf{W}^0\mathbf{A} - \mathbf{c} = \mathbf{c}^0\mathbf{B}^{-1}\mathbf{A} - \mathbf{c} = \mathbf{c}^0\bar{\mathbf{X}} - \mathbf{c} \leq \mathbf{O}$$

or

$$\mathbf{W}^0\mathbf{A} \leq \mathbf{c}$$

The vector $\mathbf{W}^0$ is a solution to the dual problem, since it satisfies the dual constraints (1.5). For this solution the corresponding value of the dual objective function (1.4) is given by $\mathbf{W}^0\mathbf{b}$, or, from (1.7) and (1.8), we have

$$\mathbf{W}^0\mathbf{b} = \mathbf{c}^0\mathbf{B}^{-1}\mathbf{b} = \mathbf{c}^0\mathbf{X}^0 = \min f(\mathbf{X}) \tag{1.10}$$

Hence, for the solution $\mathbf{W}^0$, the value of the dual objective function is equal to the minimum value of the primal objective function. Now we need only to show that $\mathbf{W}^0$ is also an optimum solution for the dual problem.

For any $n \times 1$ vector $\mathbf{X}$ which satisfies (1.2) and (1.3) and any

$1 \times m$ vector $\mathbf{W}$ satisfying (1.5), we have

$$\mathbf{WAX} = \mathbf{Wb} = g(\mathbf{W}) \tag{1.11}$$

and

$$\mathbf{WAX} \leq \mathbf{cX} = f(\mathbf{X}) \tag{1.12}$$

By (1.11) and (1.12),

$$g(\mathbf{W}) \leq f(\mathbf{X}) \tag{1.13}$$

for all feasible $\mathbf{W}$ and $\mathbf{X}$. The extreme values of (1.1) and (1.4) are related by

$$\max g(\mathbf{W}) \leq \min f(\mathbf{X}) \tag{1.14}$$

For the dual solution vector $\mathbf{W}^0$, we have from (1.10)

$$g(\mathbf{W}^0) = \mathbf{W}^0\mathbf{b} = \min f(\mathbf{X})$$

and hence, for the optimum solution to the primal problem, $\mathbf{X}^0$, and the dual solution $\mathbf{W}^0$, (1.14) becomes

$$g(\mathbf{W}^0) = f(\mathbf{X}^0)$$

i.e., for the solution $\mathbf{W}^0 = \mathbf{c}^0\mathbf{B}^{-1}$, the dual objective function takes on its maximum value. Therefore, for $\mathbf{W}^0$ and $\mathbf{X}^0$ we have the corresponding values of the objective functions related by

$$\max g(\mathbf{W}) = \min f(\mathbf{X}) \tag{1.15}$$

The above results have been shown to hold whenever the primal has a finite optimum solution. In a similar fashion, we can show that, when the dual problem has a finite optimum solution, the primal is feasible, and (1.15) holds.

To do this, transform the dual problem

$$\max g(\mathbf{W}) = \max \mathbf{Wb}$$
$$\mathbf{WA} \leq \mathbf{c}$$

to the primal format and show that its dual (which will be feasible by the above theorem) is just the original primal. We have

$$\max \mathbf{Wb} = -\min(-\mathbf{Wb})$$

or

$$-\max \mathbf{Wb} = \min(-\mathbf{Wb})$$

subject to

$$\mathbf{WA} + \mathbf{W}_3\mathbf{I} = \mathbf{c}$$
$$\mathbf{W}_3 \geq \mathbf{O}$$

where the $\mathbf{W}_3$ is a set of nonnegative slack variables. We next transform $\mathbf{W} = \mathbf{W}_1 - \mathbf{W}_2$, where $\mathbf{W}_1$ and $\mathbf{W}_2$ are nonnegative variables. Our problem is now

$$\min\,(-\mathbf{W}_1 + \mathbf{W}_2)\mathbf{b} + \mathbf{W}_3\mathbf{O} \tag{1.15a}$$

subject to

$$(\mathbf{W}_1 - \mathbf{W}_2)\mathbf{A} + \mathbf{W}_3\mathbf{I} = \mathbf{c} \tag{1.15b}$$
$$\mathbf{W}_1 \geq \mathbf{O}$$
$$\mathbf{W}_2 \geq \mathbf{O}$$
$$\mathbf{W}_3 \geq \mathbf{O}$$

and is in the standard format.

The duality theorem holds for this problem. To determine its dual in the proper format, it is convenient to multiply the constraint equation (1.15*b*) by -1 to obtain

$$-\mathbf{W}_1\mathbf{A} + \mathbf{W}_2\mathbf{A} - \mathbf{W}_3\mathbf{I} = -\mathbf{c} \tag{1.15c}$$

The dual to (1.15*a*) and (1.15*c*) is then

$$\max\,(-\mathbf{cX}) = -\min\,\mathbf{cX}$$

subject to

$$(-\mathbf{A} \quad \mathbf{A} \quad -\mathbf{I})\mathbf{X} \leq (-\mathbf{b} \quad \mathbf{b} \quad \mathbf{O})$$

which is equivalent to

$$-\max\,(-\,\mathbf{cX}) = \min\,\mathbf{cX}$$

subject to

$$-\mathbf{AX} \leq -\mathbf{b}$$
$$\mathbf{AX} \leq \mathbf{b}$$
$$-\mathbf{IX} \leq \mathbf{O}$$

or

$$\mathbf{AX} \geq \mathbf{b}$$
$$\mathbf{AX} \leq \mathbf{b}$$
$$\mathbf{X} \geq \mathbf{O}$$

or

$$\mathbf{AX} = \mathbf{b}$$
$$\mathbf{X} \geq \mathbf{O}$$

which is the original primal problem.

This completes the proof of the first part of the theorem.

To prove the second part, we note that, if the primal is unbounded, then we have by (1.13)

$$g(\mathbf{W}) \leq -\infty \tag{1.16}$$

Any solution to the dual inequalities (1.5) must have a corresponding value for the dual objective function (1.4) which is a lower bound for the primal objective function. Since this contradicts the assumption of unboundedness, we must conclude that there are no solutions to the dual problem and hence that the dual inequalities are inconsistent. A similar argument will show that, when the dual has an unbounded solution, the primal has no solutions.

Example. As an example of unsymmetric dual problems, let us look at the example that was solved by the simplex procedure in Sec. 2 of Chap. 4.

The primal problem. Minimize

$$x_2 - 3x_3 \qquad + 2x_5$$

subject to

$$\begin{aligned} x_1 + 3x_2 - x_3 \qquad\quad + 2x_5 \qquad &= 7 \\ -2x_2 + 4x_3 + x_4 \qquad\qquad\quad &= 12 \\ -4x_2 + 3x_3 \qquad\quad + 8x_5 + x_6 &= 10 \end{aligned}$$

and

$$x_j \geq 0$$

Here we have $\mathbf{c} = (0,1,-3,0,2,0)$, $\mathbf{b} = (7,12,10)$,

$$\mathbf{A} = \begin{pmatrix} 1 & 3 & -1 & 0 & 2 & 0 \\ 0 & -2 & 4 & 1 & 0 & 0 \\ 0 & -4 & 3 & 0 & 8 & 1 \end{pmatrix}$$

and

$$\mathbf{A}' = \begin{pmatrix} 1 & 0 & 0 \\ 3 & -2 & -4 \\ -1 & 4 & 3 \\ 0 & 1 & 0 \\ 2 & 0 & 8 \\ 0 & 0 & 1 \end{pmatrix}$$

The dual problem. Maximize

$$7w_1 + 12w_2 + 10w_3$$

subject to

$$
\begin{aligned}
w_1 \qquad\qquad\qquad &\leq 0 \\
3w_1 - 2w_2 - 4w_3 &\leq 1 \\
-w_1 + 4w_2 + 3w_3 &\leq -3 \\
w_2 \qquad\quad &\leq 0 \\
2w_1 \qquad + 8w_3 &\leq 2 \\
w_3 &\leq 0
\end{aligned}
$$

The final basis corresponding to an optimal solution to the primal problem, as computed by the simplex method in Sec. 2 of Chap. 4, is given by the vectors $\mathbf{P}_2$, $\mathbf{P}_3$, $\mathbf{P}_6$, that is,

$$\mathbf{B} = (\mathbf{P}_2\mathbf{P}_3\mathbf{P}_6) = \begin{pmatrix} 3 & -1 & 0 \\ -2 & 4 & 0 \\ -4 & 3 & 1 \end{pmatrix}$$

The corresponding optimum solution $\mathbf{X}^0$ is

$$\mathbf{X}^0 = \mathbf{B}^{-1}\mathbf{b} = (x_2^0,x_3^0,x_6^0) = (4,5,11)$$

$$\mathbf{c}^0 = (c_2,c_3,c_6) = (1,-3,0)$$

and the minimum value of the objective function is

$$\mathbf{c}^0\mathbf{X}^0 = (1,-3,0)\begin{pmatrix} 4 \\ 5 \\ 11 \end{pmatrix} = -11$$

From the final simplex tableau (see Tableau 4.3) we have

$$\bar{\mathbf{X}} = \begin{pmatrix} \frac{2}{5} & 1 & 0 & \frac{1}{10} & \frac{4}{5} & 0 \\ \frac{1}{5} & 0 & 1 & \frac{3}{10} & \frac{2}{5} & 0 \\ 1 & 0 & 0 & -\frac{1}{2} & 10 & 1 \end{pmatrix}$$

The vector $\mathbf{Z}$ for the optimum solution is given by

$$\mathbf{Z} = \mathbf{c}^0\bar{\mathbf{X}} - \mathbf{c} = (1,-3,0)\begin{pmatrix} \frac{2}{5} & 1 & 0 & \frac{1}{10} & \frac{4}{5} & 0 \\ \frac{1}{5} & 0 & 1 & \frac{3}{10} & \frac{2}{5} & 0 \\ 1 & 0 & 0 & -\frac{1}{2} & 10 & 1 \end{pmatrix} - (0,1,-3,0,2,0)$$

or

$$\mathbf{Z} = (-\tfrac{1}{5},0,0,-\tfrac{4}{5},-\tfrac{12}{5},0) \leq \mathbf{O}$$

The elements of $\mathbf{Z}$ are just those elements contained in the $(m + 1)$st row of the final tableau, i.e., the $z_j - c_j$ elements.

In the discussion of the elimination method described in Sec. 5 of Chap. 2, we showed that, if the original matrix of coefficients for the equations contained a unit matrix or was augmented by a unit matrix, then the solution of the equations by the elimination method also yielded the inverse of the basis that corresponds to the solution. The elimination procedure transformed the unit matrix into the desired inverse. The same property is true for a linear-programming problem that has been solved by the simplex procedure. If the original matrix of coefficients, $\mathbf{A}$, contains a unit matrix or is augmented by a unit matrix, then the inverse of the basis at each step is developed in the columns that correspond to the unit matrix. In our example, we see that $\mathbf{A}$ does contain a unit matrix whose columns correspond to $\mathbf{P}_1$, $\mathbf{P}_4$, $\mathbf{P}_6$. Hence, in our final tableau, which is the third step of the example, the columns which correspond to $\mathbf{P}_1$, $\mathbf{P}_4$, $\mathbf{P}_6$ have been transformed to the inverse of the final basis $\mathbf{B}$. These columns form the set of vectors $\mathbf{X}_1$, $\mathbf{X}_4$, $\mathbf{X}_6$, and hence

$$\mathbf{B}^{-1} = (\mathbf{X}_1\mathbf{X}_4\mathbf{X}_6) = \begin{pmatrix} 2/5 & 1/10 & 0 \\ 1/5 & 3/10 & 0 \\ 1 & -1/2 & 1 \end{pmatrix}$$

As we have shown in the proof of the duality theorem, the optimum solution $\mathbf{W}^0$ to the dual is given by

$$\mathbf{W}^0 = \mathbf{c}^0\mathbf{B}^{-1} = (1,-3,0)\begin{pmatrix} 2/5 & 1/10 & 0 \\ 1/5 & 3/10 & 0 \\ 1 & -1/2 & 1 \end{pmatrix} = (-1/5,-4/5,0)$$

We check this solution by substituting it in the dual constraints, and we have

$$\mathbf{W}^0\mathbf{A} \le \mathbf{c}$$

$$(-1/5,-4/5,0)\begin{pmatrix} 1 & 3 & -1 & 0 & 2 & 0 \\ 0 & -2 & 4 & 1 & 0 & 0 \\ 0 & 4 & 3 & 0 & 8 & 1 \end{pmatrix} \le \begin{pmatrix} 0 \\ 1 \\ -3 \\ 0 \\ 2 \\ 0 \end{pmatrix}$$

$$\begin{aligned} -1/5 &\le 0 \\ 1 &\le 1 \\ -3 &\le -3 \\ -4/5 &\le 0 \\ -2/5 &\le 2 \\ 0 &\le 0 \end{aligned}$$

We have as the value of the dual objective function

$$\mathbf{W}^0\mathbf{b} = (-\tfrac{1}{5}, -\tfrac{4}{5}, 0)\begin{pmatrix}7\\12\\10\end{pmatrix} = -11$$

The values of the variables for the optimum solution to the dual do not have to be obtained by multiplying $\mathbf{c}^0\mathbf{B}^{-1}$ if the matrix $\mathbf{A}$ contains a unit matrix. We have

$$\mathbf{W}^0 = \mathbf{c}^0\mathbf{B}^{-1} = \mathbf{c}^0(\mathbf{X}_1\mathbf{X}_4\mathbf{X}_6) = (w_1^0, w_2^0, w_3^0)$$

By definition of z_j we have $\mathbf{c}^0\mathbf{X}_1 = z_1$, $\mathbf{c}^0\mathbf{X}_4 = z_4$, and $\mathbf{c}^0\mathbf{X}_6 = z_6$. In the $(m + 1)$st row of the final tableau we have for each $\mathbf{X}_j$ the corresponding $z_j - c_j$ element. For $j = 1, 4, 6$ we note that the corresponding $c_j = 0$, and hence the elements in the $(m + 1)$st row corresponding to $j = 1, 4, 6$ are equal to the corresponding values of the dual variables, that is, $w_1^0 = z_1$, $w_2^0 = z_4$, and $w_3^0 = z_6$. If a vector which formed the unit matrix had a $c_j \neq 0$, then the value of this c_j would have to be added back to the corresponding $z_j - c_j$ in the final tableau in order to obtain the correct value for the w_i^0. We note that w_i^0 is equal to the z_j which has, for its corresponding unit vector in the initial simplex tableau, the vector whose unit element is in position i. In our example, $w_2^0 = z_4$, since $\mathbf{P}_4$ is a unit vector with its unit element in position $i = 2$.

A variation of the unsymmetric dual problems is given in Sec. 2. For both of these new problems, the symmetric dual problems, the constraints are inequalities and the variables are restricted to be nonnegative. We show that the duality theorem holds for the symmetric dual problems by transforming them to their equivalent unsymmetric primal and dual problems.

2. THE SYMMETRIC PRIMAL-DUAL PROBLEMS

The primal problem. Find a vector $\mathbf{X} = (x_1, x_2, \ldots, x_n)$ which *minimizes* the linear functional

$$f(\mathbf{X}) = \mathbf{cX} \tag{2.1}$$

subject to the conditions

$$\mathbf{AX} \geq \mathbf{b} \qquad 2.2)$$

and

$$\mathbf{X} \geq \mathbf{0} \tag{2.3}$$

The dual problem. Find a vector $\mathbf{W} = (w_1,w_2, \ldots ,w_m)$ which *maximizes* the linear functional

$$g(\mathbf{W}) = \mathbf{Wb} \tag{2.4}$$

subject to the conditions

$$\mathbf{WA} \leq \mathbf{c} \tag{2.5}$$

and

$$\mathbf{W} \geq \mathbf{O} \tag{2.6}$$

We next show that the duality theorem of Sec. 1 also applies to the symmetric dual problems. Let the m nonnegative elements of the column vector $\mathbf{Y} = (y_1,y_2, \ldots ,y_m)$ be the slack variables which transform the primal into a set of equations. The equivalent linear-programming problem in terms of partitioned matrices is to *minimize*

$$f(\mathbf{X},\mathbf{Y}) = (\mathbf{c} \mid \bar{\mathbf{O}}) \begin{pmatrix} \mathbf{X} \\ \hline \mathbf{Y} \end{pmatrix} \tag{2.7}$$

subject to the constraints

$$(\mathbf{A} \mid -\mathbf{I}) \begin{pmatrix} \mathbf{X} \\ \hline \mathbf{Y} \end{pmatrix} = \mathbf{b} \tag{2.8}$$

and

$$\mathbf{X} \geq \mathbf{O} \qquad \text{and} \qquad \mathbf{Y} \geq \mathbf{O}\dagger \tag{2.9}$$

Here $\bar{\mathbf{O}} = (0, \ldots ,0)$ is an m-component row vector and $\mathbf{I}$ is an $m \times m$ identity matrix.

The dual of this transformed primal is to find a vector $\mathbf{W}$ which maximizes

$$g(\mathbf{W}) = \mathbf{Wb} \tag{2.10}$$

subject to the constraints

$$\mathbf{W}(\mathbf{A} \mid -\mathbf{I}) \leq (\mathbf{c} \mid \bar{\mathbf{O}}) \tag{2.11}$$

We see that (2.11) decomposes to (2.5) and (2.6), respectively, i.e.,

$$\mathbf{WA} \leq \mathbf{c}$$

and

$$-\mathbf{WI} \leq \bar{\mathbf{O}}$$

The last expression is equivalent to $\mathbf{W} \geq \mathbf{O}$.

The original problems are now given as unsymmetric dual problems, and the duality theorem holds.

We next prove, for the symmetric dual problems, the following theorem which can be interpreted as establishing necessary and sufficient conditions for the optimality of feasible solutions to the primal and dual problems.

† See Sec. 5 of Chap. 2 for a discussion of the use of partitioned matrices.

Complementary slackness theorem (Dantzig and Orden [31*a*] and Dantzig [16*c*]). *For optimal feasible solutions of the primal and dual systems, whenever inequality occurs in the kth relation of either system (the corresponding slack variable is positive), then the kth variable of its dual vanishes; if the kth variable is positive in either system, the kth relation of its dual is equality (the corresponding slack variable is zero).*

Proof. We rewrite the inequalities (2.2) of the primal system as equalities by subtracting a nonnegative slack variable x_{n+i} from the ith inequality to obtain

$$\begin{aligned} a_{11}x_1 + \cdots + a_{1n}x_n - x_{n+1} \qquad\qquad &= b_1 \\ \vdots \qquad\qquad\qquad &\quad \vdots \\ a_{m1}x_1 + \cdots + a_{mn}x_n \qquad\qquad - x_{n+m} &= b_m \end{aligned} \tag{2.2'}$$

Similarly, for the dual inequalities (2.5) we add a nonnegative slack variable w_{m+j} to the jth inequality to obtain

$$\begin{aligned} a_{11}w_1 + \cdots + a_{m1}w_m + w_{m+1} \qquad\qquad &= c_1 \\ \vdots \qquad\qquad\qquad &\quad \vdots \\ a_{1n}w_1 + \cdots + a_{mn}w_m \qquad\qquad + w_{m+n} &= c_n \end{aligned} \tag{2.5'}$$

In explicit form, the primal objective function is

$$c_1x_1 + \cdots + c_nx_n = f(\mathbf{X}) \tag{2.1'}$$

and the dual objective function is

$$b_1w_1 + \cdots + b_mw_m = g(\mathbf{W}) \tag{2.4'}$$

We next multiply the ith equation of (2.2′) by the corresponding dual variable w_i, $i = 1, 2, \ldots, m$, add the resulting set of equations, and subtract this sum from (2.1′) to obtain

$$\left(c_1 - \sum_{i=1}^{m} a_{i1}w_i\right) x_1 + \left(c_2 - \sum_{i=1}^{m} a_{i2}w_i\right) x_2 + \cdots + \left(c_n - \sum_{i=1}^{m} a_{in}\right) x_n + w_1x_{n+1} + w_2x_{n+2} + \cdots + w_mx_{n+m} = f(\mathbf{X}) - \sum_{i=1}^{m} w_ib_i$$

Noting that $w_{m+j} = c_j - \sum_{i=1}^{m} a_{ij}\, w_i$ and $g(\mathbf{W}) = \sum_{i=1}^{m} w_ib_i$, we have

$$w_{m+1}x_1 + w_{m+2}x_2 + \cdots + w_{m+n}x_n + w_1x_{n+1} + w_2x_{n+2} + \cdots + w_mx_{n+m} = f(\mathbf{X}) - g(\mathbf{W}) \tag{2.12}$$

From the duality theorem, we have for an optimum solution $\mathbf{X}^0 = (x_1^0, x_2^0, \ldots, x_n^0)$ to the primal problem and

$$\mathbf{W}^0 = (w_1^0, w_2^0, \ldots, w_m^0)$$

an optimum solution to the dual problem that $f(\mathbf{X}^0) - g(\mathbf{W}^0) = 0$. Thus, for these optimum solutions and corresponding slack variables $x_{n+i}^0 \geq 0$ and $w_{m+j}^0 \geq 0$, (2.12) becomes

$$w_{m+1}^0 x_1^0 + w_{m+2}^0 x_2^0 + \cdots + w_{m+n}^0 x_n^0 + w_1^0 x_{n+1}^0 + w_2^0 x_{n+2}^0 + \cdots + w_m^0 x_{n+m}^0 = 0 \quad (2.13)$$

We note that the terms $w_{m+j}^0 x_j^0$ are the product of the jth slack variable of the dual and the jth variable of the original primal; while the terms $w_i^0 x_{n+i}^0$ are the product of the ith variable of the original dual and the ith slack variable of the primal. Since all variables are restricted to be nonnegative, all the product terms of (2.13) are nonnegative and, as the sum of these terms must be equal to zero, they individually must be equal to zero. Thus $w_{m+j}^0 x_j^0 = 0$ for all j and $w_i^0 x_{n+i}^0 = 0$ for all i. If $w_{m+k}^0 > 0$, we must have $x_k^0 = 0$; if $x_{n+k}^0 > 0$, then $w_k^0 = 0$, which establishes the first part of the theorem. If $x_k^0 > 0$, then $w_{m+k}^0 = 0$, if $w_k^0 > 0$, then $x_{n+k}^0 = 0$, which completes the proof.

In a slightly different form, the basic duality theorem can be restated as follows (Dantzig [16c]): *If feasible solutions to both the primal and dual systems exist, there exists an optimum solution to both systems and*

$$\min f(\mathbf{X}) = \max g(\mathbf{W}).$$

We note that each member of a pair of primal and dual problems can have finite optimum solutions; that one problem can have an unbounded optimum feasible solution while the other has no feasible solutions; and that both problems can have no feasible solutions.

As noted in Goldman and Tucker [51], the symmetric dual problems can be conveniently represented by the following tableau:

(≥ 0)	x_1	$\cdots$	x_j	$\cdots$	x_n	$\geq$
w_1	a_{11}	$\cdots$	a_{1j}	$\cdots$	a_{1n}	b_1
$\cdot$	$\cdot$	$\cdots$	$\cdot$	$\cdots$	$\cdot$	$\cdot$
w_i	a_{i1}	$\cdots$	a_{ij}	$\cdots$	a_{in}	b_i
$\cdot$	$\cdot$	$\cdots$	$\cdot$	$\cdots$	$\cdot$	$\cdot$
w_m	a_{m1}	$\cdots$	a_{mj}	$\cdots$	a_{mn}	b_m
$\leq$	c_1	$\cdots$	c_j	$\cdots$	c_n	max / min

By taking inner products of the rows of $\mathbf{A}$ with the row of x's, we obtain the constraints of the primal; the inner products of the columns of $\mathbf{A}$ with the column of w's yield the dual constraints.

Many linear-programming problems are originally stated in the form of either the symmetric primal or the symmetric dual. In either case, the problem is usually rewritten in equation form by adding or subtracting the necessary nonnegative slack variables. For the symmetric primal we subtract a set of slack vectors and solve the corresponding system, while for the symmetric dual we add a set of slack vectors. These slack vectors are given associated cost coefficients that are equal to zero. The resulting tableau for the primal contains a negative unit matrix, while the dual tableau contains a positive unit matrix. Hence the final simplex tableau of either will contain the optimal solution of the other. If the primal was solved to obtain the dual solution, then, since we start out with a negative unit matrix in the primal, the corresponding z_j values in the final tableau have to have their signs reversed.

As the size of a linear-programming problem that can be solved on an electronic computer is usually limited by the number of rows involved, a problem that is too large when stated in terms of the primal may be of the right dimensions when written as the corresponding dual problem, and vice versa. In either situation the optimum solution to the original problem is contained in the final tableau.

By means of the symmetric dual relationships the general linear-programming problem can be represented as a problem that is concerned with the solution of a set of inequalities in nonnegative variables. A vector **X** which satisfies the constraints

$$\begin{aligned} \mathbf{AX} &\geq \mathbf{b} \\ \mathbf{WA} &\leq \mathbf{c} \\ \mathbf{Wb} &\geq \mathbf{cX} \\ \mathbf{X} &\geq \mathbf{0} \\ \mathbf{W} &\geq \mathbf{0} \end{aligned}$$

will be an optimum solution to the problem of minimizing

$$\mathbf{cX}$$

subject to

$$\begin{aligned} \mathbf{AX} &\geq \mathbf{b} \\ \mathbf{X} &\geq \mathbf{0} \end{aligned}$$

Let us briefly investigate the economic interpretation of the activity-analysis problem of Sec. 2, Chap. 1, and its dual.[1] The primal problem

[1] See Allen [2], Baumol [3*f*], Dorfman [39] Dorfman, Samuelson, and Solow [40], Harrison [53], and Tucker [95] for further discussion. For an interpretation of the primal and dual problems (and linear programming in general) in terms of classical Lagrange multipliers, see Tucker [96].

can be written so as to maximize

$$\mathbf{cX}$$

subject to

$$\mathbf{AX} \leq \mathbf{b}$$
$$\mathbf{X} \geq \mathbf{O}$$

where the a_{ij} represent the number of units of resource i required to produce one unit of commodity j, the b_i represent the maximum number of units of resource i available, and the c_j represent the value (profit) per unit of commodity j produced. The corresponding dual problem is to minimize

$$\mathbf{Wb}$$

subject to

$$\mathbf{WA} \geq \mathbf{c}$$
$$\mathbf{W} \geq \mathbf{O}$$

Whereas the physical interpretation of the primal is straightforward, the corresponding interpretation of the dual is not so evident. Questions arise as to the meaning of the dual objective function and inequalities. These can best be answered by interpreting the primal and dual problems in terms of physical units in order to determine the meaning of the dual variables (Harrison [53]).

The primal problem is to maximize

$$\sum_{j=1}^{n} \left(\frac{\text{value}}{\text{output } j}\right) (\text{output } j) = (\text{value})$$

subject to

$$\sum_{j=1}^{n} \left(\frac{\text{input } i}{\text{output } j}\right) (\text{output } j) \leq (\text{input } i) \qquad i = 1, 2, \ldots, m$$

$$(\text{output } j) \geq 0 \qquad j = 1, 2, \ldots, n$$

and the dual is to minimize

$$\sum_{i=1}^{m} (\text{input } i) w_i = (?)$$

subject to

$$\sum_{i=1}^{m} \left(\frac{\text{input } i}{\text{output } j}\right) w_i \geq \left(\frac{\text{value}}{\text{output } j}\right) \qquad j = 1, 2, \ldots, n$$

$$w_i \geq 0 \qquad i = 1, 2, \ldots, m$$

We see that the dual constraints will be consistent if the w_i are in units of value per unit of input i. The dual problem would then be to minimize

$$\sum_{i=1}^{m} (\text{input } i)\left(\frac{\text{value}}{\text{input } i}\right) = (\text{value})$$

subject to

$$\sum_{i=1}^{m} \left(\frac{\text{input } i}{\text{output } j}\right)\left(\frac{\text{value}}{\text{input } i}\right) \geq \left(\frac{\text{value}}{\text{output } j}\right) \qquad j = 1, 2, \ldots, n$$

$$\left(\frac{\text{value}}{\text{input } i}\right) \geq 0 \qquad i = 1, 2, \ldots, m$$

Verbal descriptions of the primal and dual problems can then be stated as follows:

The primal problem. With a given unit of value of each output (c_j) and a given upper limit for the availability of each input (b_i), how much of each output (x_j) should be produced in order to maximize the value of the total output?

The dual problem. With a given availability of each input (b_i) and a given lower limit of unit value for each output (c_j), what unit values should be assigned to each input (w_i) in order to minimize the value of the total input? The variables w_i are referred to by various names, e.g., accounting, fictitious, or shadow prices.

In line with the above, let us assume that we have determined by means of the simplex method an optimum solution to the activity-analysis problem. As we are maximizing the objective function, all vectors both in and out of the basis will have an associated $z_j - c_j \geq 0$. Because of some external conditions, e.g., government regulations or marketing conditions, the problem might stipulate that the kth activity must be in the final solution at a positive level. However, the optimal basis does not include vector $\mathbf{P}_k$. We must then alter our maximum solution by forcing vector $\mathbf{P}_k$ into the basis and removing some other activity. Assuming that $\mathbf{P}_k$ does not generate a multiple solution, i.e., that we have $z_k - c_k > 0$, the introduction of $\mathbf{P}_k$ reduces the maximum value of the objective function by the amount $(z_k - c_k)\theta_0$. This value is the loss in profit forced upon the manufacturer by the external condition. The value of $z_k - c_k$ is the net cost of introducing one unit of the kth activity, where the z_k is termed the indirect cost and the c_k the direct cost. The relationship between the net costs and accounting prices can be seen by analyzing the simplex tableau corresponding to the optimal solution of the activity-analysis problem. The set of indirect costs, i.e., the $z_j - c_j$ elements that correspond to the slack vectors, represents a set of accounting prices that solve the dual problem. If all the slack variables are in the final basis, then, in terms of the primal, no outputs are generated, no inputs are used, and the maximum profit is zero. For the dual this means that the minimum value of the total input is also zero. This is evidenced by all the indirect costs corresponding to the slack vectors being equal to zero.[1] If the optimum solution of the primal does not involve any of the slack variables at a positive level, then the primal problem yields a positive profit while utilizing all the available inputs. For this case none of the indirect costs associated with the slack vectors is equal to zero, and the minimum value of the total input is positive and equal to the maximum profit.

EXERCISES

1. Write out the corresponding primal problem to the following symmetric dual problem and solve both problems by the simplex procedure: Maximize

$$w_1 + w_2 + w_3$$

subject to

$$\begin{aligned} 2w_1 + w_2 + 2w_3 &\leq 2 \\ 4w_1 + 2w_2 + w_3 &\leq 2 \\ w_1 &\geq 0 \\ w_2 &\geq 0 \\ w_3 &\geq 0 \end{aligned}$$

2. Solve the following problem by the simplex method: Minimize

$$2x_1 - 3x_2$$

subject to

$$\begin{aligned} 2x_1 - x_2 - x_3 &\geq 3 \\ x_1 - x_2 + x_3 &\geq 2 \\ x_1 &\geq 0 \\ x_2 &\geq 0 \\ x_3 &\geq 0 \end{aligned}$$

3. Write out the dual problem to the problem given in Exercise 2 and solve it by graphing the constraints.

4. Graph the constraints of the following problem and its corresponding dual problem: Minimize

$$x_1 - x_2$$

[1] A vector in the basis has its $z_j - c_j = 0$.

subject to

$$\begin{aligned} 2x_1 + x_2 &\geq 2 \\ -x_1 - x_2 &\geq 1 \\ x_1 \qquad &\geq 0 \\ x_2 &\geq 0 \end{aligned}$$

5. Write out the dual to the transportation problem of Sec. 2, Chap. 1.

6. Construct a two-variable, two-inequality primal problem which has no feasible solutions and whose corresponding dual problem also has no feasible solutions.

7. *Farkas' Lemma* (Charnes and Cooper [10*a*], Dantzig [16*c*], and Hadley [52*d*]). Prove the following: A vector $\mathbf{b}$ will satisfy $\mathbf{Wb} \geq 0$ for all $\mathbf{W}$ satisfying $\mathbf{WA} \geq \mathbf{O}$ if and only if there exists an $\mathbf{X} \geq \mathbf{O}$ such that $\mathbf{AX} = \mathbf{b}$. Hint: To prove necessity, define appropriate primal and dual problems and apply the duality theorem.

CHAPTER 6 / THE REVISED SIMPLEX METHOD

1. THE GENERAL FORM OF THE INVERSE

Upon critically analyzing the simplex computational procedure (see Chaps. 3 and 4), we see that the essential element which enables us to progress from each basic feasible solution to the optimum solution is the explicit knowledge of the representation of the vectors not in the current basis in terms of the basis vectors. Given this information, we can do the following:

1. Calculate the $z_j - c_j$ elements to determine which vector to introduce into the basis or to determine whether the current solution is optimal.
2. Determine which vector to eliminate from the basis.
3. Transform the basis and obtain the new solution.

In Sec. 5 of Chap. 2 we showed that, given a basis $\mathbf{B}$ of m-dimensional vectors, $(\mathbf{P}_1\mathbf{P}_2 \cdots \mathbf{P}_m)$, the linear combination that expresses any other m-dimensional vector $\mathbf{P}_j$ in terms of $\mathbf{B}$ is determined by computing the vector

$$\mathbf{X}_j = \mathbf{B}^{-1}\mathbf{P}_j \tag{1.1}$$

where $\mathbf{X}_j = (x_{1j},x_{2j}, \ldots ,x_{mj})$ is a column vector. We then have

$$\mathbf{P}_j = x_{1j}\mathbf{P}_1 + x_{2j}\mathbf{P}_2 + \cdots + x_{mj}\mathbf{P}_m$$

which is the desired linear combination. Consider the general linear-programming problem of minimizing

$$\mathbf{cX}$$

subject to

$$\mathbf{AX} = \mathbf{b}$$

and

$$\mathbf{X} \geq \mathbf{O}$$

If we let the matrix $\mathbf{B}$ correspond to the first m vectors of $\mathbf{A}$, such that

$$\mathbf{BX}_0 = \mathbf{b}$$
$$\mathbf{X}_0 \geq \mathbf{O}$$

where $\mathbf{X}_0 = (x_{10},x_{20}, \ldots ,x_{m0})$, we have a basic feasible solution given by

$$\mathbf{X}_0 = \mathbf{B}^{-1}\mathbf{b} \tag{1.2}$$

The linear combinations of all the vectors of $\mathbf{A}$ in terms of $\mathbf{B}$ can be determined by (1.1) for $j = 1, 2, \ldots, n$. In Sec. 1 of Chap. 4 we defined for any feasible basis the quantities

$$z_j = c_1x_{1j} + c_2x_{2j} + \cdots + c_mx_{mj} \tag{1.3}$$

for $j = 1, 2, \ldots, n$, where the c_i correspond to the cost coefficients of those vectors in the basis. For any j we have from (1.1) that (1.3) can be rewritten to read

$$z_j = \mathbf{c}_0\mathbf{X}_j = \mathbf{c}_0\mathbf{B}^{-1}\mathbf{P}_j \qquad j = 1, 2, \ldots, n$$

where $\mathbf{c}_0 = (c_1,c_2, \ldots ,c_m)$ is a row vector. Hence, given the m-dimensional row vector

$$\boldsymbol{\pi} = \mathbf{c}_0\mathbf{B}^{-1} \tag{1.4}$$

for a feasible basis $\mathbf{B}$, we can compute the corresponding z_j. The vector $\boldsymbol{\pi} = (\pi_1,\pi_2, \ldots ,\pi_m)$ is called the *pricing* or *multiplier vector* and the individual π_i are the *pricing* or *simplex multipliers.* We *price out* a vector $\mathbf{P}_j$ not in the basis by computing $\boldsymbol{\pi}\mathbf{P}_j - c_j = z_j - c_j$. From (1.1), (1.2), and (1.4) we note that the information necessary to proceed from feasible solution to feasible solution can be obtained if, for each feasible basis **B**, we have explicit knowledge of $\mathbf{B}^{-1}$ and the original data consisting of **A**, **b**, and **c**. This fact has led to the development of a computational procedure to solve the general linear-programming problem termed the *revised simplex method* (Dantzig, Orden, and Wolfe [32] and Dantzig [20, 21]).

The main difference between the original simplex method and the revised procedure is that in the former we transform all the elements of the simplex tableau by means of the elimination formulas, while in the latter we need transform only the elements of an inverse matrix by means of the same formulas.

Since its development the revised procedure and especially its variation which employs the product form of the inverse have been selected for use on the larger high-speed computers.[1] The two reasons for this are:

1. For problems whose matrix of coefficients contains a large number of zero elements the total amount of computation is reduced.[2] The revised procedure always deals with the original coefficients, and because the computer codes can be developed to multiply only nonzero elements, the total processing time is greatly reduced. Also, the original nonzero elements can be compactly stored in the computer memory. The original simplex procedure transforms the zero elements to nonzeros as the computation progresses. The total number of computations in the revised method is, in general, less than the number in the original method.[3]
2. The amount of new information the computer is required to record is, in general, reduced, since in the revised procedure we need to record only the inverse and the solution vector, while in the original method the complete simplex tableau has to be recorded. The recording is even further reduced if the product form of the inverse is used.

Before describing the systematic computational rules for the revised procedure, we wish to show how the inverse of each new basis can be obtained from the preceding basis by application of the elimination formulas. Here we are given an old basis $\mathbf{B} = (\mathbf{P}_1\mathbf{P}_2 \cdots \mathbf{P}_l \cdots \mathbf{P}_m)$, which differs from the new one in that vector $\mathbf{P}_k$ has replaced vector $\mathbf{P}_l$. Let

[1] See Dantzig and Orchard-Hays [29] and Sec. 2 of this chapter.

[2] Problems such as the caterer problem, production-scheduling problem, and interindustry problem described in Chap. 11 and many other problems are of this nature.

[3] Wagner [99] gives a complete comparison of the original and revised simplex methods. There he shows that, for $n > 3m$, the revised method is better in terms of the number of computations.

$\bar{\mathbf{B}} = (\mathbf{P}_1\mathbf{P}_2 \cdots \mathbf{P}_k \cdots \mathbf{P}_m)$ represent the new basis. We then have

$$\mathbf{B}^{-1}\mathbf{B} = \mathbf{B}^{-1}(\mathbf{P}_1\mathbf{P}_2 \cdots \mathbf{P}_l \cdots \mathbf{P}_m)$$
$$= \begin{pmatrix} 1 & 0 & \cdots & 0 & \cdots & 0 \\ 0 & 1 & \cdots & 0 & \cdots & 0 \\ \cdot & \cdot & \cdot & \cdot & \cdot & \cdot \\ 0 & 0 & \cdots & 1 & \cdots & 0 \\ \cdot & \cdot & \cdot & \cdot & \cdot & \cdot \\ 0 & 0 & \cdots & 0 & \cdots & 1 \end{pmatrix} \tag{1.5}$$

and from (1.1)

$$\mathbf{B}^{-1}\bar{\mathbf{B}} = \mathbf{B}^{-1}(\mathbf{P}_1\mathbf{P}_2 \cdots \mathbf{P}_k \cdots \mathbf{P}_m)$$
$$= \begin{pmatrix} 1 & 0 & \cdots & x_{1k} & \cdots & 0 \\ 0 & 1 & \cdots & x_{2k} & \cdots & 0 \\ \cdot & \cdot & \cdot & \cdot & \cdot & \cdot \\ 0 & 0 & \cdots & x_{lk} & \cdots & 0 \\ \cdot & \cdot & \cdot & \cdot & \cdot & \cdot \\ 0 & 0 & \cdots & x_{mk} & \cdots & 1 \end{pmatrix} \tag{1.5a}$$

If we let b_{ij} equal the element in the ith row and jth column of $\mathbf{B}^{-1}$ and let $\bar{b}_{ij}$ be the corresponding element of $\bar{\mathbf{B}}^{-1}$, we have the $\bar{b}_{ij}$ given by the elimination formulas

$$\begin{aligned} \bar{b}_{ij} &= b_{ij} - \frac{b_{lj}}{x_{lk}} x_{ik} \qquad \text{for } i \neq l \\ \bar{b}_{lj} &= \frac{b_{lj}}{x_{lk}} \end{aligned} \tag{1.6}$$

The validity of this transformation can be verified by direct multiplication of $(\bar{\mathbf{B}}^{-1}\bar{\mathbf{B}})$ to determine whether the product yields the identity matrix.[1] For example, we have the inner product of the first row of $\bar{\mathbf{B}}^{-1}$ and the first column of $\bar{\mathbf{B}}$ given by[2]

$$\left(b_{11} - \frac{b_{l1}}{x_{lk}} x_{1k}\right) a_{11} + \left(b_{12} - \frac{b_{l2}}{x_{lk}} x_{1k}\right) a_{21} + \cdots + \left(b_{1m} - \frac{b_{lm}}{x_{lk}} x_{1k}\right) a_{m1}$$

or

$$(b_{11}a_{11} + b_{12}a_{21} + \cdots + b_{1m}a_{m1}) - (b_{l1}a_{11} + b_{l2}a_{21} + \cdots + b_{lm}a_{m1}) \frac{x_{1k}}{x_{lk}} \tag{1.7}$$

[1] From (1.5a), $\bar{\mathbf{B}}^{-1}$ in terms of $\mathbf{B}^{-1}$ is given by $\mathbf{B}^{-1}\bar{\mathbf{B}} = \mathbf{I}_l$, where $\mathbf{I}_l$ is the right-hand side of (1.5a). Then $\bar{\mathbf{B}} = \mathbf{B}\mathbf{I}_l$, $\bar{\mathbf{B}}^{-1} = (\mathbf{B}\mathbf{I}_l)^{-1}$, and finally, $\bar{\mathbf{B}}^{-1} = \mathbf{I}_l^{-1}\mathbf{B}^{-1}$. In Sec. 2, we show that $\mathbf{I}_l^{-1} = \mathbf{E}^l$, where $\mathbf{E}^l$ is an identity matrix except for the lth column.

[2] This product is obtained by the rule of matrix multiplication described in Sec. 1 of Chap. 2.

The first term of (1.7) is the inner product of the first row of $\mathbf{B}^{-1}$ and $\mathbf{P}_1$, which from (1.5) is equal to 1; and the second term of (1.7) contains the inner product of the lth row of $\mathbf{B}^{-1}$ and $\mathbf{P}_1$, which equals 0. Hence (1.7) is equal to 1. In a similar manner we can show that $(\bar{\mathbf{B}}^{-1}\bar{\mathbf{B}}) = \mathbf{I}$, and hence formulas (1.6) do generate $\bar{\mathbf{B}}^{-1}$.†

We next describe the computational format for the revised simplex method as developed by Dantzig [19] and Orchard-Hays [81]. We shall not attempt to develop the revised method fully, in the way the original simplex method was presented in Chaps. 3 and 4, but the reader will be able to determine the relationship between these methods.

In order to facilitate the use of the revised procedure, we shall introduce and define below two new variables x_{n+m+1} and x_{n+m+2}.

For the revised procedure we write the general linear-programming problem as follows: Maximize

$$x_{n+m+1} \tag{1.8}$$

subject to

$$\begin{aligned} a_{11}x_1 + a_{12}x_2 + \cdots + a_{1n}x_n &= b_1 \\ \cdots\cdots\cdots\cdots\cdots\cdots\cdots &\cdots \\ a_{m1}x_1 + a_{m2}x_2 + \cdots + a_{mn}x_n &= b_m \end{aligned} \tag{1.9}$$

$$\begin{aligned} c_1x_1 + c_2x_2 + \cdots + c_nx_n + x_{n+m+1} &= 0 \\ x_1 &\geq 0 \\ x_2 &\geq 0 \\ &\vdots \\ x_n &\geq 0 \end{aligned} \tag{1.10}$$

Here we assume all $b_i \geq 0$. Since

$$x_{n+m+1} = -c_1x_1 - c_2x_2 - \cdots - c_nx_n$$

and since the linear-programming problem that minimizes an objective function is basically the same as the one which maximizes the negative of this objective function subject to the same constraints, the above problem is equivalent to the problem of minimizing

$$c_1x_1 + c_2x_2 + \cdots + c_nx_n \tag{1.8a}$$

† The matrices formed by vectors $\mathbf{P}_1$, $\mathbf{P}_4$, and $\mathbf{P}_6$ in the example of Sec. 2, Chap. 4, illustrate this transformation.

subject to

$$
\begin{aligned}
a_{11}x_1 + a_{12}x_2 + \cdots + a_{1n}x_n &= b_1 \\
\cdots\cdots\cdots\cdots\cdots\cdots & \\
a_{m1}x_1 + a_{m2}x_2 + \cdots + a_{mn}x_n &= b_m
\end{aligned}
\tag{1.9a}
$$

$$
\begin{aligned}
x_1 \qquad\qquad\qquad &\geq 0 \\
x_2 \qquad\qquad &\geq 0 \\
\cdot \qquad &\;\; \cdot \\
\cdot \quad &\;\; \cdot \\
\cdot \;\; &\;\; \cdot \\
x_n &\geq 0
\end{aligned}
\tag{1.10a}
$$

Since the objective function in the revised problem consists of only x_{n+m+1}, that is, the value of x_{n+m+1} is the negative value of the minimizing objective function, we do not restrict the sign of x_{n+m+1} in (1.8) and (1.9).

As in the original method, the revised procedure has to begin with a basis consisting of an identity matrix whose unit vectors correspond to either real or artificial vectors. If the revised procedure starts with artificial vectors, we again have the problem of determining a first basic feasible solution (if one exists) and then continuing on to the optimum solution. We shall let Phase I refer to the part of the computation that determines feasibility and Phase II to the part that develops the optimum solution. In order to generalize the concept of artificial vectors and to facilitate the computation of Phase I, we calculate a "redundant" equation defined as

$$a_{m+2,\,1}x_1 + a_{m+2,\,2}x_2 + \cdots + a_{m+2,\,n}x_n + x_{m+n+2} = b_{m+2}$$

where

$$
\begin{aligned}
a_{m+2,j} &= -\sum_{i=1}^{m} a_{ij} \qquad j = 1, 2, \ldots, n \\
b_{m+2} &= -\sum_{i=1}^{m} b_i
\end{aligned}
\tag{1.11}
$$

We see that $a_{m+2,\,j}$ is just the negative sum of the coefficients in the jth column of the coefficient matrix $\mathbf{A}$ and that b_{m+2} is the negative sum of all the b_i. The significance of the "redundant" variable x_{m+n+2} will be described below. If we let $c_j = a_{m+1,\,j}$, we then rewrite the revised problem (1.8) to (1.10) to maximize

$$x_{n+m+1} \tag{1.12}$$

subject to

$$\begin{array}{lr}
a_{11}x_1 + a_{12}x_2 + \cdots + a_{1n}x_n + x_{n+1} = b_1 & \\
a_{21}x_1 + a_{22}x_2 + \cdots + a_{2n}x_n + x_{n+2} = b_2 & \\
\cdots\cdots\cdots\cdots\cdots\cdots\cdots\cdots\cdots\cdots\cdots\cdots & \\
a_{m1}x_1 + a_{m2}x_2 + \cdots + a_{mn}x_n + x_{n+m} = b_m & (1.13) \\
a_{m+1,\,1}x_1 + a_{m+1,\,2}x_2 + \cdots + a_{m+1,\,n}x_n + x_{n+m+1} = 0 & \\
a_{m+2,\,1}x_1 + a_{m+2,\,2}x_2 + \cdots + a_{m+2,\,n}x_n + x_{n+m+2} = b_{m+2} &
\end{array}$$

$$\begin{array}{lr}
x_1 \geq 0 & \\
x_2 \geq 0 & \\
\cdots & \\
x_n \geq 0 & (1.14) \\
x_{n+1} \geq 0 & \\
\cdots & \\
x_{n+m} \geq 0 &
\end{array}$$

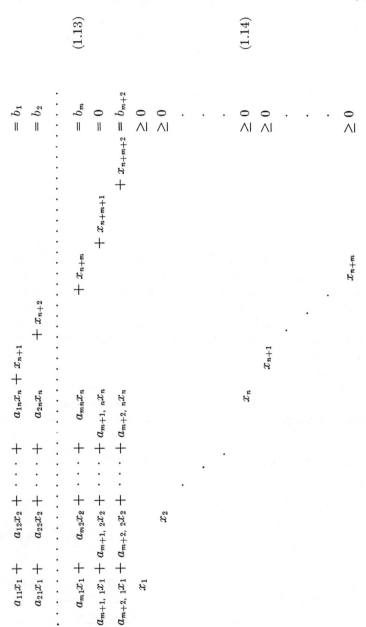

where the nonnegative artificial variables x_{n+i} for $i = 1, 2, \ldots, m$ have been added to the corresponding equations. The artificial variables can be interpreted as the measure of the error between the sides of the original equations of (1.9) evaluated for a vector $\mathbf{X} = (x_1, x_2, \ldots, x_n)$ which does not satisfy all the constraints of (1.9). From (1.11) and (1.13) we have

$$x_{n+1} + x_{n+2} + \cdots + x_{n+m} + x_{n+m+2} = 0$$

The variable x_{n+m+2} can then be thought of as representing the negative of the absolute sum of errors of an approximate nonnegative solution to the equations. Since $x_{n+i} \geq 0$ for $i = 1, 2, \ldots, m$, it is clear that x_{n+m+2} cannot be positive.

As written in (1.12) to (1.14), the revised problem consists of $m + 2$ equations in $n + m + 2$ variables. Here a basic feasible solution will contain $m + 2$ variables from the set $(x_1, x_2, \ldots, x_n, x_{n+m+1}, x_{n+m+2})$. The signs of the last two variables are unrestricted, and they will always be in the solution. The variables from the set $(x_1, x_2, \ldots, x_n)$ that are in the optimum solution represent a corresponding basic maximum feasible solution to the problem of (1.8) to (1.10) with x_{n+m+1} the value of the objective function. This optimal solution is also a basic minimum feasible solution to (1.8*a*) to (1.10*a*), where

$$-x_{n+m+1} = c_1 x_1 + \cdots + c_n x_n$$

is the corresponding value of the objective function. The computational procedure for the revised problem starts out with the m artificial variables $x_{n+1}, \ldots, x_{n+m}$ and variables x_{n+m+1} and x_{n+m+2} in the solution. We must employ the Phase I procedure to find a first basic feasible solution.

In Phase I we first consider solving the problem of maximizing x_{n+m+2}, subject to (1.13) and (1.14), with both x_{n+m+1} and x_{n+m+2} unrestricted as to sign. If the maximum value of x_{n+m+2} is zero, then all the x_{n+i} for $i = 1, 2, \ldots, m$ must also equal zero, and the x_j for $j = 1, 2, \ldots, n$ of this "preliminary maximum solution" represent a basic feasible solution to the problems (1.9), (1.10) and (1.9*a*), (1.10*a*). If the maximum of x_{n+m+2} is less than zero, then this implies that at least one artificial variable is in the Phase I solution with a nonzero value and hence that no feasible solutions exist for the original problem. In the former case we then go on to Phase II, where we maximize x_{n+m+1}, subject to (1.13) and (1.14), while keeping x_{n+m+2} equal to zero. It should be noted that the set of basic feasible solutions found in Phase II might contain artificial variables with zero values. The final Phase II solution is the desired optimum solution.

We shall next describe and illustrate the complete computational procedure for the revised problem which starts with a full artificial basis. In the computational tableau for the procedure we need to keep record of only the following information: the variables in the solution, the corresponding values of the variables, and the inverse of the current basis. In developing new solutions, we must have a means of calculating the z_j terms. We shall do this by having the vector defined by (1.4) available for the current basis. We can then determine which vector is to be introduced into the basis and calculate (1.1), where $\mathbf{P}_j$ corresponds to the vector being introduced. We next find out which vector is to be eliminated, using the method described in Chap. 4, and then transform the inverse of the current basis by the elimination formulas to obtain the inverse of the new basis.

Let us detach the required coefficients from (1.13) and arrange them in an $m + 2 \times n$ matrix as follows:

$$\bar{\mathbf{A}} = \begin{pmatrix} a_{11} & a_{12} & \cdots & a_{1k} & \cdots & a_{1n} \\ a_{21} & a_{22} & \cdots & a_{2k} & \cdots & a_{2n} \\ \cdots & \cdots & \cdots & \cdots & \cdots & \cdots \\ a_{l1} & a_{l2} & \cdots & a_{lk} & \cdots & a_{ln} \\ \cdots & \cdots & \cdots & \cdots & \cdots & \cdots \\ a_{m1} & a_{m2} & \cdots & a_{mk} & \cdots & a_{mn} \\ a_{m+1,\,1} & a_{m+1,\,2} & \cdots & a_{m+1,\,k} & \cdots & a_{m+1,\,n} \\ a_{m+2,\,1} & a_{m+2,\,2} & \cdots & a_{m+2,\,k} & \cdots & a_{m+2,\,n} \end{pmatrix}$$

We shall denote each column vector of $\bar{\mathbf{A}}$ by $\bar{\mathbf{A}}_j$. Let us compare the information in matrix $\bar{\mathbf{A}}$ with the initial artificial-basis computational tableau for the original procedure for the minimization problem described in Chap. 4 (see Tableau 4.4). We note that the matrix formed by rows $1, 2, \ldots, m, m + 1, m + 2$ and the columns $\mathbf{P}_1, \mathbf{P}_2, \ldots, \mathbf{P}_n$ of the tableau is equal to matrix $\bar{\mathbf{A}}$, except that rows $m + 1$ and $m + 2$ of $\bar{\mathbf{A}}$ are the negative of the corresponding rows of the tableau. As in the original method, row $m + 2$ of $\bar{\mathbf{A}}$ will be used in computing the $z_j - c_j$ elements when artificial variables are still in the solution, and row $m + 1$ will be used when they have been eliminated.[1]

Since our starting basis for the revised procedure consists of an identity matrix, its inverse is also an identity matrix, and this information is recorded in the following $m + 2 \times m + 2$ matrix:

$$\mathbf{U} = \left(\begin{array}{cccccc|cc} 1 & 0 & \cdots & 0 & \cdots & 0 & 0 & 0 \\ 0 & 1 & \cdots & 0 & \cdots & 0 & 0 & 0 \\ \cdots & \cdots & \cdots & \cdots & \cdots & \cdots & \cdots & \cdots \\ 0 & 0 & \cdots & 1 & \cdots & 0 & 0 & 0 \\ \cdots & \cdots & \cdots & \cdots & \cdots & \cdots & \cdots & \cdots \\ 0 & 0 & \cdots & 0 & \cdots & 1 & 0 & 0 \\ \hline 0 & 0 & \cdots & 0 & \cdots & 0 & 1 & 0 \\ 0 & 0 & \cdots & 0 & \cdots & 0 & 0 & 1 \end{array}\right)$$

[1] Dantzig and Orchard-Hays [29] write the revised problem with the equations containing variables x_{n+m+1} and x_{n+m+2} as the zeroth and first equations, respectively. They renumber the equations to make $x_{n+m+1} = x_0$ and $x_{n+m+2} = x_{n+1}$.

The first m rows and m columns of **U** represent the inverse of the starting basis **B**. The last two rows of **U** will be used to determine which vector should be introduced into the basis, with row $m + 2$ generating the information in Phase I and row $m + 1$ in Phase II. We will embed the matrix **U** in the starting computational tableau and transform its elements as described below. Let us designate the elements of **U** by u_{ij} and its rows by the row vectors $\mathbf{U}_i = (u_{i1}, u_{i2}, \ldots, u_{i,m+2})$. For convenience in describing the computational procedure, we shall let $\mathbf{U}_i$ represent both the original and transformed versions of the ith row of **U**. The matrix **U** will be the inverse of an $m + 2 \times m + 2$ matrix whose columns are vectors of (1.13).

We arrange the matrix **U** and the initial solution of $x_{n+i} = b_i$ for $i = 1, 2, \ldots, m$; $x_{n+m+1} = 0$; and $x_{n+m+2} = b_{m+2}$ as shown in the starting tableau of Tableau 6.1. The computational steps for the phases are as follows:

Phase I. Artificial variables in the solution with positive values

STEP 1. If $x_{m+n+2} < 0$, compute

$$\delta_j = \mathbf{U}_{m+2}\bar{\mathbf{A}}_j\dagger$$
$$= u_{m+2,\,1}a_{1j} + u_{m+2,\,2}a_{2j} + \cdots + u_{m+2,\,m+2}a_{m+2,\,j} \qquad j = 1, 2, \ldots, n$$

and continue to Step 2.

If $x_{m+n+2} = 0$, go to Step 1 of Phase II.

STEP 2. If all $\delta_j \geq 0$, then x_{n+m+2} is at its maximum, and hence no feasible solution exists for the problem (1.8) to (1.10).

If at least one $\delta_j < 0$, then the variable to be introduced into the solution, x_k, corresponds to

$$\delta_k = \min \delta_j$$

If more than one of the δ_j are equal to the minimum, select δ_k such that k is the smallest index. This is an arbitrary, but computationally sound, rule for breaking ties.

† In Phase I for the revised problem the objective function to be maximized is x_{m+n+2}. Hence, all cost coefficients except $c_{m+n+2} = 1$ for this maximizing problem are equal to zero. Similarly, in Phase II the objective function to be maximized is x_{m+n+1}, and the cost coefficients are all equal to zero except $c_{m+n+1} = 1$. The quantities δ_j of Phase I and γ_j defined in Phase II can be interpreted in two ways. They are equal to, respectively, the artificial and real $z_j - c_j$ elements of the original simplex procedure adapted to maximizing the objective functions of the revised problem. The criterion then states that the optimal solution has been reached when all $z_j - c_j \geq 0$. For the corresponding iteration, the quantities δ_j and γ_j are also equivalent to the $c_j - z_j$ elements that would be computed in solving the minimization problem defined by (1.8*a*) to (1.10*a*). Here the criterion states that the optimal solution has been reached when all $c_j - z_j \geq 0$. In setting up the revised problem, we shall always write the original problem as one to be minimized and then determine matrix $\bar{\mathbf{A}}$, where $a_{m+1,\,j} = c_j$ of the minimization problem.

STEP 3. Compute

$$\begin{aligned} x_{ik} &= \mathbf{U}_i\bar{\mathbf{A}}_k \\ &= u_{i1}a_{1k} + u_{i2}a_{2k} + \cdots + u_{i,\,m+2}a_{m+2,\,k} \end{aligned}$$

for $i = 1, 2, \ldots, m, m+1, m+2$. The variable x_l to be eliminated from the solution corresponds to the ratio

$$\theta_0 = \min_i \frac{x_{i0}}{x_{ik}} = \frac{x_{l0}}{x_{lk}} \qquad i = 1, 2, \ldots, m$$

where the ratio is formed only for those $x_{ik} > 0$. If there are ties for the minimum, we select l to be the smallest index.[1] [For discussion purposes we have taken the ratios for the values of index $i = 1, 2, \ldots, m$. The ratios x_{i0}/x_{ik} are actually taken for those i whose corresponding x_{i0} are in the basic solution to (1.13).]

STEP 4. The new values of the variables in the basic solution are obtained by the formulas

$$\begin{aligned} x'_{i0} &= x_{i0} - \frac{x_{l0}}{x_{lk}}\, x_{ik} \qquad \text{for } i \neq k \\ x'_{k0} &= \frac{x_{l0}}{x_{lk}} \end{aligned}$$

The new elements of the matrix $\mathbf{U}$ are transformed by

$$\begin{aligned} u'_{ij} &= u_{ij} - \frac{u_{lj}}{x_{lk}}\, x_{ik} \qquad \text{for } i \neq l \\ u'_{lj} &= \frac{u_{lj}}{x_{lk}} \end{aligned}$$

Under this transformation the $(m+1)$st and $(m+2)$nd columns of $\mathbf{U}$ will never change. These unit vectors enable us to add the correct c_j when we compute the $c_j - z_j$. Using the transformed rows of $\mathbf{U}$ as shown in the transformed section of Tableau 6.1, the steps of Phase 1 are repeated until it is determined either that no feasible solutions exist or that the value of $x_{n+m+2} = 0$. In the latter case we go on to Phase II.

Phase II. No artificial variables in the solution with positive values

STEP 1. Here $x_{m+n+2} = 0$. Compute

$$\begin{aligned} \gamma_j &= \mathbf{U}_{m+1}\bar{\mathbf{A}}_j \\ &= u_{m+1,\,1}a_{1j} + u_{m+1,\,2}a_{2j} + \\ &\qquad \cdots + u_{m+1,\,m+2}a_{m+2,\,j} \qquad j = 1, 2 \ldots n \end{aligned}$$

[1] In this situation the resultant basic feasible solution will be degenerate. We shall discuss in Chap. 7 the theoretical computational rule for the degenerate case given by Dantzig [19]. It might be noted that the originators of the revised procedure do not incorporate the "degeneracy routine" in the computer codes developed at their research center (Orchard-Hays [82]).

STEP 2. If all $\gamma_j \geq 0$, then x_{n+m+1} is at its maximum value and the corresponding basic feasible solution is an optimum solution. The negative value of x_{n+m+1} is the true value of the objective function that was to be minimized.

If at least one $\gamma_j < 0$, let

$$\gamma_k = \min \gamma_j$$

Variable x_k is selected to be introduced into the solution. Ties in selecting γ_k are broken by selecting the variable with the smallest index.

STEP 3. Compute

$$\begin{aligned} x_{ik} &= \mathbf{U}_i\bar{\mathbf{A}}_k \\ &= u_{i1}a_{1k} + u_{i2}a_{2k} + \cdots + u_{i,\, m+2}a_{m+2,\, k} \end{aligned}$$

for $i = 1, 2, \ldots, m, m+1, m+2$. The variable x_{l0} to be eliminated from the solution is determined by finding the ratio

$$\theta_0 = \min_i \frac{x_{i0}}{x_{ik}} = \frac{x_{l0}}{x_{lk}} \qquad i = 1, 2, \ldots, m$$

for those $x_{ik} > 0$. (See Phase I, Step 3, for actual range of index i.) Ties are broken in the usual fashion. If all $x_{ik} \leq 0$, then the procedure has yielded a solution whose value of the objective function can be made arbitrarily large.

STEP 4. The new values of the variables are obtained by the formulas

$$x'_{i0} = x_{i0} - \frac{x_{l0}}{x_{lk}} x_{ik} \qquad \text{for } i \neq k$$

$$x'_{k0} = \frac{x_{l0}}{x_{lk}}$$

The elements of the matrix $\mathbf{U}$ are transformed by

$$u'_{ij} = u_{ij} - \frac{u_{lj}}{x_{lk}} x_{ik} \qquad \text{for } i \neq l$$

$$u'_{lj} = \frac{u_{lj}}{x_{lk}}$$

The steps are repeated until an optimum solution with either a finite or an infinite value of the objective function is determined.

The initial solution to (1.12) to (1.14) and the above steps can be arranged and carried out by the computational scheme shown in Tableau 6.1. A column has been set aside for the recording of the x_{ik} associated with each inverse. The inverse matrix $\mathbf{U}$ consists of the $m + 2 \times m + 2$ array enclosed in heavy lines. The last two columns of the transformed $\mathbf{U}$ do not change. The inverse of the m-dimensional basis selected from the $\mathbf{P}_j$ of (1.13) without the $(m + 1)$st and $(m + 2)$nd rows is enclosed in heavy and dashed lines.

Tableau 6.1

Row index of tableau	Index of variables in solution	Value of variables	Matrix $\mathbf{U}$: Inverse of basis, $\mathbf{B}^{-1}$ / $\mathbf{U}_{m+1}$, $\mathbf{U}_{m+2}$	$\mathbf{U}_i\bar{\mathbf{A}}_k$ $(\mathbf{B}^{-1}\mathbf{P}_k)$ / $\mathbf{U}_{m+1}\bar{\mathbf{A}}_k$, $\mathbf{U}_{m+2}\bar{\mathbf{A}}_k$

1. Starting Tableau

1	$n+1$	$x_{n+1} = b_1$	1	0	⋯	0	⋯	0	0	0	x_{1k}
2	$n+2$	$x_{n+2} = b_2$	0	1	⋯	0	⋯	0	0	0	x_{2k}
⋮	⋮	⋮	⋮	⋮	⋯	⋮	⋯	⋮	⋮	⋮	⋮
l	$n+l$	$x_{n+l} = b_l$	0	0	⋯	1	⋯	0	0	0	x_{lk}
⋮	⋮	⋮	⋮	⋮	⋯	⋮	⋯	⋮	⋮	⋮	⋮
m	$n+m$	$x_{n+m} = b_m$	0	0	⋯	0	⋯	1	0	0	x_{mk}
$m+1$	$n+m+1$	$x_{n+m+1} = 0$	0	0	⋯	0	⋯	0	1	0	$x_{m+1,\,k}$
$m+2$	$n+m+2$	$x_{n+m+2} = b_{m+2}$	0	0	⋯	0	⋯	0	0	1	$x_{m+2,\,k}$

2. Transformed Tableau

1	$n+1$	x'_{n+1}	u'_{11}	u'_{12}	⋯	u'_{1l}	⋯	u'_{1m}	0	0	
2	$n+2$	x'_{n+2}	u'_{21}	u'_{22}	⋯	u'_{2l}	⋯	u'_{2m}	0	0	
⋮	⋮	⋮	⋮	⋮	⋯	⋮	⋯	⋮	⋮	⋮	
l	k	x'_k	u'_{l1}	u'_{l2}	⋯	u'_{ll}	⋯	u'_{lm}	0	0	
⋮	⋮	⋮	⋮	⋮	⋯	⋮	⋯	⋮	⋮	⋮	
m	$n+m$	x'_{n+m}	u'_{m1}	u'_{m2}	⋯	u'_{ml}	⋯	u'_{mn}	0	0	
$m+1$	$n+m+1$	x'_{n+m+1}	$u'_{m+1,\,1}$	$u'_{m+1,\,2}$	⋯	$u'_{m+1,\,l}$	⋯	$u'_{m+1,\,m}$	1	0	
$m+2$	$n+m+2$	x'_{n+m+2}	$u'_{m+2,\,1}$	$u'_{m+2,\,2}$	⋯	$u'_{m+2,\,l}$	⋯	$u'_{m+2,\,m}$	0	1	

If the revised problem has unit vectors in its explicit mathematical statement, these vectors can be used in the starting basis for the revised procedure if their cost coefficients are equal to zero; for example, slack vectors. If these vectors are used, formulas (1.11) must be changed to read, respectively,

$$a_{m+2,\,j} = -\sum_{i \neq \mathbf{B}} a_{ij} \tag{1.15}$$

and

$$b_{m+2} = -\sum_{i \neq \mathbf{B}} b_i \tag{1.16}$$

where $i \neq \mathbf{B}$ represents the set of indices of the rows that still require an artificial vector. If m appropriate natural unit vectors are available, then $b_{m+2} = 0$ and the procedure starts with Phase II.

To illustrate the revised procedure and to help point out the relationships between the two simplex methods, we shall next solve the example of Sec. 3, Chap. 4, using the revised procedure.

Example. Maximize

$$x_1 + 2x_2 + 3x_3 - x_4$$

subject to

$$\begin{aligned} x_1 + 2x_2 + 3x_3 \qquad &= 15 \\ 2x_1 + x_2 + 5x_3 \qquad &= 20 \\ x_1 + 2x_2 + x_3 + x_4 &= 10 \end{aligned}$$

and $x_j \geq 0$.

Here $m = 3$ and $n = 4$. The objective function for the corresponding minimization problem is $-x_1 - 2x_2 - 3x_3 + x_4$.

For the revised procedure with a full artificial basis we rewrite the example to read: Maximize

$$x_8$$

subject to

$$\begin{aligned} x_1 + 2x_2 + 3x_3 \qquad + x_5 \qquad\qquad\qquad &= 15 \\ 2x_1 + x_2 + 5x_3 \qquad\qquad + x_6 \qquad\qquad &= 20 \\ x_1 + 2x_2 + x_3 + x_4 \qquad\qquad + x_7 \qquad &= 10 \\ -x_1 - 2x_2 - 3x_3 + x_4 \qquad\qquad\qquad + x_8 \quad &= 0 \\ -4x_1 - 5x_2 - 9x_3 - x_4 \qquad\qquad\qquad\qquad + x_9 &= -45 \end{aligned}$$

The coefficients of the first four variables for the fourth row are equal to the corresponding c_j as written in the original objective function to be minimized. The coefficients of variables x_1, x_2, x_3, x_4 and the right side of the fifth equation were obtained by formulas (1.11).

The $\bar{\mathbf{A}}$ and $\mathbf{U}$ matrices for the example are

$$\bar{\mathbf{A}} = \begin{pmatrix} 1 & 2 & 3 & 0 \\ 2 & 1 & 5 & 0 \\ 1 & 2 & 1 & 1 \\ -1 & -2 & -3 & 1 \\ -4 & -5 & -9 & -1 \end{pmatrix}$$

and

$$\mathbf{U} = \begin{pmatrix} 1 & 0 & 0 & 0 & 0 \\ 0 & 1 & 0 & 0 & 0 \\ 0 & 0 & 1 & 0 & 0 \\ 0 & 0 & 0 & 1 & 0 \\ 0 & 0 & 0 & 0 & 1 \end{pmatrix}$$

The starting tableau and the sequence of iterations are shown in Tableau 6.2.

2. THE PRODUCT FORM OF THE INVERSE

In Sec. 1 we developed the computational formulas (1.6) that enabled us to compute the inverse of a basis $\mathbf{\bar{B}}$ which differs by one vector from a basis $\mathbf{B}$ with known inverse $\mathbf{B}^{-1}$. Making use of the notation of Sec. 1, we have

$$\mathbf{B} = (\mathbf{P}_1\mathbf{P}_2 \cdots \mathbf{P}_l \cdots \mathbf{P}_m)$$

$$\mathbf{\bar{B}} = (\mathbf{P}_1\mathbf{P}_2 \cdots \mathbf{P}_k \cdots \mathbf{P}_m)$$

$$\mathbf{B}^{-1} = \begin{pmatrix} b_{11} & b_{12} & \cdots & b_{1l} & \cdots & b_{1m} \\ \cdots & \cdots & \cdots & \cdots & \cdots & \cdots \\ b_{l1} & b_{l2} & \cdots & b_{ll} & \cdots & b_{lm} \\ \cdots & \cdots & \cdots & \cdots & \cdots & \cdots \\ b_{m1} & b_{m2} & \cdots & b_{ml} & \cdots & b_{mm} \end{pmatrix}$$

and

$$\mathbf{B}^{-1}\mathbf{P}_k = \mathbf{X}_k = \begin{pmatrix} x_{1k} \\ \cdot \\ \cdot \\ \cdot \\ x_{lk} \\ \cdot \\ \cdot \\ \cdot \\ x_{mk} \end{pmatrix}$$

Also

$$\mathbf{\bar{B}}^{-1} = \begin{pmatrix} \bar{b}_{11} & \bar{b}_{12} & \cdots & \bar{b}_{1l} & \cdots & \bar{b}_{1m} \\ \cdots & \cdots & \cdots & \cdots & \cdots & \cdots \\ \bar{b}_{l1} & \bar{b}_{l2} & \cdots & \bar{b}_{ll} & \cdots & \bar{b}_{lm} \\ \cdots & \cdots & \cdots & \cdots & \cdots & \cdots \\ \bar{b}_{m1} & \bar{b}_{m2} & \cdots & \bar{b}_{ml} & \cdots & \bar{b}_{mm} \end{pmatrix}$$

Tableau 6.2

Row index of tableau	Index of variables in solution	Value of variables	Matrix U	x_{ik}

1. Starting Tableau

Phase I

Row index of tableau	Index of variables in solution	Value of variables	U					x_{ik}
1	5	15	1	0	0	0	0	3
2	6	20	0	1	0	0	0	5 (circled)
3	7	10	0	0	1	0	0	1
4	8	0	0	0	0	1	0	−3
5	9	−45	0	0	0	0	1	−9

$\delta_k = \delta_3 = \mathbf{U}_{m+2}\bar{\mathbf{A}}_3 = -9$

$\theta_0 = \frac{x_6}{x_{63}} = \frac{20}{5} = 4$

2. Second Iteration

Row index of tableau	Index of variables in solution	Value of variables	U					x_{ik}
1	5	3	1	−3/5	0	0	0	7/5 (circled)
2	3	4	0	1/5	0	0	0	1/5
3	7	6	0	−1/5	1	0	0	9/5
4	8	+12	0	3/5	0	1	0	−7/5
5	9	−9	0	9/5	0	0	1	−16/5

$\delta_k = \delta_2 = \mathbf{U}_{m+2}\bar{\mathbf{A}}_2 = -16/5$

$\theta_0 = \frac{x_5}{x_{52}} = \frac{15}{7}$

3. Third Iteration

Row index of tableau	Index of variables in solution	Value of variables	U					x_{ik}
1	2	15/7	5/7	−3/7	0	0	0	0
2	3	25/7	−1/7	3/7	0	0	0	0
3	7	15/7	−9/7	4/7	1	0	0	1 (circled)
4	8	+15	1	0	0	1	0	1
5	9	−15/7	16/7	3/7	0	0	1	−1

$\delta_k = \delta_4 = \mathbf{U}_{m+2}\bar{\mathbf{A}}_4 = -1$

$\theta_0 = \frac{x_7}{x_{74}} = \frac{15}{7}$

4. Fourth Iteration

Row index of tableau	Index of variables in solution	Value of variables	U					x_{ik}
1	2	15/7	5/7	−3/7	0	0	0	−1/7
2	3	25/7	−1/7	2/7	0	0	0	3/7
3	4	15/7	−9/7	4/7	1	0	0	6/7 (circled)
4	8	+90/7	16/7	−4/7	−1	1	0	−6/7
5	9	0	1	1	1	0	1	0

$x_9 = 0$

$\gamma_k = \gamma_1 = \mathbf{U}_{m+1}\bar{\mathbf{A}}_1 = -6/7$

$\theta_0 = \frac{x_4}{x_{41}} = \frac{5}{2}$

5. Fifth Iteration

Row index of tableau	Index of variables in solution	Value of variables	U					x_{ik}
1	2	5/2	3/6	−2/6	1/6	0	0	
2	3	5/2	3/6	0	−3/6	0	0	
3	1	5/2	−9/6	4/6	7/6	0	0	
4	8	+15	1	0	0	1	0	
5	9	0	1	1	1	0	1	

All $\gamma_i \geq 0$

Optimum solution:

$x_1 = x_2 = x_3 = 5/2,\ x_4 = 0$

Value of objective function:

$x_8 = 15$

where

$$\bar{b}_{ij} = b_{ij} - \frac{x_{ik}}{x_{lk}} b_{lj} \qquad \text{for } i \neq l$$

$$\bar{b}_{lj} = b_{lj} \frac{1}{x_{lk}}$$

Let us form the $m \times m$ matrix

$$\mathbf{E}^l = \begin{pmatrix} 1 & 0 & \cdots & -\dfrac{x_{1k}}{x_{lk}} & \cdots & 0 \\ 0 & 1 & \cdots & -\dfrac{x_{2k}}{x_{lk}} & \cdots & 0 \\ \cdot & \cdot & \cdot & \cdot & \cdot & \cdot \\ 0 & 0 & \cdots & \dfrac{1}{x_{lk}} & \cdots & 0 \\ \cdot & \cdot & \cdot & \cdot & \cdot & \cdot \\ 0 & 0 & \cdots & -\dfrac{x_{mk}}{x_{lk}} & \cdots & 1 \end{pmatrix}$$

The matrix $\mathbf{E}^l$ is called an elementary matrix, i.e., an identity matrix except for the lth column. The reader can readily verify by direct multiplication that the matrix product

$$\mathbf{E}^l\mathbf{B}^{-1} = \bar{\mathbf{B}}^{-1}$$

Hence the premultiplication of $\mathbf{B}^{-1}$ by $\mathbf{E}^l$ is equivalent to the application of formulas (1.6). As in (1.6), this is accomplished with the knowledge of only $\mathbf{X}_k$ and $\mathbf{B}^{-1}$. If we let

$$y_{il} = -\frac{x_{ik}}{x_{lk}} \qquad \text{for } i \neq l \tag{2.1}$$

$$y_{ll} = \frac{1}{x_{lk}} \tag{2.2}$$

then $\mathbf{E}^l$ can be written as

$$\mathbf{E}^l = \begin{pmatrix} 1 & \cdots & y_{1l} & \cdots & 0 \\ \cdot & \cdot & \cdot & \cdot & \cdot \\ 0 & \cdots & y_{ll} & \cdots & 0 \\ \cdot & \cdot & \cdot & \cdot & \cdot \\ 0 & \cdots & y_{ml} & \cdots & 1 \end{pmatrix}$$

In fact, to write out the explicit representation of the elementary matrix, we need to know only the column index l and the set of m elements y_{il}.

Using a shorthand notation, we can let

$$\mathbf{E}^l = (l; y_{1l}, \ldots, y_{ll}, \ldots, y_{ml})$$

In the original simplex method we usually initiate the computation with either an artificial or a real basis $\mathbf{B}_0$ which consists of an m-dimensional unit matrix. The inverse of $\mathbf{B}_0$ is, of course, an identity matrix; let us denote $\mathbf{B}_0^{-1} = \mathbf{I} = \mathbf{E}_0$. The inverse of the next basis $\mathbf{B}_1$, which differs from $\mathbf{B}_0$ in that vector $\mathbf{P}_k$ has replaced vector $\mathbf{P}_l$, can be computed by multiplying

$$\mathbf{E}_1{}^l\mathbf{E}_0 = \mathbf{B}_1^{-1}$$

where $\mathbf{E}_1{}^l$ is the elementary matrix obtained in making the first change of basis. In general, given the associated set of elementary matrices, the inverse of the pth basis can be obtained from

$$\mathbf{E}_p{}^l\mathbf{E}_{p-1}{}^l \cdots \mathbf{E}_1{}^l\mathbf{E}_0 = \mathbf{B}_p^{-1}$$

We should note that, in each $\mathbf{E}_p{}^l$, the index l corresponds to the positional index of the column in the preceding basis that is being replaced. We see that, for any basis $\mathbf{B}_p$, given the corresponding set of elementary matrices, we can obtain $\mathbf{B}_p^{-1}$, and hence the methodology of the revised simplex procedure should apply.

In the product form of the revised simplex method the computation begins with the $m + 2 \times m + 2$ identity matrix $\mathbf{U}$. $\mathbf{U}$ is interpreted as being the inverse of the initial basis to the problem stated by (1.12) to (1.14). The associated elementary matrices are of order $m + 2$, and the rules of Phase I and Phase II of Sec. 1 apply, with the following modifications: Since, in the general form of the revised method, we always have the explicit knowledge of the "inverse matrix" $\mathbf{U}$, we can immediately calculate in Phase I, Step 1, the necessary $\delta_j = \mathbf{U}_{m+2}\bar{\mathbf{A}}_j$. In the product form, however, we have to determine first the current $\mathbf{U}_{m+2}$, and this is best done by computing

$$\mathbf{V}_{m+2}\mathbf{E}_p{}^l\mathbf{E}_{p-1}{}^l \cdots \mathbf{E}_1{}^l = \mathbf{U}_{m+2}$$

where

$$\mathbf{V}_{m+2} = (0,0, \ldots, 0,1)$$

is a unit row vector of dimension $m + 2$. The calculation can be efficiently carried out by interpreting the product to be

$$\{[(\mathbf{V}_{m+2}\mathbf{E}_p{}^l)\mathbf{E}_{p-1}{}^l] \cdots \mathbf{E}_1{}^l\}$$

We should note that, in the product of any row vector

$$\mathbf{V} = (v_1, \ldots, v_{m+2})$$

and an elementary matrix $\mathbf{E}_p{}^l$, that is, in the product $\mathbf{W} = \mathbf{V}\mathbf{E}_p{}^l$, the elements can be computed by the formulas

$$w_i = v_i \qquad \text{for } i \neq l$$

$$w_l = \sum_i v_i y_{il}$$

In Step 3, the vector $\mathbf{X}_k$ is best computed by

$$\mathbf{X}_k = \{\mathbf{E}_p{}^l[\mathbf{E}_{p-1}{}^l \cdots (\mathbf{E}_1{}^l\bar{\mathbf{A}}_k)]\}$$

In computing these products, we note that, in the product of a column vector $\mathbf{A}_j = (a_{1j}, \ldots, a_{m+2,\,j})$ and an elementary matrix $\mathbf{E}_p{}^l$, that is, in $\mathbf{E}_p{}^l\mathbf{A}_j = \mathbf{D}_j$, the elements can be computed by the formulas

$$d_{ij} = a_{ij} + y_{il}a_{lj} \qquad \text{for } i \neq l$$

$$d_{lj} = y_{ll}a_{lj}$$

The new elementary matrix and the current solution are developed in Step 4 by formulas (2.1) and (2.2), and

$$x'_{i0} = x_{i0} + y_{il}x_{l0} \qquad \text{for } i \neq k$$

$$x'_{k0} = y_{ll}x_{l0}$$

respectively.

The above modifications apply to Phase II, except that in Step 1 we compute

$$\{[(\mathbf{V}_{m+1}\mathbf{E}_p{}^l)\mathbf{E}_{p-1}{}^l] \cdots \mathbf{E}_1{}^l\} = \mathbf{U}_{m+1}$$

where

$$\mathbf{V}_{m+1} = (0,0, \ldots ,0,1,0)$$

is a unit row vector of dimension $m + 2$.

The major advantage in generating the inverse for each basis by means of elementary matrices is that only a minimum amount of information, namely, $\mathbf{E}_p{}^l$, need be recorded. This is extremely important when one must use a computer which has a limited high-speed storage. Here we need only record in the shorthand notation that $\mathbf{E}_p{}^l = (l;y_{1l}, \ldots ,y_{m+2,l})_p$ instead of recording the full inverse $\mathbf{U}$ after each iteration. For the more recent computer codes, the revised simplex procedure, using a product form of the inverse, has been employed.

REMARKS

The material in this chapter is from Orchard-Hays [81] and Dantzig and Orchard-Hays [29]. For additional remarks on the computational elements of the general form of the revised procedure, the reader is referred to Dantzig [21]; for the product form of the inverse, see Dantzig, Orchard-Hays, and Waters [30], and, in general, Orchard-Hays [83*a*].

EXERCISES

1. Solve the example of Sec. 1 by means of the product form of the inverse.

2. Solve the following problems by both forms of the revised procedure:

a. Maximize

$$5x_1 - x_2 + x_3 - 10x_4 + 7x_5$$

subject to

$$\begin{aligned} 3x_1 - x_2 - x_3 \qquad\qquad &= 4 \\ x_1 - x_2 + x_3 + x_4 \qquad &= 1 \\ 2x_1 + x_2 + 2x_3 \qquad + x_5 &= 7 \end{aligned}$$

and

$$x_j \geq 0$$

b. Minimize

$$-x_1 + 2x_2$$

subject to

$$\begin{aligned} 5x_1 - 2x_2 &\leq 3 \\ x_1 + x_2 &\geq 1 \\ -3x_1 + x_2 &\leq 3 \\ -3x_1 - 3x_2 &\leq 2 \\ x_1 \qquad &\geq 0 \\ x_2 &\geq 0 \end{aligned}$$

3. Solve Exercise 2*b* by graphical methods.

4. Solve the following problem by the original simplex method and by both forms of the revised procedure: Minimize

$$x_1 + x_2 + x_3$$

subject to

$$\begin{aligned} x_1 \qquad - x_4 \qquad - 2x_6 &= 5 \\ x_2 \quad + 2x_4 - 3x_5 + x_6 &= 3 \\ x_3 + 2x_4 - 5x_5 + 6x_6 &= 5 \end{aligned}$$

and

$$x_j \geq 0$$

CHAPTER 7 / DEGENERACY PROCEDURES

A degenerate basic feasible solution to a linear-programming problem is one in which some x_{i0}, where the index i corresponds to a vector of the admissible basis, is equal to zero. That is, if the vectors $(\mathbf{P}_1\mathbf{P}_2 \cdots \mathbf{P}_m)$ form an admissible basis, then, in the degenerate case, the corresponding nonnegative linear combination

$$x_{10}\mathbf{P}_1 + x_{20}\mathbf{P}_2 + \cdots + x_{m0}\mathbf{P}_m = \mathbf{P}_0$$

has at least one $x_{i0} = 0$. In Chap. 4 we assumed in proving the validity of the simplex procedure that all basic feasible solutions were nondegenerate. This assumption was necessary in order to demonstrate that, for each successive admissible basis, the associated value of the objective function is smaller than those that precede it. Hence, we will reach the minimum solution in a finite number of solutions, since there are only a finite number of possible bases. This proof breaks down if we admit the existence of degenerate basic feasible solutions. The latter is, of course, the more realistic situation. For a degenerate solution, we have the possibility of computing a $\theta_0 = x_{l0}/x_{lk}$ [see Eqs. (1.6) and (1.7) of Chap. 4] for which $\theta_0 = 0$. This choice of vector $\mathbf{P}_l$ to be eliminated and $\mathbf{P}_k$ to be introduced into the basis will give a new feasible solution whose value of the objective function is equal to the preceding one.[1] It is then theoretically possible to select a sequence of admissible bases that cycles, i.e., a sequence of bases that is repeatedly selected without ever satisfying the optimality criteria and hence never reaches a minimum solution. The possibility of cycling is crucial only if the current basic

[1] Here we should distinguish between a multiple solution and a degenerate solution. A multiple solution has the same value of the objective function as the preceding solution because $z_k - c_k = 0$ with $\mathbf{P}_k$ not in the basis and $\theta_0 > 0$, while a degenerate solution will yield a new solution with the same value of the objective function because $\theta_0 = 0$.

feasible solution has more than one $x_{i0} = 0$ (see Exercise 2, Chap. 4). With at least two $x_{i0} = 0$, it could be possible to have a tie with $\theta_0 = 0$ in the selection of the vector to be eliminated from the basis. It is also possible to have a tie in computing θ_0 even if the current solution is not degenerate. Here, of course, $\theta_0 > 0$, and the new solution will have an improved value for the objective function. However, because of this tie in the old solution, the new solution will be degenerate.

From the above discussion, we see that any device developed to overcome the degeneracy restriction must be concerned with the determination of a unique θ_0 and hence of the index l of the vector to be eliminated. A number of such techniques have been developed (Dantzig [17], Charnes [9], Wolfe [103], and Dantzig, Orden, and Wolfe [32]). Computational experience on digital computers has minimized the importance of these techniques, since there have not been any practical problems that have been known to cycle. In other words, the successful solution of hundreds of problems has not hinged on the development of these techniques. For this reason, these procedures have not been incorporated in most computer codes. What is important, however, is that these devices "make the simplex method available, without blemish, as a crisp tool for proving pure theorems" (Hoffman [59]).

1. PERTURBATION TECHNIQUES

Geometrically, a degenerate situation is one in which the vector $\mathbf{P}_0$ lies on a bounding hyperplane or edge of the convex cone determined by its basis vectors. For example, in Fig. 7.1 vector $\mathbf{P}_0$ can be expressed as a positive combination of $\mathbf{P}_1$ and $\mathbf{P}_2$ but as a nonnegative combination of $\mathbf{P}_1$, $\mathbf{P}_2$, and $\mathbf{P}_3$ with $x_{30} = 0$. However, if we move, i.e., perturb, the vector $\mathbf{P}_0$ in such a manner that it lies inside the convex cone determined by vectors $\mathbf{P}_1$, $\mathbf{P}_2$, and $\mathbf{P}_3$, then the corresponding solution will be nondegenerate. We can do this by taking a positive linear combination of

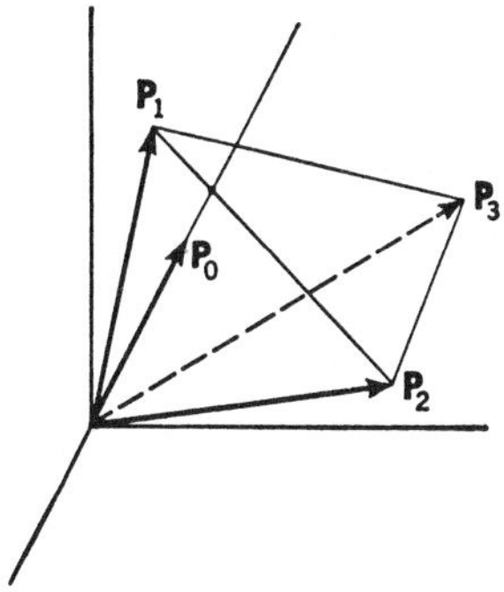

FIGURE 7.1

these vectors and adding it to $\mathbf{P}_0$. However, we wish to do this so as not to destroy or hide the original problem. Hence we take our positive combination small, and, as suggested by Charnes [9], we would let the positive combination for our example be

$$\epsilon\mathbf{P}_1 + \epsilon^2\mathbf{P}_2 + \epsilon^3\mathbf{P}_3$$

where ϵ is some small positive number. The constraints of the new problem would be

$$x_{10}\mathbf{P}_1 + x_{20}\mathbf{P}_2 + x_{30}\mathbf{P}_3 = \mathbf{P}_0 + \epsilon\mathbf{P}_1 + \epsilon^2\mathbf{P}_2 + \epsilon^3\mathbf{P}_3 = \mathbf{P}_0(\epsilon)$$

Its geometric interpretation is given in Fig. 7.2.

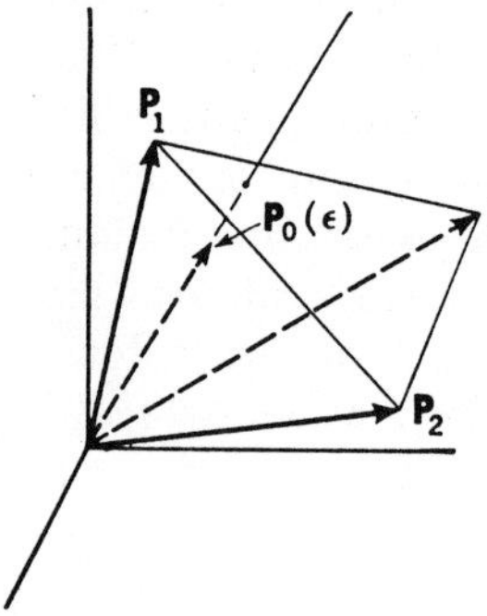

FIGURE 7.2

For the general linear-programming problem we wish to perturb the problem in a similar manner so as to ensure that, for any possible admissible basis, the corresponding solution will be nondegenerate. (Hence the perturbation technique and other schemes are procedures for justifying the nondegeneracy assumption of the simplex method.) To accomplish this for the general problem, we rewrite the constraints

$$x_1\mathbf{P}_1 + x_2\mathbf{P}_2 + \cdots + x_m\mathbf{P}_m + \cdots + x_n\mathbf{P}_n = \mathbf{P}_0 \tag{1.1}$$

to read

$$x_1\mathbf{P}_1 + x_2\mathbf{P}_2 + \cdots + x_m\mathbf{P}_m + \cdots + x_n\mathbf{P}_n = \mathbf{P}_0 + \epsilon\mathbf{P}_1 + \epsilon^2\mathbf{P}_2 + \cdots + \epsilon^m\mathbf{P}_m + \cdots + \epsilon^n\mathbf{P}_n = \mathbf{P}_0(\epsilon) \tag{1.2}$$

Let us assume that the vectors $\mathbf{P}_1, \mathbf{P}_2, \ldots, \mathbf{P}_m$ form an admissible basis $\mathbf{B}$. Then a solution to (1.1) is

$$\bar{\mathbf{X}}_0 = \mathbf{B}^{-1}\mathbf{P}_0 \geq 0 \tag{1.3}$$

and a solution to (1.2) is

$$\bar{\mathbf{X}}_0(\epsilon) = \mathbf{B}^{-1}\mathbf{P}_0(\epsilon) \tag{1.4}$$

Let

$$\mathbf{X}_j = \mathbf{B}^{-1}\mathbf{P}_j \tag{1.5}$$

Then (1.4) is given by

$$\begin{aligned}\bar{\mathbf{X}}_0(\epsilon) &= \mathbf{B}^{-1}\mathbf{P}_0 + \epsilon\mathbf{B}^{-1}\mathbf{P}_1 + \epsilon^2\mathbf{B}^{-1}\mathbf{P}_2 + \\ &\qquad \cdots + \epsilon^m\mathbf{B}^{-1}\mathbf{P}_m + \cdots + \epsilon^n\mathbf{B}^{-1}\mathbf{P}_n \\ &= \bar{\mathbf{X}}_0 + \epsilon\mathbf{X}_1 + \epsilon^2\mathbf{X}_2 + \cdots + \epsilon^m\mathbf{X}_m + \cdots + \epsilon^n\mathbf{X}_n \qquad (1.6)\end{aligned}$$

We note that, since $\mathbf{B}$ consists of vectors $\mathbf{P}_1, \mathbf{P}_2, \ldots, \mathbf{P}_m$, $\mathbf{X}_j = \mathbf{B}^{-1}\mathbf{P}_j$ is an identity vector with the unit element in position j for $j = 1, 2, \ldots, m$. Hence the $\bar{x}_{i0}(\epsilon)$ are given by

$$\bar{x}_{i0}(\epsilon) = \bar{x}_{i0} + \sum_{j=1}^{n} \epsilon^j x_{ij} \qquad (1.7)$$

or

$$\bar{x}_{i0}(\epsilon) = \bar{x}_{i0} + \epsilon^i + \sum_{j=m+1}^{n} \epsilon^j x_{ij} \qquad (1.8)$$

From (1.8), and by taking $\epsilon > 0$ but sufficiently small, we can make all $\bar{x}_{i0}(\epsilon) > 0$ for $i = 1, 2, \ldots, m$. (We can always rearrange the problem to make the basis vectors the first m.) In determining the vector $\mathbf{P}_l$ to be eliminated from the basis for the problem of (1.2), the criterion that ensures the feasibility of the new solution is the selection of

$$\theta_0 = \frac{\bar{x}_{l0}(\epsilon)}{x_{lk}} = \min_i \frac{\bar{x}_{i0}(\epsilon)}{x_{ik}} = \min_i \frac{\bar{x}_{i0} + e^i + \sum_{j=m+1}^{n} \epsilon^j x_{ij}}{x_{ik}} > 0 \qquad (1.9)$$

for $x_{ik} > 0$. We see from (1.8) that ties cannot occur since $\bar{x}_{l0}(\epsilon)$ is the only variable that contains ϵ^l.

Once a solution to the linear-programming problem with the constraints (1.2) has been obtained, by letting ϵ equal zero, we shall have the corresponding extreme-point solution to (1.1). In actual practice, it is not necessary to select an ϵ and rewrite the problem as was done in (1.2). From (1.6) and (1.9) we note that the information required to determine θ_0 is the $\bar{x}_{i0}$ of problem (1.1) and the coefficients of the ϵ^j. This information is all contained in the simplex tableau of the unperturbed problem. For ϵ small enough, we note that the significant coefficients of ϵ are the first ones, starting with $j = 1, 2, \ldots, n$. The procedure can then be routinized as follows: If, for all $\bar{x}_{i0}$ in the basic solution, the ratios $\bar{x}_{i0}/x_{ik}$ for $x_{ik} > 0$ yield a unique $\theta_0 = \min_i (\bar{x}_{i0}/x_{ik})$, then vector $\mathbf{P}_l$ is uniquely determined. If, however, there are ties for the minimum for some set of indices, we compute starting with $j = 1$ the ratios x_{ij}/x_{ik} for all rows i in this set and compare these ratios. The index that corresponds to the algebraically smallest ratio determines the index of the vector to be eliminated. If there are still ties in computing the minimum, we form the next ratios for the tied set of indices for column $j + 1$ and repeat the analysis. For example, if for the admissible basis $(\mathbf{P}_1\mathbf{P}_2 \cdots \mathbf{P}_m)$ we have

$$\theta_0 = \frac{\bar{x}_{10}}{x_{1k}} = \frac{\bar{x}_{20}}{x_{2k}}$$

we then compute x_{11}/x_{1k} and x_{21}/x_{2k} and compare. If

$$\min_i \frac{x_{i1}}{x_{ik}} = \frac{x_{11}}{x_{1k}} \qquad \text{for } i = 1, 2$$

then vector $\mathbf{P}_1$ is eliminated. If

$$\min_i \frac{x_{i1}}{x_{ik}} = \frac{x_{21}}{x_{2k}} \qquad \text{for } i = 1, 2$$

then $\mathbf{P}_2$ is eliminated. In both cases $\mathbf{P}_k$ is the vector introduced. (The selection of the new vector does not depend in any way on the degenerate situation.) If $x_{11}/x_{1k} = x_{21}/x_{2k}$, we form the ratios x_{12}/x_{1k} and x_{22}/x_{2k} and repeat the comparison until the tie situation is broken. We know from (**1**.8) that this will happen for some j. The simplex tableau is transformed in the usual manner with the selected x_{lk} as pivot element. The new $\bar{\mathbf{X}}_0'(\epsilon)$ solution to (1.2) will be nondegenerate for ϵ sufficiently small, and the above process is repeated until a minimum solution is found.

The above perturbation scheme is ideally adapted to the original simplex method since all the necessary information to execute the scheme is readily available. For the revised procedure a different approach to the problem also yields an efficient technique that requires knowledge of the inverse of the current basis (Dantzig, Orden, and Wolfe [32]). We shall not discuss the development of the method except to describe the rule to be used when breaking ties in the computation of θ_0. This rule is given in Dantzig [21] and is: "If two or more indices $l_1, l_2, \ldots$ are tied for the minimum, divide the corresponding entries in the *first column of the inverse* by $x_{l_1,k}, x_{l_2,k}, \ldots$, respectively, and take the index of the row with the minimizing ratio for l. If there still remain ties (for the minimum), repeat for those indices that are still tied, using as ratios the corresponding entries in the *second column of the inverse* divided by their respective $x_{l_i,k}$. Ratios formed from successive columns of the inverse are used until all ties are resolved. (Since no two columns of an inverse can be proportional, a *unique* l will be chosen by the last column.)"

2. EXAMPLE OF CYCLING

The literature contains very few examples of problems that cycle when they are solved by the original simplex method (Hoffman [58] and Beale

Tableau 7.1

	Basis	c	P_0	P_1	P_2	P_3	P_4	P_5	P_6	P_7
I.		c		−3/4	150	−1/50	6	0	0	0
	P_5	0	0	1/4	−60	−1/25	9	1	0	0
	P_6	0	0	1/2	−90	−1/50	3	0	1	0
	P_7	0	1	0	0	1	0	0	0	1
			0	3/4	−150	1/50	−6	0	0	0
II.	P_1	−3/4	0	1	−240	−4/25	36	4	0	0
	P_6	0	0	0	30	3/50	−15	−2	1	0
	P_7	0	1	0	0	1	0	0	0	1
			0	0	30	7/50	−33	−3	0	0
III.	P_1	−3/4	0	1	0	8/25	−84	−12	8	0
	P_2	150	0	0	1	1/500	−1/2	−1/15	1/30	0
	P_7	0	1	0	0	1	0	0	0	1
			0	0	0	2/25	−18	−1	−1	0
IV.	P_3	−1/50	0	25/8	0	1	−525/2	−75/2	25	0
	P_2	150	0	−1/160	1	0	1/40	1/120	−1/60	0
	P_7	0	1	−25/8	0	0	525/2	75/2	−25	1
			0	−1/4	0	0	3	2	−3	0
V.	P_3	−1/50	0	−125/2	10,500	1	0	50	−150	0
	P_4	6	0	−1/4	40	0	1	1/3	−2/3	0
	P_7	0	1	125/2	−10,500	0	0	−50	150	1
			0	1/2	−120	0	0	1	−1	0
VI.	P_5	0	0	−5/4	210	1/50	0	1	−3	0
	P_4	6	0	1/6	−30	−1/150	1	0	1/3	0
	P_7	0	1	0	0	1	0	0	0	1
			0	7/4	−330	−1/50	0	0	2	0
VII.	P_5	0	0	1/4	−60	−1/25	9	1	0	0
	P_6	0	0	1/2	−90	−1/50	3	0	1	0
	P_7	0	1	0	0	1	0	0	0	1
			0	3/4	−150	1/50	−6	0	0	0

Tableau 7.2

	Basis		P_0	P_1	P_2	P_3	P_4	P_5	P_6	P_7
		c		$-3/4$	150	$-1/50$	6	0	0	0
I.	P_5	0	0	(1/4)	-60	$-1/25$	9	1	0	0
	P_6	0	0	1/2	-90	$-1/50$	3	0	1	0
	P_7	0	1	0	0	1	0	0	0	1
			0	3/4	-150	1/50	-6	0	0	0
II.	P_1	$-3/4$	0	1	-240	$-4/25$	36	4	0	0
	P_6	0	0	0	(30)	3/50	-15	-2	1	0
	P_7	0	1	0	0	1	0	0	0	1
			0	0	30	7/50	-33	-3	0	0
III.	P_1	$-3/4$	0	1	0	8/25	-84	-12	8	0
	P_2	150	0	0	1	(1/500)	$-1/2$	$-1/15$	1/30	0
	P_7	0	1	0	0	1	0	0	0	1
			0	0	0	2/25	-18	-1	-1	0
IV.	P_1	$-3/4$	0	1	-160	0	-4	$-4/3$	8/3	0
	P_3	$-1/50$	0	0	500	1	-250	$-100/3$	50/3	0
	P_7	0	1	0	-500	0	(250)	100/3	$-50/3$	1
			0	0	-40	0	2	5/3	$-7/3$	0
V.	P_1	$-3/4$	2/125	1	-168	0	0	$-4/5$	12/5	2/125
	P_3	$-1/50$	1	0	0	1	0	0	0	1
	P_4	6	1/250	0	-2	0	1	(2/15)	$-1/15$	1/250
			$-1/125$	0	-36	0	0	7/5	$-11/5$	$-1/125$
VI.	P_1	$-3/4$	1/25	1	-180	0	6	0	2	1/25
	P_3	$-1/50$	1	0	0	1	0	0	0	1
	P_5	0	3/100	0	-15	0	15/2	1	$-1/2$	3/100
			$-1/20$	0	-15	0	$-21/2$	0	$-3/2$	$-1/20$

[4]). The problem of Beale [4] is discussed in terms of its dual. We shall next illustrate the cycling phenomenon in terms of the primal by a previously unpublished example due to Beale.

The problem is to minimize

$$-\tfrac{3}{4}x_1 + 150x_2 - \tfrac{1}{50}x_3 + 6x_4$$

subject to

$$\begin{aligned}
\tfrac{1}{4}x_1 - 60x_2 - \tfrac{1}{25}x_3 + 9x_4 + x_5 \qquad\qquad &= 0 \\
\tfrac{1}{2}x_1 - 90x_2 - \tfrac{1}{50}x_3 + 3x_4 \qquad + x_6 \qquad &= 0 \\
x_3 \qquad\qquad + x_7 &= 1
\end{aligned}$$

and

$$x_j \geq 0$$

In showing how this problem cycles, we shall select for the vector to be eliminated the one whose row index is the smallest. (This is the rule used in Chap. 4 for breaking ties.) In computing θ_0, ties arose in the first, third, and fifth solutions. The steps in the process are shown in Tableau 7.1, in which it will be noted that the seventh solution is identical with the first solution. If we should continue the process with the seventh tableau, we should only repeat the previous solutions and never reach the minimum solution.

However, by applying the degeneracy procedure described in Sec. 1, we select a different sequence of solutions and do determine the minimum solution. The steps in this computation are shown in Tableau 7.2, where the sixth solution gives the final answer. There is no recurrence of any solution. Ties occur in computing θ_0 in the first and third solutions. The first three solutions in both Tableaus 7.1 and 7.2 are the same. The main difference occurs in going from the third to the fourth solution, where in the second case we eliminated $\mathbf{P}_2$ instead of $\mathbf{P}_1$. The minimum value of the objective function is $-\tfrac{1}{20}$.

EXERCISES

1. Show, using the trigonometric functions, that the following problem (Hoffman [58]) will cycle after ten iterations if the standard rules of the simplex method are applied. Note that the cycle is started by introducing $\mathbf{P}_4$ followed by $\mathbf{P}_5$. Determine an appropriate range of w and a value of ϕ.

Minimize

$$\frac{\cos\phi - 1}{\cos\phi}x_4 + wx_5 + 2wx_7 + 4\sin^2\phi\, x_8 + w(2 - 4\cos^2\phi)x_9 + 4\sin^2\phi\, x_{10} + w(1 - 2\cos\phi)x_{11}$$

subject to

$$x_1 = 1$$

$$x_2 + \cos\phi\, x_4 - w\cos\phi\, x_5 + \cos 2\phi\, x_6 - 2w\cos^2\phi\, x_7 + \cos 2\phi\, x_8 + 2w\cos^2\phi\, x_9 + \cos\phi\, x_{10} + w\cos\phi\, x_{11} = 0$$

$$x_3 + \frac{\tan\phi\sin\phi}{w}x_4 + \cos\phi\, x_5 + \frac{\tan\phi\sin 2\phi}{w}x_6 + \cos 2\phi\, x_7 - 2\frac{\sin^2\phi}{w}x_8 + \cos 2\phi\, x_9 - \frac{\tan\phi\sin\phi}{w}x_{10} + \cos\phi\, x_{11} = 0$$

2. Solve the problem of Exercise 1 by using Charnes' rule for resolving ties.

3. Solve the cycling example of Sec. 2 using the revised simplex method and the corresponding rule for breaking ties.

4. In Exercise 1, let the intersection of the second and third rows and the fourth and fifth columns be denoted by the 2×2 matrix $\mathbf{A}$. Note that $\mathbf{A}^2$ is the corresponding 2×2 matrix associated with the sixth and seventh columns, $\mathbf{A}^3$ with the eighth and ninth columns, $\mathbf{A}^4$ with the tenth and eleventh columns, and $\mathbf{A}^5 = \mathbf{I}$ with the first and second columns. Attempt to construct a new linear-programming problem with a similar set of matrices of finite order that will cycle.

5. Starting with vectors $\mathbf{P}_1$, $\mathbf{P}_2$, $\mathbf{P}_7$ as the initial basis, will the following problem cycle when the regular simplex rules are applied (Yudin and Gol'shtein [106]):

Maximize

$$x_3 - x_4 + x_5 - x_6$$

subject to

$$\begin{aligned} x_1 \quad + 2x_3 - 3x_4 - 5x_5 + 6x_6 \quad &= 0 \\ x_2 + 6x_3 - 5x_4 - 3x_5 + 2x_6 \quad &= 0 \\ 3x_3 + x_4 + 2x_5 + 4x_6 + x_7 &= 1 \\ x_j &\geq 0 \end{aligned}$$

6. Prove that the minimum length of a cycle in a linear-programming problem is six iterations (Yudin and Gol'shtein [106]).

CHAPTER 8 / PARAMETRIC LINEAR PROGRAMMING

1. THE PARAMETRIC OBJECTIVE FUNCTION

A major task in the development of realistic linear-programming models is the gathering of accurate and reliable numerical values for the coefficients. Hence it is important to study the behavior of solutions to a linear-programming problem when the coefficients of that problem are allowed to vary. This type of investigation is the function of *parametric linear programming*. We can vary the coefficients of the **A** matrix, the objective-function cost coefficients, or the constants of the right-hand side of the equations. Not much work has been done on the first of these alternatives, and only a special case of the last two types of variation has been studied in some detail.[1] There are still many problems to be solved in this area. In what follows, we shall be concerned only with the material contained in the references cited in the footnote below.

The investigation of parametric programming as applied to the variation of the cost coefficients originated in the study of the production-scheduling problem described in Sec. 1 of Chap. 11.[2] There we introduced the parameter λ, which measures the cost of a unit increase in output relative to that of storing a unit for 1 month. Before selecting a particular production schedule that is optimal for some specified value of λ, the manufacturer might find it advantageous to investigate a set of optimal solutions corresponding to ranges of values of λ. In this manner he can more accurately select the solution that not only best fits his production and storage capabilities, but also suits those factors that have not been incorporated into the linear-programming model. The procedure that enables one to compute efficiently the associated optimal solutions for ranges of λ is described below. This computational scheme is basically a variation of the standard simplex method.

[1] See Gass and Saaty [48] and Manne [74]. Credit should be given to W. W. Jacobs for initiating the early work in parametric programming and establishing the underlying concepts.

[2] The reader not familiar with this problem should first review the cited section.

Mathematically, the problem can be stated as follows: Let $\delta \leq \lambda \leq \phi$, where δ may be an arbitrary, algebraically small, but finite, number and ϕ may be an arbitrary, algebraically large, but finite number. For each λ in this interval, find a vector $\mathbf{X} = (x_1, x_2, \ldots, x_n)$ which minimizes

$$\sum_{j=1}^{n} (d_j + \lambda d_j')x_j \tag{1.1}$$

subject to

$$\begin{aligned} \sum_{j=1}^{n} a_{ij}x_j &= b_i \qquad i = 1, \ldots, m \\ x_j &\geq 0 \qquad j = 1, \ldots, n \end{aligned} \tag{1.2}$$

where d_j, d_j', a_{ij}, and b_i are given constants.

We assume that our problem is nondegenerate and that we have available a basic feasible solution of (1.2). Using the simplex method, we can solve our problem for $\lambda = \delta$ and obtain either a solution (Case A) or the information that the objective function (1.1) with $\lambda = \delta$ has no finite minimum on the convex set defined by (1.2)(Case B).

Case A. Since in our problem the cost coefficients $c_j = d_j + \lambda d_j'$, we can for any basis represent the terms $z_j - c_j$ as a linear function of λ. Let us write these linear functions as $z_j - c_j = \alpha_j + \lambda\beta_j$. Then, since we are in Case A,

$$\alpha_j + \delta\beta_j \leq 0 \qquad j = 1, \ldots, n$$

which means that the inequalities

$$\alpha_j + \lambda\beta_j \leq 0 \qquad j = 1, \ldots, n \tag{1.3}$$

are consistent. For all $\beta_j < 0$, we have

$$\lambda \geq -\frac{\alpha_j}{\beta_j}$$

and for all $\beta_j > 0$,

$$\lambda \leq -\frac{\alpha_j}{\beta_j}$$

Let us define

$$\underline{\lambda} = \begin{cases} \max\limits_{\beta_j<0} -\dfrac{\alpha_j}{\beta_j} \\ \text{or} \\ -\infty \qquad \text{if all } \beta_j \geq 0 \end{cases}$$

$$\bar{\lambda} = \begin{cases} \min\limits_{\beta_j>0} -\dfrac{\alpha_j}{\beta_j} \\ \text{or} \\ +\infty \qquad \text{if all } \beta_j \leq 0 \end{cases}$$

Hence, by the simplex method, the current solution will yield the minimum for all λ satisfying (1.3), i.e., for all λ such that

$$\underline{\lambda} \leq \lambda \leq \bar{\lambda}$$

If $\bar{\lambda} = +\infty$, our problem is over. Assume then that $\bar{\lambda}$ is finite, $\bar{\lambda} = -\alpha_k/\beta_k$ for $\beta_k > 0$. If all the corresponding $x_{ik} \leq 0$, then we know from the simplex method and the definition of $\bar{\lambda}$ that our problem has no minimum for $\lambda > \bar{\lambda}$; thus we are finished.[1] If at least one $x_{ik} > 0$, then the simplex process introduces the vector $\mathbf{P}_k$ into the basis and eliminates a vector $\mathbf{P}_l$ in the usual manner. Note that

$$x_{lk} > 0 \tag{1.4}$$

Theorem 1. *The new basis yields a minimum for at least one value of λ. If $\underline{\lambda}' \leq \lambda \leq \bar{\lambda}'$ is the entire set of values of λ for which the new basis yields a minimum, then $\underline{\lambda}' = \bar{\lambda}$.*

Proof. The basis is certainly feasible, since we have followed the simplex prescription. Further, the inequalities (for the new basis)

$$\alpha_j' + \lambda\beta_j' \leq 0 \qquad j = 1, \ldots, n \tag{1.5}$$

are consistent. For if we imagine $\lambda = \bar{\lambda}$, we have inserted into the basis the vector $\mathbf{P}_k$ whose $z_k - c_k = \alpha_k + \bar{\lambda}\beta_k = 0$. The new basis will still be a minimum for $\lambda = \bar{\lambda}$, since $\bar{\lambda}$ satisfies (1.5).

All that remains to be shown is that $\lambda < \bar{\lambda}$ does not satisfy (1.5). We have

$$\begin{aligned} \alpha_l' &= -\frac{\alpha_k}{x_{lk}} \\ \beta_l' &= -\frac{\beta_k}{x_{lk}} \end{aligned} \tag{1.6}$$

In order to satisfy the inequality from the set (1.5) which corresponds to the eliminated vector $\mathbf{P}_l$, we must have

$$\alpha_l' + \lambda\beta_l' \leq 0$$

Or, by (1.4) and (1.6),

$$-\alpha_k - \lambda\beta_k \leq 0$$

Since $\beta_k > 0$, this last inequality forces the new range of λ to be

$$\lambda \geq -\frac{\alpha_k}{\beta_k} = \bar{\lambda}$$

[1] As in the general simplex method, the column vector $\mathbf{X}_k = (x_{1k}, \ldots, x_{mk})$, where $\mathbf{X}_k = \mathbf{B}^{-1}\mathbf{P}_k$ for the current basis $\mathbf{B}$.

In this manner, we proceed from one range of values of λ to the next until we include $\lambda = \phi$. To prove that this process is valid, we have only to assure ourselves that no basis is repeated. This assurance is provided by the following remark: If $\bar{\lambda}$ is replaced by $\bar{\lambda} + \epsilon$, where ϵ is any (conceptually small) positive number, the steps followed would be precisely the same as we have already described. The nondegeneracy assumption guarantees that we will either solve the problem for $\lambda = \bar{\lambda} + \epsilon$ or obtain the information that there is no minimum for that value of λ. Hence we cannot remain indefinitely at a value λ such that $\underline{\lambda} = \lambda = \bar{\lambda}$. After we leave a basis, we cannot return to it or to any basis corresponding to lower values of λ. By our theorem, these intervals of λ will not overlap. It is entirely possible (indeed, it has frequently occurred) that, corresponding to some solutions, $\underline{\lambda} = \bar{\lambda}$. But, as we have seen, this cannot persist indefinitely through the various changes of basis. The various $\underline{\lambda}$ and $\bar{\lambda}$ that arise are called *characteristic values of* λ, and the minimizing solutions corresponding to the various values of λ are called *characteristic solutions.*

Case B. Here, as we attempt to find a minimum feasible solution for $\lambda = \delta$, a vector chosen to go into the basis cannot do so because all of its elements x_{ik} are nonpositive. There are two possible situations:

1. Let $\mathbf{P}_k$ be the vector that cannot be introduced into the basis. Here we are given $\alpha_k + \delta\beta_k > 0$ and all $x_{ik} \leq 0$. If $\beta_k \geq 0$, then the problem has no finite minimum solutions for any λ.

2. If $\beta_k < 0$, the inequality $\alpha_k + \lambda\beta_k > 0$ will hold for all

$$\lambda < \lambda_1' = -(\alpha_k/\beta_k)$$

and hence there will be no finite minimum feasible solutions for λ in the region $\delta \leq \lambda < \lambda_1'$. At this stage, we do not know whether a finite minimum solution exists for λ_1'. If all $\alpha_j + \lambda_1'\beta_j \leq 0$, we have a minimum feasible solution for λ_1', and λ_1 can be determined by

$$\lambda_1 = \min_{\beta_j > 0} (-\alpha_j/\beta_j)$$

The characteristic solution holds for $\lambda_1' \leq \lambda \leq \lambda_1$, and the procedure of Case *A* can now be applied. If not all $\alpha_j + \lambda_1'\beta_j \leq 0$, any vector having $\alpha_j + \lambda_1'\beta_j > 0$ can be introduced into the basis. This criterion is used in the following iterations until all the transformed $\alpha_j + \lambda_1'\beta_j \leq 0$ or until a vector with $\alpha_t + \lambda_1'\beta_t > 0$ cannot be introduced because all its transformed elements are nonpositive. The former condition can be handled by the method of Case *A*. In the latter condition, if $\beta_t \geq 0$, no finite minimum solutions exist. For $\beta_t < 0$, we know there are no finite solutions for $\lambda < \lambda_2' = -(\alpha_t/\beta_t)$, where $\lambda_2' > \lambda_1'$. We now attempt to determine whether a finite solution exists for $\lambda = \lambda_2'$. Successive applications of the above procedure either will lead us to a finite minimum for some λ (and then Case *A* can be used) or will show that there are no λ for which a finite minimum exists.

If we are given a characteristic solution for an arbitrary range

$$\lambda_i \leq \lambda \leq \lambda_{i+1}$$

we can proceed to the right of λ_{i+1} by the procedure of Case *A*. We can proceed to the left of λ_i by introducing the vector $\mathbf{P}_q$ which has

$$\lambda_i = \max_{\beta_j<0} -\frac{\alpha_j}{\beta_j} = -\frac{\alpha_q}{\beta_q}$$

If all $x_{iq} \leq 0$, then there are no finite solutions for $\lambda < \lambda_i$.

Summarizing, we have seen that

1. By a modification of the general simplex procedure, it is possible to investigate systematically and solve the one-parameter objective-function problem.

2. Given any finite minimum solution, we can determine a set of characteristic solutions and the associated characteristic values for all possible values of the parameter.

3. A solution is minimum over a closed interval of λ.

4. The set of λ for which minimum solutions exist is closed and connected.

Many large-scale linear-programming problems with a parametric objective function have been solved using an electronic computer. A typical problem consisting of 33 equations and 65 variables was solved for all positive values of the parameter in 53 iterations. The complete solution included 23 characteristic solutions and corresponding characteristic values. Multiple solutions were generated but not tabulated. All problems considered had a finite minimum for $\lambda = 0$, and the range of interest was $0 \leq \lambda < \infty$. As a computational aid, it was found advisable to solve for the $\delta = 0$ solution using the regular simplex procedure and to use this solution to initiate the parametric process.

Example. Minimize

$$\lambda x - y$$

subject to

$$\begin{aligned} 3x - y &\geq 5 \\ 2x + y &\leq 3 \end{aligned} \tag{1.7}$$

where

$$-\infty < \delta \leq \lambda \leq \phi < +\infty$$

We transform system (1.7) to a set of equalities with nonnegative variables, with one artificial variable v with associated large positive cost coefficient w, to obtain the system (1.8):

$$\begin{aligned} 3(x_1 - x_2) - (y_1 - y_2) - u_1 \quad\quad + \quad v &= 5 \\ 2(x_1 - x_2) + (y_1 - y_2) \quad\quad + u_2 \quad\quad &= 3 \end{aligned} \tag{1.8}$$

with the corresponding objective function of

$$\lambda(x_1 - x_2) - (y_1 - y_2) \qquad\qquad + wv$$

Step I of Tableau 8.1 is the system (1.8) in the usual simplex tableau. Since there is a natural unit vector, $\mathbf{P}_6$, in the system, we need to employ only one artificial vector, $\mathbf{P}_7$, with large positive weight, w. The first basis is $(\mathbf{P}_7\mathbf{P}_6)$. The $z_j - c_j$ elements are written as $\alpha_j + \lambda\beta_j + w\gamma_j$ and are entered in the $(m + 1)$st, $(m + 2)$nd, and $(m + 3)$rd rows, respectively. For example, vector $\mathbf{P}_1$ has $z_1 - c_1 = -\lambda + 3w$; hence we enter 0, -1, 3 in the corresponding rows under $\mathbf{P}_1$. The value of the objective function is $z = 5w$. Since $\mathbf{P}_1$ has the largest positive value in the $(m + 3)$rd row, it is introduced into the basis, and $\mathbf{P}_6$ is eliminated (Step II). Similarly we eliminate $\mathbf{P}_7$ by $\mathbf{P}_4$ to reach Step III. Now, since all the elements in the $(m + 3)$rd row equal zero, we have our first feasible solution. The basis vectors are $\mathbf{P}_1$ and $\mathbf{P}_4$ and the solution is $x_1 = 8/5$, $y_2 = 1/5$; $x_2 = y_1 = u_1 = u_2 = 0$. The value of the objective function is $z = 1/5 + 8/5\lambda$.

Since we first wish to determine whether a finite solution for $\delta = \lambda$ exists, we should next introduce a vector with a *negative* element in row $m + 2$. The only one is $\mathbf{P}_5$. However, all $x_{i5} \leq 0$, and hence there are no finite solutions for $\lambda < \lambda_1' = -\alpha_5/\beta_5 = -2$. Applying the procedure of Case B, we see that all $\alpha_j + \lambda_1'\beta_j \leq 0$, and Step III gives a minimum for $-2 \leq \lambda \leq 3$ $(\lambda_1 = -\alpha_6/\beta_6)$. Introducing $\mathbf{P}_k = \mathbf{P}_6$ into the basis (we should expect a solution for $\lambda \geq 3$), we eliminate $\mathbf{P}_1$ and have a new feasible solution with $\mathbf{P}_4$ and $\mathbf{P}_6$ as basis vectors. The solution is $y_2 = 5$, $u_2 = 8$, $z = 5$; $x_1 = x_2 = y_1 = u_1 = 0$. (Note that here the value of the objective function is independent of λ.) Applying the algorithm of Case A, we have $\mathbf{P}_k = \mathbf{P}_2$ and $\lambda_1 = \lambda_2 = 3$. $\mathbf{P}_2$ cannot be introduced, since all $x_{i2} \leq 0$, and we have no bounded solutions for $\lambda > 3$. Therefore, the problem has only one characteristic solution for $-2 \leq \lambda < 3$ and a multiple solution for $\lambda = 3$.

Tableau 8.1

I. Equations (1.8) in Simplex Tableau (with Artificial Vector $\mathbf{P}_7$)

i	Basis	c	$\mathbf{P}_0$	λ $\mathbf{P}_1$	$-\lambda$ $\mathbf{P}_2$	-1 $\mathbf{P}_3$	1 $\mathbf{P}_4$	0 $\mathbf{P}_5$	0 $\mathbf{P}_6$	w $\mathbf{P}_7$
1	$\mathbf{P}_7$	w	5	3	−3	−1	1	−1	0	1
2	$\mathbf{P}_6$	0	3	(2)	−2	1	−1	0	1	0
$m+1$			0	0	0	1	−1	0	0	0
$m+2$			0	−1	1	0	0	0	0	0
$m+3$			5	(3)	−3	−1	1	−1	0	0

II. Vector $\mathbf{P}_1$ Eliminated Vector $\mathbf{P}_6$

$\mathbf{P}_7$	w	1/2	0	0	−5/2	(5/2)	−1	−3/2	1
$\mathbf{P}_1$	λ	3/2	1	−1	1/2	−1/2	0	1/2	0
		0	0	0	1	−1	0	0	0
		3/2	0	0	1/2	−1/2	0	1/2	0
		1/2	0	0	−5/2	(5/2)	−1	−3/2	0

III. Vector $\mathbf{P}_4$ Eliminated the Artificial Vector $\mathbf{P}_7$

$\mathbf{P}_4$	1	1/5	0	0	−1	1	−2/5	−3/5
$\mathbf{P}_1$	λ	8/5	1	−1	0	0	−1/5	(1/5)
		1/5	0	0	0	0	−2/5	−3/5
		8/5	0	0	0	0	−1/5	1/5
		0	0	0	0	0	0	0

IV. Vector $\mathbf{P}_6$ Eliminated Vector $\mathbf{P}_1$

$\mathbf{P}_4$	1	5	3	−3	−1	1	−1	0
$\mathbf{P}_6$	0	8	5	−5	0	0	−1	1
		5	3	−3	0	0	−1	0
		0	−1	1	0	0	0	0

An application of the above technique to a problem from the theory of games is given in Chap. 12.

2. THE PARAMETRIC DUAL PROBLEM

Associated with the parametric objective-function problem is the following general dual interpretation, i.e., where the parameter is contained in the right-hand side of the equations: Let $\sigma \leq \theta \leq \rho$. For each θ in this interval find a vector

$$\mathbf{X} = (x_1, x_2, \ldots, x_n)$$

which minimizes

$$\sum_{j=1}^{n} c_j x_j$$

subject to

$$\sum_{j=1}^{n} a_{ij} x_j = b_i + \theta b_i' \qquad i = 1, 2, \ldots, m$$

and

$$x_j \geq 0 \qquad j = 1, 2, \ldots, n$$

Assume that we have found a minimum feasible solution $\bar{\mathbf{X}} = (\bar{x}_1, \bar{x}_2, \ldots, \bar{x}_n)$ for the problem when $\theta = \sigma$. We can write each $\bar{x}_i$ as a linear combination of σ and have

$$\bar{x}_i = q_i + \sigma p_i \geq 0$$

which implies that the inequalities

$$\bar{x}_i = q_i + \theta p_i \geq 0 \tag{2.1}$$

are consistent.

If all $p_i = 0$, then the solution $\bar{\mathbf{X}}$ is a minimum feasible solution for all θ; if all $p_i \geq 0$, then the solution $\bar{\mathbf{X}}$ is a minimum feasible solution for all $\theta \geq \sigma$; and if all $p_i \leq 0$, the solution $\bar{\mathbf{X}}$ is a minimum feasible solution for all $\theta \leq \sigma$. In general, however, the p_i will be both positive and negative, and in order to determine for what range of values of θ the given solution $\bar{\mathbf{X}}$ is a minimum, we must perform an analysis similar to that performed for the parametric objective-function problem.

For $p_i > 0$, we have

$$\theta \geq -\frac{q_i}{p_i}$$

and define

$$\underline{\theta} = \begin{cases} \max\limits_{p_i > 0} -\dfrac{q_i}{p_i} \\ \text{or} \\ -\infty & \text{if all } p_i \leq 0 \end{cases}$$

For $p_i < 0$, we have

$$\theta \leq -\frac{q_i}{p_i}$$

and define

$$\bar{\theta} = \begin{cases} \min\limits_{p_i<0} -\dfrac{q_i}{p_i} \\ \text{or} \\ +\infty \qquad \text{if all } p_i \geq 0 \end{cases}$$

The given solution (2.1) is then a minimum for

$$\underline{\theta} \leq \theta \leq \bar{\theta}$$

We shall assume that the upper bound $\bar{\theta}$ is not $+\infty$. As θ is increased, the solution remains optimal, i.e., each $z_j - c_j \leq 0$, but it need not stay feasible.

For a sufficiently large increase in θ, one of the values

$$\bar{x}_i = q_i + \theta p_i$$

will be forced negative. If $\bar{x}_l$ is the first of these variables to go negative, then

$$\bar{\theta} = -\frac{q_l}{p_l}$$

where $p_l < 0$. We next wish to determine a new minimum feasible solution for a range of $\theta \geq \bar{\theta}$. We have to determine a vector to be introduced into the basis and a vector to be eliminated that will keep the right-hand elements of the transformed equations nonnegative and the transformed $z_j - c_j$ nonpositive. The method of selection is summarized as follows:

Theorem 2. *If the vector $\mathbf{P}_l$ corresponding to $\bar{\theta} = -q_l/p_l$ is eliminated from the basis and a vector $\mathbf{P}_k$ with*

$$\frac{z_k - c_k}{x_{lk}} = \min_{x_{lj}<0} \frac{z_j - c_j}{x_{lj}}$$

is introduced into the basis, the new solution is a minimum for at least one value of θ. If $\underline{\theta}' \leq \theta \leq \bar{\theta}'$ is the entire set of θ for which the new basis yields a minimum, then $\underline{\theta}' = \bar{\theta}$.

Proof. The new solution $\bar{\mathbf{X}}' = (\bar{x}'_1, \bar{x}'_2, \ldots, \bar{x}'_n)$ is given by

$$\bar{x}'_i = q'_i + \theta p'_i = q_i + \theta p_i - \frac{x_{ik}}{x_{lk}}(q_l + \theta p_l) \qquad \text{for } i \neq l$$

$$\bar{x}'_l = q'_l + \theta p'_l = \frac{q_l + \theta p_l}{x_{lk}}$$

The new solution is certainly feasible for $\theta = \bar{\theta} = -q_l/p_l$. If $\bar{\mathbf{X}}'$ is feasible for any other θ, then it must be for $\theta \geq \bar{\theta}$, since $x_{lk} < 0$, $p_l < 0$, and

$$\frac{q_l + \theta p_l}{x_{lk}} \geq 0$$

implies

$$\theta \geq -\frac{q_l}{p_l} = \bar{\theta}$$

To show that the new basis is optimal, we have

$$(z_j - c_j)' = z_j - c_j - \frac{x_{lj}}{x_{lk}}(z_k - c_k)$$

Since all $z_j - c_j \leq 0$ and $x_{lk} < 0$, all $(z_j - c_j)'$ with an $x_{lj} \geq 0$ are also nonpositive. In order to have

$$z_j - c_j - \frac{x_{lj}}{x_{lk}}(z_k - c_k) \leq 0$$

for $x_{lj} < 0$, we must have

$$z_j - c_j \leq \frac{x_{lj}}{x_{lk}}(z_k - c_k)$$

or

$$\frac{z_j - c_j}{x_{lj}} \geq \frac{z_k - c_k}{x_{lk}}$$

Hence $\mathbf{P}_k$ must be the vector not in the basis that corresponds to

$$\min_{x_{lj}<0} \frac{z_j - c_j}{x_{lj}}$$

If all $x_{lj} \geq 0$, then there are no feasible solutions for $\theta > \bar{\theta}$. The proof is left to the reader (see Sec. 2 of Chap. 9).

3. SENSITIVITY ANALYSIS

Except for the discussions in Secs. 1 and 2, we have assumed that all the coefficients of a linear-programming problem are given. However, for many problems either these constants, i.e., the a_{ij}, b_i, and c_j, are estimates or they vary over time. For example, in a diet problem such as the chicken-feed problem, the cost of any individual feed will vary from week to week, and it is quite important to know the range of cost for which the solution remains optimal. Similar situations may arise for other elements of a problem. Investigations that deal with the changes in the optimum solution due to changes in the data are termed *sensitivity analyses*. In this section, we shall be concerned with the analysis that determines the range of a given element for which the solution, as originally stated, remains optimal, and we conclude with a discussion of the sensitivity of the objective function with regard to an implicit parameter in the neighborhood of the optimal vertex. Additional work in sensitivity and related areas is given in Shetty [90*a*], Courtillot [16*a*], Garvin [45*a*], Madansky [72*a*], Wagner [99*a*], Saaty [88*a*], and Webb [100*a*].

a. Variation of the c_j. For an optimum solution we have the set of inequalities $z_j - c_j \leq 0$ holding for all j.

Let Δc_j be the amount to be added to the corresponding c_j. For those variables not in the final basis, we must have $z_j - (c_j + \Delta c_j) \leq 0$; hence, $z_j - c_j \leq \Delta c_j$, with Δc_j having no upper bound. An appropriate change in c_j does not change the value of the objective function since $x_j = 0$. For the example of Tableau 4.3, we have for those variables not in the final solution that $-1/5 \leq \Delta c_1$, $-4/5 \leq \Delta c_4$, and $-12/5 \leq \Delta c_5$.

For a variable in the final basis, a Δc_j affects all z_j, for all j not in the basis, since

$$z_j - c_j = \sum_{i \text{ in basis}} x_{ij}c_i - c_j \leq 0$$

Let the change Δc_k occur for some basis variable x_k. Then

$$\sum_{i \text{ in basis}} x_{ij}c_i + x_{kj}\Delta c_k - c_j \leq 0$$

or

$$x_{kj}\Delta c_k \leq -(z_j - c_j)$$

For those $x_{kj} > 0$

$$\Delta c_k \leq \frac{-(z_j - c_j)}{x_{kj}}$$

and for those $x_{kj} < 0$

$$\Delta c_k \geq \frac{-(z_j - c_j)}{x_{kj}}$$

Hence

$$\max_{x_{kj}<0} \frac{-(z_j - c_j)}{x_{kj}} \leq \Delta c_k \leq \min_{x_{kj}>0} \frac{-(z_j - c_j)}{x_{kj}}$$

for all j not in the basis. If no $x_{kj} > 0$, there is no upper bound, and if no $x_{kj} < 0$, there is no lower bound. For the example of Tableau 4.3, let $\Delta c_k = \Delta c_6$. We have

$$\max\left(\frac{4/5}{-1/2}\right) \leq \Delta c_6 \leq \min\left(\frac{1/5}{1}, \frac{12/5}{10}\right)$$

$$-8/5 \leq \Delta c_6 \leq 1/5$$

For $\Delta c_k = \Delta c_2$, we have

$$-\infty \leq \Delta c_2 \leq \min\left(\frac{1/5}{2/5}, \frac{4/5}{1/10}, \frac{12/5}{4/5}\right)$$

$$-\infty \leq \Delta c_2 \leq 1/2$$

With Δc_k restricted by the appropriate bounds, the basis remains optimal but the value of the objective function changes by $\Delta c_k x_k$.

b. Variation of the b_i. A change in a b_i must be of the magnitude that preserves the feasibility of the basis. For the optimum solution, $\mathbf{X}^0 = \mathbf{B}^{-1}\mathbf{b} \geq \mathbf{O}$, and for a change Δb_l in b_l, we must have, letting $\bar{\mathbf{b}}$ be the new right-hand side,

$$\bar{\mathbf{X}}^0 = \mathbf{B}^{-1}\bar{\mathbf{b}} = (x_i + b_{il}\Delta b_l) \geq \mathbf{O} \tag{3.1}$$

for all i in the basis and where b_{il} is the element in the ith row and lth column of $\mathbf{B}^{-1}$.

For $b_{il} > 0$, we have

$$\Delta b_l \geq \frac{-x_i}{b_{il}}$$

and for $b_{il} < 0$

$$\Delta b_l \leq \frac{-x_i}{b_{il}}$$

Hence,

$$\max_{b_{il}>0} \frac{-x_i}{b_{il}} \leq \Delta b_l \leq \min_{b_{il}<0} \frac{-x_i}{b_{il}}$$

For the example of Tableau 4.3, $\mathbf{B} = (\mathbf{P}_2\mathbf{P}_3\mathbf{P}_6)$, $\mathbf{X}^0 = (4,5,11)$, $\mathbf{b} = (7,12,10)$, and

$$\mathbf{B}^{-1} = \begin{pmatrix} 2/5 & 1/10 & 0 \\ 1/5 & 3/10 & 0 \\ 1 & -1/2 & 1 \end{pmatrix}$$

Letting $\Delta b_l = \Delta b_2$ yields

$$\max\left(\frac{-4}{1/10}, \frac{-5}{3/10}\right) \leq \Delta b_2 \leq \frac{-11}{-1/2}$$

$$\frac{-50}{3} \leq \Delta b_2 \leq 22$$

The new solution is given by (3.1), and the change in the value of the objective function is given by $\sum_{i \text{ in basis}} b_{il}\,\Delta b_l c_i$.

c. Variation in a_{ij}. Let a_{lk}, the element to be varied, be the element of the lth row of column $\mathbf{P}_k$. Assume that $\mathbf{P}_k$ is a vector of the optimum basis and that the optimum solution is $\mathbf{X}^0 = \mathbf{B}^{-1}\mathbf{b}$. The new matrix is given by $\bar{\mathbf{B}} = \mathbf{B} + \Delta a_{lk}\,\mathbf{O}_{lk}$, where $\mathbf{O}_{lk}$ is a null matrix except for element (l,k), which equals unity. To preserve feasibility (assuming that $\bar{\mathbf{B}}$ is a nonsingular matrix), we must have

$$\bar{\mathbf{B}}^{-1}\mathbf{b} = (\mathbf{B} + \Delta a_{lk}\,\mathbf{O}_{lk})^{-1}\mathbf{b} \geq \mathbf{O} \tag{3.2}$$

and to preserve optimality

$$z_j - c_j = \mathbf{c}_0\bar{\mathbf{B}}^{-1}\mathbf{P}_j - c_j \leq 0 \tag{3.3}$$

for all j, where $\mathbf{c}_0$ is the set of cost coefficients for the basis variables. We have

$$\bar{\mathbf{B}} = \mathbf{B}(\mathbf{I} + \mathbf{B}^{-1}\,\Delta a_{lk}\,\mathbf{O}_{lk})$$

and, by Sec. 1 of Chap. 2,

$$\bar{\mathbf{B}}^{-1} = (\mathbf{I} + \mathbf{B}^{-1}\,\Delta a_{lk}\,\mathbf{O}_{lk})^{-1}\mathbf{B}^{-1}$$

Equation (3.2) can then be rewritten as

$$\begin{aligned}\bar{\mathbf{B}}^{-1}\mathbf{b} &= (\mathbf{I} + \mathbf{B}^{-1}\,\Delta a_{lk}\,\mathbf{O}_{lk})^{-1}\mathbf{B}^{-1}\mathbf{b} \\ &= (\mathbf{I} + \mathbf{B}^{-1}\,\Delta a_{lk}\,\mathbf{O}_{lk})^{-1}\mathbf{X}^0 \end{aligned}\tag{3.4}$$

We now must determine for what conditions on Δa_{lk} the inverse $(\mathbf{I} + \mathbf{B}^{-1}\,\Delta a_{lk}\,\mathbf{O}_{lk})^{-1}$ exists and restricts (3.4) to be nonnegative.

The term $\mathbf{B}^{-1}\,\Delta a_{lk}\,\mathbf{O}_{lk}$ can be written as

$$\mathbf{B}^{-1}\,\Delta a_{lk}\,\mathbf{O}_{lk} =$$

$$\mathbf{B}^{-1}\begin{pmatrix} 0 & \cdots & 0 & \cdots & 0 \\ \cdot & \cdot & \cdot & \cdot & \cdot \\ 0 & \cdots & \Delta a_{lk} & \cdots & 0 \\ \cdot & \cdot & \cdot & \cdot & \cdot \\ 0 & \cdots & 0 & \cdots & 0 \end{pmatrix} = \begin{pmatrix} 0 & \cdots & b_{1l}\,\Delta a_{lk} & \cdots & 0 \\ \cdot & \cdot & \cdot & \cdot & \cdot \\ 0 & \cdots & b_{kl}\,\Delta a_{lk} & \cdots & 0 \\ \cdot & \cdot & \cdot & \cdot & \cdot \\ 0 & \cdots & b_{ml}\,\Delta a_{lk} & \cdots & 0 \end{pmatrix}$$

$$\qquad\qquad\uparrow\ k\text{th column} \qquad\qquad\qquad\qquad \uparrow\ k\text{th column}$$

where the b_{il} are the corresponding elements of the lth column of $\mathbf{B}^{-1}$. Hence,

$$(\mathbf{I} + \mathbf{B}^{-1}\,\Delta a_{lk}\,\mathbf{O}_{lk}) =$$

$$\begin{pmatrix} 1 & \cdots & b_{1l}\,\Delta a_{lk} & \cdots & 0 \\ \cdot & \cdot & \cdot & \cdot & \cdot \\ 0 & \cdots & 1 + b_{kl}\,\Delta a_{lk} & \cdots & 0 \\ \cdot & \cdot & \cdot & \cdot & \cdot \\ 0 & \cdots & b_{ml}\,\Delta a_{lk} & \cdots & 1 \end{pmatrix} \leftarrow k\text{th row} \tag{3.5}$$

The inverse of (3.5) is given by

$$(\mathbf{I} + \mathbf{B}^{-1}\,\Delta a_{lk}\,\mathbf{O}_{lk})^{-1} = \begin{pmatrix} 1 & \cdots & \dfrac{-b_{1l}\,\Delta a_{lk}}{1 + b_{kl}\,\Delta a_{lk}} & \cdots & 0 \\ \cdots & \cdots & \cdots & \cdots & \cdots \\ 0 & \cdots & \dfrac{1}{1 + b_{kl}\,\Delta a_{lk}} & \cdots & 0 \\ \cdots & \cdots & \cdots & \cdots & \cdots \\ 0 & \cdots & \dfrac{-b_{ml}\,\Delta a_{lk}}{1 + b_{kl}\,\Delta a_{lk}} & \cdots & 1 \end{pmatrix} \tag{3.6}$$

Substituting (3.6) in (3.4), we obtain

$$\bar{\mathbf{B}}^{-1}\mathbf{b} = (\mathbf{I} + \mathbf{B}^{-1}\,\Delta a_{lk}\,\mathbf{O}_{lk})^{-1}\mathbf{X}^0$$

$$\bar{\mathbf{B}}^{-1}\mathbf{b} = \begin{pmatrix} x_1 - \dfrac{b_{1l}\,\Delta a_{lk}}{1 + b_{kl}\,\Delta a_{lk}}\,x_k \\ \cdot \\ \cdot \\ \cdot \\ \dfrac{1}{1 + b_{kl}\,\Delta a_{lk}}\,x_k \\ \cdot \\ \cdot \\ \cdot \\ x_m - \dfrac{b_{ml}\,\Delta a_{lk}}{1 + b_{kl}\,\Delta a_{lk}}\,x_k \end{pmatrix} \tag{3.7}$$

In order for (3.7) to exist and to be nonnegative, we must have the following:

$$1 + b_{kl}\,\Delta a_{lk} > 0 \tag{3.8}$$

$$x_i - \frac{b_{il}\,\Delta a_{lk}}{1 + b_{kl}\,\Delta a_{lk}}\,x_k \geq 0 \tag{3.9}$$

Solving for Δa_{lk}, (3.9) yields (assuming that $1 + b_{kl}\,\Delta a_{lk} > 0$) the conditions for all $i \neq k$:

$$\Delta a_{lk} \leq \frac{x_i}{b_{il}x_k - b_{kl}x_i} \qquad \text{for } b_{il}x_k - b_{kl}x_i > 0$$

$$\frac{x_i}{b_{il}x_k - b_{kl}x_i} \leq \Delta a_{lk} \qquad \text{for } b_{il}x_k - b_{kl}x_i < 0$$

The upper and lower bounds do not exist if corresponding denominators do not exist. If the change Δa_{lk} is for a vector in the basis, the feasibility conditions impose the restriction (3.8) and

$$\max_{i \neq k} \frac{x_i}{(b_{il}x_k - b_{kl}x_i) < 0} \leq \Delta a_{lk} \leq \min_{i \neq k} \frac{x_i}{(b_{il}x_k - b_{kl}x_i) > 0} \tag{3.10}$$

For those vectors $\mathbf{P}_j$ not in the basis, the optimality conditions (3.3) must be simultaneously satisfied. Since $\bar{\mathbf{B}}^{-1} = (\mathbf{I} + \mathbf{B}^{-1}\,\Delta a_{lk}\,\mathbf{O}_{lk})^{-1}\mathbf{B}^{-1}$, the above analysis yields for (3.3)

$$\mathbf{c}_0(\mathbf{I} + \mathbf{B}^{-1}\,\Delta a_{lk}\,\mathbf{O}_{lk})^{-1}\mathbf{B}^{-1}\mathbf{P}_j - c_j \leq 0 \tag{3.11}$$

or

$$\mathbf{c}_0(\mathbf{I} + \mathbf{B}^{-1}\,\Delta a_{lk}\,\mathbf{O}_{lk})^{-1}\mathbf{X}_j - c_j \leq 0 \tag{3.12}$$

where $\mathbf{X}_j = \mathbf{B}^{-1}\mathbf{P}_j$. From (3.6), inequalities (3.12) become

$$\mathbf{c}_0 \begin{pmatrix} x_{1j} - \dfrac{b_{1l}\,\Delta a_{lk}}{1 + b_{kl}\,\Delta a_{lk}}\, x_{kj} \\ \cdot \\ \cdot \\ \cdot \\ \dfrac{1}{1 + b_{kl}\,\Delta a_{lk}}\, x_{kj} \\ \cdot \\ \cdot \\ \cdot \\ x_{mj} - \dfrac{b_{ml}\,\Delta a_{lk}}{1 + b_{kl}\,\Delta a_{lk}}\, x_{kj} \end{pmatrix} - c_j \leq 0$$

This yields

$$c_1\left(x_{1j} - \frac{b_{1l}\,\Delta a_{lk}}{1 + b_{kl}\,\Delta a_{lk}}\, x_{kj}\right) + \cdots + c_k \frac{1}{1 + b_{kl}\,\Delta a_{lk}}\, x_{kj} + \cdots$$
$$+ c_m\left(x_{mj} - \frac{b_{ml}\,\Delta a_{lk}}{1 + b_{kl}\,\Delta a_{lk}}\, x_{kj}\right) + c_k x_{kj} - c_k x_{kj} - c_j \leq 0$$

where the term $c_k x_{kj} - c_k x_{kj}$ has been added to the left-hand side.

Collecting terms, we have

$$\sum_{i \text{ in basis}} c_i x_{ij} - \frac{b_{1l}\,\Delta a_{lk}}{1 + b_{kl}\,\Delta a_{lk}}\, c_1 x_{kj} - \cdots + \frac{1}{1 + b_{kl}\,\Delta a_{lk}}\, c_k x_{kj}$$
$$- \cdots - \frac{b_{ml}\,\Delta a_{lk}}{1 + b_{kl}\,\Delta a_{lk}}\, c_m x_{kj} - c_k x_{kj} - c_j \leq 0 \tag{3.13}$$

By multiplying $c_k x_{kj}$ by $1 + b_{kl}\,\Delta a_{lk}/(1 + b_{kl}\,\Delta a_{lk})$, substituting for z_j, and collecting terms in (3.13), we have

$$(z_j - c_j) - \sum_{i \text{ in basis}} \frac{b_{il}\,\Delta a_{lk}}{1 + b_{kl}\,\Delta a_{lk}}\, c_i x_{kj} \leq 0$$

and from (3.8),

$$(z_j - c_j)(1 + b_{kl}\,\Delta a_{lk}) - \Delta a_{lk} \sum_i b_{il} c_i x_{kj} \leq 0$$

$$(z_j - c_j) + (z_j - c_j) b_{kl}\,\Delta a_{lk} - \Delta a_{lk} \sum_i b_{il} c_i x_{kj} \leq 0$$

$$(z_j - c_j) \leq \Delta a_{lk} \left[\sum_i b_{il} c_i x_{kj} - (z_j - c_j) b_{kl}\right]$$

If the term in brackets is positive,

$$\frac{z_j - c_j}{\sum_i b_{il} c_i x_{kj} - (z_j - c_j) b_{kl}} \leq \Delta a_{lk}$$

and if it is negative,

$$\Delta a_{lk} \leq \frac{z_j - c_j}{\sum_i b_{il} c_i x_{kj} - (z_j - c_j) b_{kl}}$$

or, for those j not in the basis,

$$\max_j \frac{z_j - c_j}{\left[x_{kj} \sum_i b_{il} c_i - (z_j - c_j) b_{kl}\right] > 0} \leq \Delta a_{lk} \leq \min_j \frac{z_j - c_j}{\left[x_{kj} \sum_i b_{il} c_i - (z_j - c_j) b_{kl}\right] < 0} \tag{3.14}$$

The Δa_{lk} is unbounded if corresponding denominators do not exist. We note that the summation term is the lth element of the row vector $\mathbf{c}_0\mathbf{B}^{-1}$. In order to apply a change Δa_{lk} to the lth component of a vector $\mathbf{P}_k$ in the basis, Δa_{lk} must satisfy (3.8), (3.10), and (3.14). The corresponding changes in the value of the variables are given by (3.7). The total change in the value of the objective function is determined by subtracting the value of the old solution $\mathbf{X}^0$ from the value of the solution (3.7), i.e.,

$$\left(x_1 - \frac{b_{1l}\,\Delta a_{lk}}{1 + b_{kl}\,\Delta a_{lk}} x_k\right) c_1 + \cdots + \frac{c_k x_k}{1 + b_{kl}\,\Delta a_{lk}} + \cdots + \left(x_m - \frac{b_{ml}\,\Delta a_{lk}}{1 + b_{kl}\,\Delta a_{lk}} x_k\right) c_m - c_1 x_1 - \cdots - c_k x_k - \cdots - c_m x_m = z$$

Adding the term

$$\frac{b_{kl}\,\Delta a_{lk}}{1 + b_{kl}\,\Delta a_{lk}} c_k x_k - \frac{b_{kl}\,\Delta a_{lk}}{1 + b_{kl}\,\Delta a_{lk}} c_k x_k$$

and collecting terms, we have

$$z = -\sum_i b_{il} c_i \frac{\Delta a_{lk} x_k}{1 + b_{kl}\,\Delta a_{lk}}$$

For a change to an element a_{lj} of a vector $\mathbf{P}_j$ not in the basis, a_{lj} must be such that

$$\mathbf{c}_0\mathbf{B}^{-1}\mathbf{P}_j - c_j \leq 0$$

For a given Δa_{lj}, we have

$$\mathbf{c}_0(\mathbf{X}_j + \Delta a_{lj}\,\mathbf{B}^l) - c_j \leq 0 \tag{3.15}$$

where $\mathbf{B}^l$ is the lth column of $\mathbf{B}^{-1}$. Then (3.15) is

$$(z_j - c_j) + \Delta a_{lj} \sum_i b_{il}c_i \leq 0$$

Hence, for the vector $\mathbf{P}_j$ not in the basis, Δa_{lj} must lie between the limits

$$\frac{-(z_j - c_j)}{\left(\sum_i b_{il}c_i\right) < 0} \leq \Delta a_{lj} \leq \frac{-(z_j - c_j)}{\left(\sum_i b_{il}c_i\right) > 0}$$

That is,

$$\Delta a_{lj} \leq \begin{cases} \dfrac{-(z_j - c_j)}{\sum_i b_{il}c_i} & \text{if } \sum_i b_{il}c_i > 0 \\ +\infty & \text{if } \sum_i b_{il}c_i \leq 0 \end{cases}$$

$$\Delta a_{lj} \geq \begin{cases} \dfrac{-(z_j - c_j)}{\sum_i b_{il}c_i} & \text{if } \sum_i b_{il}c_i < 0 \\ -\infty & \text{if } \sum_i b_{il}c_i \geq 0 \end{cases}$$

Another approach to investigating the sensitivity of an optimum solution to changes in the constant terms is given in Saaty [88*a*] and Webb [100*a*]. From the statement of the symmetric primal and dual linear-programming problems (Chap. 5, Sec. 2), we have

$$\begin{aligned} \sum_j a_{ij}x_j &\geq b_i \qquad i = 1, 2, \ldots, m \\ x_j &\geq 0 \qquad j = 1, 2, \ldots, n \\ f = \sum_j c_jx_j &= \min \end{aligned}$$

and

$$\begin{aligned} \sum_i a_{ij}w_i &\leq c_j \qquad j = 1, 2, \ldots, n \\ w_i &\geq 0 \qquad i = 1, 2, \ldots, m \\ g = \sum_i b_iw_i &= \max \end{aligned}$$

We note that $f = \mathbf{c}_0\mathbf{B}^{-1}\mathbf{b}$, where $\mathbf{B}$ is the optimal basis and $\mathbf{c}_0$ is the set of cost coefficients for the optimal-basis vectors.

When f is considered as a function of a_{ij}, b_i, and c_j, and these elements are considered as functions of an implicit parameter t, the total derivative of f with respect to t is given by

$$\frac{df}{dt} = \sum_{i,j=1}^{m,n} \frac{\partial f}{\partial a_{ij}} \frac{da_{ij}}{dt} + \sum_{j=1}^{n} \frac{\partial f}{\partial c_j} \frac{dc_j}{dt} + \sum_{i=1}^{m} \frac{\partial f}{\partial b_i} \frac{db_i}{dt}$$

From [88*a*] the total derivative can then be written in a computable form as

$$\frac{df}{dt} = -\sum_{i,j=1}^{m,n} \frac{\partial f}{\partial b_i} \frac{\partial f}{\partial c_j} \frac{da_{ij}}{dt} + \sum_{j=1}^{n} \frac{\partial f}{\partial c_j} \frac{dc_j}{dt} + \sum_{i=1}^{m} \frac{\partial f}{\partial b_i} \frac{db_i}{dt} \tag{3.16}$$

Letting $\mathbf{X}^0$ be the optimal vector for the primal and $\mathbf{W}^0$ be the optimal vector of the dual, we have, based on the duality theorem,

$$\frac{\partial f}{\partial \mathbf{c}} = \mathbf{X}^0 \qquad \frac{\partial f}{\partial \mathbf{b}} = \mathbf{W}^0$$

In (3.16) all the partials and derivatives can be solved explicitly. The formula provides us with the sensitivity of the objective function with regard to an implicit parameter in the neighborhood of the optimal vertex.

As an example, consider the linear-programming problem of Tableau 4.3. We shall assume that the coefficients are multiplied by powers of t as follows: $a_{ij}t^2$, b_it, c_jt^3. We are interested in the rate of change of the objective function with respect to t in the neighborhood of the optimal point. Computing all the partial derivatives and derivatives, we have

$$\frac{\partial f}{\partial b_1} = -\tfrac{1}{5} \qquad \frac{\partial f}{\partial b_2} = -\tfrac{4}{5}$$

$$\frac{\partial f}{\partial c_2} = 4 \qquad \frac{\partial f}{\partial c_3} = 5 \qquad \frac{\partial f}{\partial c_6} = 11$$

$$\frac{dc_2}{dt} = 3t^2 \qquad \frac{dc_3}{dt} = -9t^2 \qquad \frac{dc_5}{dt} = 6t^2$$

$$\frac{db_1}{dt} = 7 \qquad \frac{db_2}{dt} = 12 \qquad \frac{db_3}{dt} = 10$$

$$\frac{da_{11}}{dt} = 2t \qquad \frac{da_{12}}{dt} = 6t \qquad \frac{da_{13}}{dt} = -2t \qquad \frac{da_{15}}{dt} = 4t$$

$$\frac{da_{22}}{dt} = -4t \qquad \frac{da_{23}}{dt} = 8t \qquad \frac{da_{24}}{dt} = 2t$$

$$\frac{da_{32}}{dt} = -8t \qquad \frac{da_{33}}{dt} = 6t \qquad \frac{da_{35}}{dt} = 16t \qquad \frac{da_{36}}{dt} = 2t$$

and the total derivative

$$\begin{aligned}\frac{df}{dt} &= -(-\tfrac{1}{5})(4)(6t) - (-\tfrac{1}{5})(5)(-2t) \\ &\quad -(-\tfrac{4}{5})(4)(-4t) - (-\tfrac{4}{5})(5)(8t) \\ &\quad +(4)(3t^2) + (5)(-9t^2) \\ &\quad +(-\tfrac{1}{5})(7) + (-\tfrac{4}{5})(12) \\ &= -33t^2 + 22t - 11\end{aligned}$$

The last expression indicates the rate of change of the objective function with respect to a change in t. We note that for a small increment of t we can estimate the effect on the objective function, i.e.,

$$\Delta f = -33t^2\,\Delta t + 22t\,\Delta t - 11\Delta t$$

For instance, if we evaluate the effect of Δt in this example (where $t = 1$), we have

$$\Delta f = -22\Delta t$$

One must be sure that Δt is chosen small enough so that the old optimal basis is still the solution.

In Webb [100*a*] an extension based on the derivation of df/dt is given. It is shown that

$$\frac{\partial f}{\partial a_{ij}} = -x_j{}^0 w_i{}^0$$

In addition, by letting $t = c_j$ in the derivative df/dt, we have

$$\frac{df}{dc_j} = x_j{}^0$$

and by allowing $t = b_i$

$$\frac{df}{db_i} = w_i{}^0$$

Two additional important relationships can also be developed. The variation of the primal solution with respect to changes in a_{ij}, when evaluated for a particular $x_j{}^0$, is given by

$$\left.\frac{\partial x_j}{\partial a_{ij}}\right]_{x_j{}^0} = -\frac{x_j{}^0 w_i{}^0}{c_i}$$

and the variation of the dual solution with respect to changes in a_{ij}, when evaluated for a particular $w_i{}^0$, is given by

$$\left.\frac{\partial w_i}{\partial a_{ij}}\right]_{w_i{}^0} = -\frac{x_j{}^0 w_i{}^0}{b_i}$$

Extensions of this technique into statistical distributions over a multiplicity of optimal solutions are given in Webb [100*a*].

REMARKS

For additional reading the reader is referred to Saaty and Gass [89] and Barnett [3*e*].

EXERCISES

1. Solve the following parametric objective-function problem for all values of λ: Minimize

$$2\lambda x_1 + (1 - \lambda)x_2 - 3x_3 + \lambda x_4 + 2x_5 - 3\lambda x_6$$

subject to

$$\begin{aligned} x_1 + 3x_2 - x_3 + 2x_5 &= 7 \\ -2x_2 + 4x_3 + x_4 &= 12 \\ -4x_2 + 3x_3 + 8x_5 + x_6 &= 10 \end{aligned}$$

and

$$x_j \geq 0$$

2. Write the dual to the problem in Exercise 1 and solve for all values of the parameter.

3. Solve the production-scheduling problem (Exercise 1, Chap. 11) with the objective function $16s_0 + \sum_{t=1}^{4} s_t + \lambda \sum_{t=1}^{5} y_t$ to be minimum for all values of $\lambda \geq 0$.

4. Modify the computational procedure of the revised simplex method to handle a parametric objective function.

5. By means of the simplex method and a two-dimensional graph of the parameter space, solve the following two-parameter problem:
Minimize

$$x_1 + \lambda x_2 + \mu x_3 + x_4 + x_5 + x_6 + x_7 + \lambda x_8 + \mu x_9$$

subject to

$$\begin{aligned} x_1 - x_4 - 2x_6 + 4x_7 + 2x_8 + x_9 &= 4 \\ x_2 + 2x_4 - 2x_5 + x_6 + 2x_7 + 4x_8 &= 2 \\ x_3 + x_5 + x_6 + x_7 + 2x_9 &= 1 \\ x_j &\geq 0 \end{aligned}$$

For this problem a feasible solution exists for all values and combinations of λ and μ. Construct an example that does not have this property. What is the dual analogue for the two-parameter problem? Can an efficient computational scheme be developed for the n-parameter ($n \geq 2$) problem?

6. Describe procedures that will allow for the addition or deletion of a constraint.

7. Generalize the formulas for varying a single b_i to include variations in all b_i.

8. Generalize the formulas for varying a single c_j to include variations in all c_j for the set of basis variables.

9. Apply formula (3.16) to Exercise 1 when $\lambda = t^2$, each a_{ij} is multiplied by $t - 1$, and each b_i is multiplied by $t^2 - t + 1$.

10. In the corresponding optimal solutions, determine the range of the variation for the elements in the following problems:

a. In Exercise 1*a* of Chap. 4:

1. The cost coefficient c_1.
2. The cost coefficient c_3.
3. The right-hand-side coefficient b_1.

b. In Exercise 1*c* of Chap. 4:

1. The cost coefficient c_1.
2. The coefficient a_{24}.
3. The coefficient a_{32}.
4. The right-hand-side coefficient b_3.
5. The cost coefficient c_4.

11. Let the optimum basis to a linear-programming problem be denoted by $\mathbf{B} = (\mathbf{P}_1 \cdots \mathbf{P}_s \cdots \mathbf{P}_m)$. Let $\mathbf{P}_s$ be changed (perturbed) to $\mathbf{P}_s'$. Determine conditions on $\mathbf{P}_s'$ such that the $\mathbf{B}' = (\mathbf{P}_1 \cdots \mathbf{P}_s' \cdots \mathbf{P}_m)$ is the optimal basis (Barnett [3*e*]). Similarly, for a vector $\mathbf{P}_k$ not in the optimum basis, let it be perturbed to $\mathbf{P}_k'$. What are the conditions on $\mathbf{P}_k'$ such that $\mathbf{B}$ remains the optimal basis?

12. For the linear-programming problem minimize $\mathbf{cX}$ subject to $\mathbf{AX} = \mathbf{b}, \mathbf{X} \geq \mathbf{O}$, we have determined the optimal solution with corresponding basis $\mathbf{B}$. Consider the new problem $\mathbf{AX} = \bar{\mathbf{b}}$ with $\bar{b}_i = b_i + \theta$ for all i and all values of θ. Develop a formula which tells for what range of θ the basis $\mathbf{B}$ will be an optimum basis for the new problem. Note that the basis $\mathbf{B}$ is optimum for $\theta = 0$ and $\bar{\mathbf{b}} = \mathbf{b} + \boldsymbol{\theta}$ where $\boldsymbol{\theta}$ is a column vector with elements all equal to θ.

13. For the parametric dual problem of maximizing $\sum_{j=1}^{n} c_j x_j$ subject to

$$\sum_{j=1}^{n} a_{ij}x_j = b_i + \theta b_i' \qquad i = 1, 2, \ldots, m$$

$$x_j \geq 0 \qquad \text{all } j$$

prove that the objective function is a concave function of θ.† Also, show for the parametric primal problem with the objective of minimizing $\sum_{j=1}^{n} (d_j + \lambda d_j')x_j$, that the objective function is a convex function of λ. Hint for dual: Let $z_0(\theta) = \max \sum_{j=1}^{n} c_j x_j$ and show that $\alpha z_0(\theta_1) + (1 - \alpha)z_0(\theta_2) \leq z_0(\theta_3)$, where $0 \leq \alpha \leq 1$, $\theta_1 < \theta_2$, and $\theta_3 = \alpha\theta_1 + (1 - \alpha)\theta_2$.

† See Chap. 13 for definitions of concave and convex functions.

CHAPTER 9 / **ADDITIONAL COMPUTATIONAL TECHNIQUES**

In this chapter we shall consider the question faced by the investigator in the field who has formulated a linear-programming problem and determined the necessary coefficients and is ready to solve the associated numerical example. *What preliminary analysis and computational devices should he use in order to solve the problem with the minimum amount of computing?*

One reason that the above query is an open question is that the mathematical model of a specific application may lead to a simplified computational scheme. This is the case, for example, for the transportation and the production-scheduling problems discussed in Chaps. 10 and 11, respectively. In the former, a computationally simple variation of the simplex method is devised, while in the latter we could either reduce the number of equations in the system and solve it by the general simplex procedure or solve it by even simpler techniques. However, in the final analysis, it may prove to be more efficient not to worry about such schemes, since investigations of such possible variations, although interesting, are in themselves time-consuming and may be unfruitful. Hence we shall preface our discussion by emphasizing that *any* linear-programming problem can be solved by the tried and proved techniques of the standard original and revised simplex procedures described in Chaps. 4 and 6, respectively. Since the development of reduction and new computational methods is dependent upon the specific mathematical equations under investigation, we are here necessarily limited to discussing techniques for reducing the amount of computation that are applicable to most linear-programming problems. These schemes are all based on overcoming certain peculiarities of the simplex procedure and are designed to reduce the total number of iterations required to reach the minimum solution. Some of these are described in Sec. 1 below, but let us first review a few pertinent points of the simplex method.

The general simplex method is divided into two distinct computational phases: Phase I is concerned with determining a first basic feasible solution, while Phase II, which starts with this first solution, is concerned with obtaining a minimum feasible solution. Computational experience has shown that it takes approximately m iterations (where m is the number of equations in the final model) to solve the problem if the computation starts with an explicit basis of m unit vectors; approximately $2m$ iterations are required if a full artificial basis is used. Hence it appears to be advantageous to eliminate Phase I altogether, or at least to start with as few artificial vectors as possible. Further, if we are in Phase I, the selection of a new vector to be introduced into the basis depends only on a criterion that consistently reduces the contribution of the artificial variables to the value of the objective function. This criterion is used until the "artificial" part of the objective function is made zero. The criterion has no control over that part of the objective function that must eventually be reduced to a minimum by Phase II. This "real" part is allowed to fluctuate, and usually the value of the objective function for the first feasible solution is far from the minimum value. It seems reasonable to expect that the number of iterations required in Phase II depends on how close the first feasible solution is to the minimum. That is, the closer the first solution, the fewer the iterations required. What is more important is that the artificial-basis technique does not allow us to use certain information inherent in many programming problems. For example, from our knowledge of a particular problem, we may expect a specific set of m vectors to yield a solution that is close to the minimum. However, since there is no assurance that this set of vectors forms a feasible basis, it cannot, in general, be efficiently employed in the standard computational procedures. Again, it is felt that the total number of iterations will be reduced if we utilize some of our knowledge about the "expected solution."[1] The majority of the schemes developed for speeding the computation are designed to offset the above points and to vary the initial simplex computational scheme in order to obtain a "good" first feasible solution (Gass [46], Orchard-Hays [82], and Vajda [97]).

Experimental work has been conducted comparing the efficacy of proposed Phase I methods, alternative rules for selecting the vector to enter the basis, and the alternative forms for carrying out the simplex algorithm—standard, revised, and revised-product form of the inverse (Wolfe and Cutler [105*g*] and Quandt and Kuhn [85*c*]). It appears as if

[1] The cautious wording of the above sentences is due to the existence of the ever-present "counter-examples." Problems have been constructed that start with a full artificial basis and reach the minimum in exactly m iterations; problems also exist that start with a high value of the objective function and rapidly reach the minimum or, conversely, start close to the minimum and "creep" to the optimum.

the product form using the pivot-selection rule of Dickson and Frederick [34*a*][1] offers the best combination for achieving computational efficiency. Here, efficiency is measured in terms of total number of iterations, number of computations, and computational time.

1. DETERMINING A FIRST FEASIBLE SOLUTION

A set of $m \times n$ constraints to a linear-programming problem can take on three basic configurations. Let us first assume that the final set of constraints of the model is given by

$$\mathbf{AX} = \mathbf{b} \qquad \text{with } \mathbf{b} \geq \mathbf{O} \qquad m < n \tag{1.1}$$

We assume at least one $b_i > 0$.

For the set (1.1) we have three situations:

1. The set contains no unit vectors, and the computation starts with a set of m artificial vectors.
2. The set contains k distinct unit vectors, and the Phase I procedure starts with $m - k$ artificial vectors.
3. The set contains a basis of m unit vectors, and the computation begins with Phase II.

Next, if the model is of the form

$$\mathbf{AX} \leq \mathbf{b} \qquad \text{with } \mathbf{b} \geq \mathbf{O} \tag{1.2}$$

we rewrite (1.2) as equalities in terms of nonnegative variables (as described in Sec. 4 of Chap. 2 and Sec. 4 of Chap. 4). The resulting set of equations contains a basis of m unit vectors, and the computation begins with Phase II.

A third possibility is that the constraints are of the form

$$\mathbf{AX} \geq \mathbf{b} \qquad \text{with } \mathbf{b} \geq \mathbf{O} \tag{1.3}$$

[1] This rule is as follows: Define $d_j = (z_j - c_j)^2/[(z_j - c_j)^2 + \sum_i x_{ij}^2(+)]$, where the $x_{ij}(+)$ are the positive elements in jth column of the simplex tableau. The vector $\mathbf{P}_k$ to be introduced into the basis corresponds to $d_k = \max_j d_j$ for which $(z_j - c_j) > 0$. Modifications of this rule which are computationally simpler have also proved effective. This rule is based on the angle, and thus the projections, the vectors $\mathbf{P}_j$ make with cost axis (see Figs. 4.1 and 4.2 of Chap. 4). Dickson and Frederick noted that, in general, the rule required a 10 per cent increase in computation time per iteration, but solved their test problems in 30 to 70 per cent less iterations when compared with the standard selection rule.

Here we also rewrite (1.3) as equations in nonnegative variables, but the resulting system contains a set of m negative unit vectors. Let this system be given by

$$\mathbf{AX} - \bar{\mathbf{X}} = \mathbf{b} \tag{1.4}$$

where $\bar{\mathbf{X}} = (x_{n+1}, x_{n+2}, \ldots, x_{n+m})$ is a nonnegative column vector. By applying a simple transformation to the coefficients of $\mathbf{X}$, $\bar{\mathbf{X}}$, and the column vector $\mathbf{b}$, we can start the computation with only one artificial vector. The scheme calls for determining $\max_i b_i = b_s$ and adding the sth row to the negative of every other row of (1.4). The resulting set of m equations will contain $m - 1$ distinct positive unit vectors and will require one artificial vector, which corresponds to the sth variable. The transformation that is applied to (1.4) is given by

$$\begin{aligned}
a'_{ij} &= -a_{ij} + a_{sj} \qquad \text{for } i \neq s \qquad j = 1, 2, \ldots, n+m \\
a'_{sj} &= a_{sj} \qquad j = 1, 2, \ldots, n+m \\
b'_i &= -b_i + b_s \qquad \text{for } i \neq s \\
b'_s &= b_s
\end{aligned} \tag{1.5}$$

To illustrate this scheme, let us consider the inequalities

$$\begin{aligned}
x_1 - x_2 + 3x_3 &\geq 1 \\
2x_1 - x_2 + x_3 &\geq 0 \\
x_1 + x_2 - x_3 &\geq 2
\end{aligned} \tag{1.6}$$

We rewrite (1.6) as equalities to obtain[1]

$$\begin{aligned}
x_1 - x_2 + 3x_3 - x_4 \qquad\qquad &= 1 \\
2x_1 - x_2 + x_3 \qquad - x_5 \qquad &= 0 \\
x_1 + x_2 - x_3 \qquad\qquad - x_6 &= 2
\end{aligned} \tag{1.7}$$

Here $\max_i b_i = b_3 = 2$, and applying formulas (1.5) to (1.7) gives us

$$\begin{aligned}
2x_2 - 4x_3 + x_4 \qquad\qquad - x_6 &= 1 \\
-x_1 + 2x_2 - 2x_3 \qquad + x_5 - x_6 &= 2 \\
x_1 + x_2 - x_3 \qquad\qquad - x_6 &= 2
\end{aligned}$$

[1] The transformation does not have to be applied to those rows which have their corresponding $b_i = 0$, as in the second equation, but here we illustrate the general application of the transformation.

We can, of course, formulate a model that contains a mixture of inequalities and equalities, that is, $\mathbf{AX} \gtreqless \mathbf{b}$, where $\mathbf{b}$ is unrestricted. Here, by suitable manipulation of the appropriate constraints, we can, as above, introduce as many unit vectors as possible.

Up to now we have been assuming that the computation must begin with an explicit unit matrix for the first feasible basis. In Sec. 2 of Chap. 4 we briefly mentioned the possibility of selecting an arbitrary set of m vectors for the first admissible basis. Here the plan calls for the selection of m vectors to form an initial "expected solution." This selection should be based on a careful, but not time-consuming, analysis of the program objectives. We next attempt to compute the inverse of the corresponding $m \times m$ matrix. The following possibilities may occur:

1. The set of m vectors is linearly independent (the inverse exists).
   a. The corresponding solution is feasible.
   For this case the simplex procedure starts Phase II with a *preferred* solution.
   b. The corresponding solution is not feasible.
   Here, as will be described below, we can start Phase I with only one artificial vector.

2. The set of m vectors is linearly dependent and is of rank $k < m$. For this situation we can apply a transformation to the tableau to initiate Phase I with either $m - k$ or $m - k + 1$ artificial vectors (see Gass [46] for further discussion). We could also continue to choose other vectors for our initial basis from a secondary set of preferred vectors until we have chosen a linearly independent set of m vectors (see Orchard-Hays [82] for additional comments along these lines).

For Case 1*b* above, the following simple transformation, which is similar to the one applied to the set of inequalities $\mathbf{AX} \geq \mathbf{b}$, will enable us to begin Phase I with one artificial vector. For ease of discussion we shall describe the transformation in terms of the original simplex tableau of Chap. 4 (see Tableau 4.1). Let us assume that the first m vectors, $\mathbf{P}_1, \mathbf{P}_2, \ldots, \mathbf{P}_m$ have been selected as the preferred set and are linearly independent but do not form an admissible basis. Let x_{ij} for $i = 1, 2, \ldots, m + 1$ and $j = 0, 1, \ldots, n$ be the transformed element in the ith row and jth column after the matrix $(\mathbf{P}_1\mathbf{P}_2 \cdots \mathbf{P}_m)$ has been inverted, and let $x_{ii} = 1$ for $i = 1, 2, \ldots, m$. Here we assume some $x_{i0} < 0$. Let $x_{s0} = \min\limits_i x_{i0}$. Next, transform all the elements of the tableau by the formulas

$$
\begin{aligned}
x'_{ij} &= x_{ij} - x_{sj} \qquad \text{for } i \neq s,\ m+1 \\
x'_{sj} &= -x_{sj}
\end{aligned}
\tag{1.8}
$$

Formulas (1.8) hold for $j = 0, 1, \ldots, n$ and represent the subtraction of the sth row from all the other rows and the multiplication of the sth row by (-1). This transformation yields a basis consisting of $m - 1$ vectors of the selected m and one artificial vector whose corresponding variable $x'_{s0} > 0$. The new elements of the $(m + 1)$st row are

$$
\begin{aligned}
x'_{m+1,j} &= x_{m+1,j} - x_{sj} \sum_1^m c_i \qquad \text{for } j \neq s \\
x'_{m+1,s} &= -\sum_1^m c_i
\end{aligned}
\tag{1.8'}
$$

The elements in row s of the tableau guide the succeeding iterations until the artificial variable is reduced to zero or until it is determined that the problem is not feasible. Row s corresponds to the $(m + 2)$nd row of the artificial-basis tableau with one artificial vector. Since we wish to control the "real" part of the value of the objective function, we can modify the criterion that selects the new vector for the basis as follows: For the tableau given by (1.8) and (1.8′), let vector $\mathbf{P}_k$ correspond to

$$
x'_{m+1,k} = \max_{x_{sj}' > 0} x'_{m+1,j} \tag{1.9}
$$

Formula (1.9) states that the vector selected has the "artificial" part of its $z_j - c_j$ equal to $x'_{sj} > 0$ and the "real" part equal to $x'_{m+1,j}$ as algebraically large as possible. This criterion will consistently reduce the artificial part and tend to reduce the real part of the objective function. If

$$
\min_i \frac{x'_{i0}}{x'_{ik}} = \frac{x'_{s0}}{x'_{sk}} \qquad \text{for } x'_{ik} > 0
$$

then the artificial vector will be eliminated, and Phase II begins. If x'_{sk} is not the pivot element, then the next iteration introduces a vector $\mathbf{P}_q$ with $x''_{m+1,q} = \max_{x_{sj}'' > 0} x''_{m+1,j}$ etc., until the artificial vector is eliminated or until it is determined that no feasible solutions exist.

For some linear-programming models the problem might call for the minimization of different objectives over the same set of linear constraints. This problem of multiple objective functions can be conveniently handled by setting up the basic constraints and solving the problem for the first objective function. Once the basic minimum feasible solution for this problem has been computed, it is a simple matter to determine whether this solution is also a minimum for any of the other objective functions. If we were using the original simplex procedure, we would include in the tableau an extra row for each objective function which transforms under the usual elimination formulas. By looking at the corresponding $z_j - c_j$ elements, we could determine the additional optimized objective functions. If we were using the revised procedure, then the optimality of the objective functions could be determined by means of the inverse of the current basis and the corresponding new cost coefficients. If the optimum solution for the first objective function were not the minimum for one of the other objective functions, then this first optimum solution would be used as the first feasible solution for the new objective function.

The dual of this problem, the problem of multiple constant vectors, may also arise, e.g., solutions to economic models for a number of different bills of goods might be desired (see Chap. 11, Sec. 2). This problem can be resolved by the following techniques: Let the original problem be that of minimizing

$$\mathbf{cX} \tag{1.10}$$

subject to

$$\mathbf{AX} = \mathbf{P}_0 \tag{1.11}$$

$$\mathbf{X} \geq \mathbf{O} \tag{1.12}$$

In addition to (1.10) to (1.12), we also wish to solve the following set of problems: To minimize

$$\mathbf{cX}$$

subject to

$$\mathbf{AX} = \mathbf{P}_{0q} \qquad q = 1, 2, \ldots, r$$

$$\mathbf{X} \geq \mathbf{O}$$

The problem is first solved for (1.10) to (1.12). In the original simplex method the additional $\mathbf{P}_{0q}$ are attached to the tableau and transformed in the usual manner. If, after the minimum feasible solution for $\mathbf{P}_0$ is obtained, all components of all the $\mathbf{P}_{0q}$ are nonnegative, then this solution is a minimum for all the $\mathbf{P}_{0q}$. Suppose that, for some $q = r$, not all the elements of the transformed vector $\mathbf{P}_{0r}$ are greater than or equal to zero. To develop a starting solution for this problem, we find the minimum element of the transformed vector $\mathbf{P}_{0r}$ and apply (1.8) and (1.8′). The resulting solution will consist of one artificial vector and $m - 1$ vectors of the previous minimum feasible solution. The problem is then solved in the usual manner for $\mathbf{P}_{0r}$, and similarly for any of the remaining $\mathbf{P}_{0q}$. However, instead of applying the procedure associated with Eqs. (1.8), we might look at the situation in the light of its dual interpretation.

Here we have selected a basis—let us say $\mathbf{B} = (\mathbf{P}_1\mathbf{P}_2 \cdots \mathbf{P}_m)$—computed its inverse, and determined that $\mathbf{B}^{-1}\mathbf{P}_0 \geq \mathbf{O}$ and that

$$\mathbf{c}^0\mathbf{B}^{-1}\mathbf{P}_j - c_j = z_j - c_j \leq 0 \tag{1.13}$$

for all j, where $\mathbf{c}^0 = (c_1,c_2, \ldots ,c_m)$ is a row vector. Hence **B** is an admissible basis for (1.10) to (1.12). We have also determined that **B** is not an admissible basis for the constraints

$$\mathbf{AX} = \mathbf{P}_{0r} \tag{1.14}$$

in that

$$\mathbf{B}^{-1}\mathbf{P}_{0r} \tag{1.15}$$

is not a nonnegative vector. We say, for the problem determined by (1.14), that the solution (1.15) is optimal but not feasible; i.e., all $z_j - c_j \leq 0$, but not all elements of $\mathbf{B}^{-1}\mathbf{P}_{0r}$ are nonnegative. The dual problem to (1.10), (1.14), and (1.12) is to maximize

$$\mathbf{WP}_{0r} \tag{1.16}$$

subject to

$$\mathbf{WA} \leq \mathbf{c}\dagger \tag{1.17}$$

For the dual problem we see from (1.13) that the basis **B** yields a solution to (1.17), in that

$$\mathbf{W}^0\mathbf{P}_j - c_j \leq 0$$

for $\mathbf{W}^0 = \mathbf{c}^0\mathbf{B}^{-1}$. We also have that, for the dual, this solution is feasible but not optimal. A computational technique which enables us to employ efficiently the information concerning feasibility and optimality of the primal and dual contained in the above problem was first proposed by Lemke [70] and is called the *dual simplex method.*

2. THE DUAL SIMPLEX METHOD

To consider the general procedure, let us write our problem as follows: To minimize

$$\mathbf{cX}$$

subject to

$$\mathbf{AX} = \mathbf{b}$$
$$\mathbf{X} \geq \mathbf{O}$$

with its dual problem: To maximize

$$\mathbf{Wb}$$

† The dual problem is described in Chap. 5.

subject to

$$\mathbf{WA} \leq \mathbf{c}$$

As we did in the case of multiple-constant vectors, let us assume we have selected a basis $\mathbf{B} = (\mathbf{P}_1\mathbf{P}_2 \cdots \mathbf{P}_m)$ such that at least one element of $\mathbf{B}^{-1}\mathbf{b}$ is negative and $\mathbf{c}^0\mathbf{B}^{-1}\mathbf{P}_j \leq c_j$ for all j. A solution to the dual constraints is given by $\mathbf{W}^0 = \mathbf{c}^0\mathbf{B}^{-1}$ with its corresponding value of the objective function being

$$\mathbf{c}^0\mathbf{B}^{-1}\mathbf{b} \tag{2.1}$$

We wish to develop a computational procedure for the dual which will yield a maximizing solution, and hence, by the duality theorems of Chap. 5, a minimizing solution to the primal. This procedure must then determine a new basis for which

1. The dual inequalities will still be satisfied.
2. The value of the dual objective function will increase (or remain the same)[1] until the maximum or unbounded solution is reached.

In this fashion we will always preserve optimality of the primal and, in a finite number of steps, determine a feasible and optimum solution to the primal.

Let us denote the rows of $\mathbf{B}^{-1}$ by $\mathbf{B}_i$. Hence the m components of the *nonfeasible solution* of the primal are given by $x_{i0} = \mathbf{B}_i\mathbf{b}$ for $i = 1, 2, \ldots, m$. Let $x_{l0} = \mathbf{B}_l\mathbf{b} = \min_i \mathbf{B}_i\mathbf{b} < 0$. The vector $\mathbf{P}_l$, as we shall see, will be the one eliminated from the basis. For those vectors not in the basis, i.e., for those having $\mathbf{W}^0\mathbf{P}_j < c_j$, compute $x_{lj} = \mathbf{B}_l\mathbf{P}_j$. Let us assume that at least one $x_{lj} < 0$. For the set of $x_{lj} < 0$ form the ratios $(z_j - c_j)/x_{lj}$. Let

$$\theta = \min_{x_{lj}<0} \frac{z_j - c_j}{x_{lj}} = \frac{z_k - c_k}{x_{lk}}^{\dagger} > 0 \tag{2.2}$$

The vector $\mathbf{P}_k$ is selected to replace $\mathbf{P}_l$, and the new basis will yield a solution to the dual constraints. The new basis is

$$\bar{\mathbf{B}} = (\mathbf{P}_1 \cdots \mathbf{P}_{l-1}\mathbf{P}_k\mathbf{P}_{l+1} \cdots \mathbf{P}_m)$$

[1] The value of the dual objective for the new basis may not increase if the dual solution is degenerate. This is manifested by more than m of the quantities $\mathbf{c}^0\mathbf{B}^{-1}\mathbf{P}_j = c_j$.

† This criterion for selecting a vector to be introduced into the basis is equivalent to the one used in the parametric dual problem (Theorem 2 of Sec. 2, Chap. 8). The proof of that theorem also applies to this situation.

and $\bar{\mathbf{B}}^{-1}$ is obtained by application of the elimination formulas on $\mathbf{B}^{-1}$ as described in Sec. 1 of Chap. 6. The reader can readily verify that the new solution to the dual constraints is

$$\bar{\mathbf{W}} = \mathbf{W}^0 - \theta \mathbf{B}_l \tag{2.3}$$

and the corresponding value of the objective function is

$$\bar{\mathbf{W}}\mathbf{b} = \mathbf{W}^0\mathbf{b} - \theta x_{l0} \tag{2.4}$$

The new solution to the primal constraints can be computed by the usual elimination formulas or directly from $\bar{\mathbf{X}} = \bar{\mathbf{B}}^{-1}\mathbf{b}$. If all $\bar{\mathbf{X}} \geq 0$, then we have determined an optimum feasible solution to the primal. If not, then we know we have made at least $\bar{x}_{l0} = \bar{\mathbf{B}}_l\mathbf{b} > 0$. For this situation, we repeat the above dual simplex process of selecting the vector to be eliminated and then the vector to be introduced into the basis, until we find a basis that solves the dual and is also an admissible basis for the primal or until we have determined that the dual has an unbounded solution and hence that there are no feasible solutions to the primal. The latter case arises when, in computing the $x_{lj} = \mathbf{B}_l\mathbf{P}_j$, all $x_{lj} \geq 0$. If this is true, we have from (2.2) and (2.3) that we can construct a solution to the dual constraints for any $\theta > 0$, since

$$(\mathbf{W}^0 - \theta\mathbf{B}_l)\mathbf{P}_j = \mathbf{W}^0\mathbf{P}_j - \theta\mathbf{B}_l\mathbf{P}_j \leq \mathbf{W}^0\mathbf{P}_j < c_j$$

From (2.4) the corresponding value of the objective function can be made as large as possible, since $x_{l0} < 0$. This situation is revealed in the primal tableau, in that we would have all $x_{lj} \geq 0$, $x_{l0} < 0$, with the implication that the lth equation, which has been transformed to a nonnegative sum of nonnegative variables, is equal to a negative number. The dual simplex procedure can be employed as a variation of the original simplex procedure or of the revised procedure, since it uses exactly the information contained in their respective tableaus.

To handle degeneracy in the dual problem, Lemke [70] sets up the perturbed dual

$$\mathbf{W}\mathbf{P}_j \leq c_j + \epsilon^j \qquad j = 1, 2, \ldots, n$$

and shows that a scheme similar to the degeneracy method for the primal may be established. In the dual, we must have a procedure for selecting a unique vector $\mathbf{P}_k$ to be introduced into the basis. The situation is not unique when the value of θ in Eq. (2.2) assumes the minimum for more than one value of j.† For these j, we first compute the set of ratios x_{1j}/x_{lj}. If there is a unique minimum for these ratios, the index j of the minimum corresponds to the vector to be introduced. If not, we next

† In practice, if there are ties for the minimum, then the ratio that corresponds to the smallest index is selected for θ.

compute for the tied columns the ratios x_{2j}/x_{lj} and repeat the analysis, etc. In this manner we select a unique $\mathbf{P}_k$, and the problem will not cycle.

In the above presentation of the dual procedure we assumed knowledge of an optimal but not feasible solution to the primal, i.e., a solution to the dual constraints. This is equivalent to starting the simplex procedure with a known feasible but not optimal solution and hence eliminating the artificial Phase I computations. The main advantage to using the dual simplex method is that the dual statements of many problems have explicit solutions. These problems are characterized by having a set of nonnegative cost coefficients, e.g., the diet problem. The dual constraints are immediately solved by the vector $\mathbf{W} = \mathbf{O}$. Here we take the identity matrix as the corresponding artificial basis. Then, with only minor changes, the dual method can be applied to determine the optimum solution to the primal (Dantzig [22]).

For those problems which do not have explicit solutions, i.e., which do not have all $c_j \geq 0$, a number of schemes have been proposed (Dantzig [22] and Vajda [97]). The first technique sets all negative costs equal to zero and starts with the identity matrix as an initial basis. Here the dual method, with the appropriate changes, is applied until the optimum is reached. The corresponding basis will be an admissible one for the primal, and the usual simplex procedure can proceed to optimize the original objective function. The second scheme constructs the dual analogue of the artificial-variable technique and solves the associated problem in two phases by the dual simplex method.

As an example of the dual method, consider the following problem: Minimize

$$x_3 + x_4 + x_5$$

subject to

$$\begin{aligned} x_1 \quad - x_3 + x_4 - x_5 &= -2 \\ x_2 - x_3 - x_4 + x_5 &= 1 \\ x_j &\geq 0 \end{aligned}$$

The dual objective is to maximize

$$-2w_1 + w_2$$

and the constraints are

$$\begin{aligned} w_1 \qquad &\leq 0 \\ w_2 &\leq 0 \\ -w_1 - w_2 &\leq 1 \\ w_1 - w_2 &\leq 1 \\ -w_1 + w_2 &\leq 1 \end{aligned}$$

An initial basis is given by

$$\mathbf{B} = (\mathbf{P}_1\mathbf{P}_2) = \begin{pmatrix} 1 & 0 \\ 0 & 1 \end{pmatrix}$$

Setting up the primal in the usual simplex tableau, we have the following:

i	Basis	c		0	0	1	1	1
			$\mathbf{P}_0$	$\mathbf{P}_1$	$\mathbf{P}_2$	$\mathbf{P}_3$	$\mathbf{P}_4$	$\mathbf{P}_5$
1	$\mathbf{P}_1$	0	−2	1	0	(−1)	1	−1
2	$\mathbf{P}_2$	0	1	0	1	−1	−1	1
3			0	0	0	−1	−1	−1

Since all the $z_j - c_j$ elements are nonpositive, the basis **B** is a feasible basis for the dual, i.e., optimal but not feasible for the primal. The dual solution is

$$\mathbf{W}^0 = \mathbf{c}^0\mathbf{B}^{-1} = (0 \quad 0)\begin{pmatrix} 1 & 0 \\ 0 & 1 \end{pmatrix} = (0 \quad 0)$$

Applying the dual algorithm, we see that $\mathbf{P}_1$ is to be eliminated

$$(x_l = x_1 = -2)$$

and vector $\mathbf{P}_3$ is to be introduced into the basis, because

$$\theta = \frac{z_3 - c_3}{x_{13}} = \frac{z_5 - c_5}{x_{15}} = 1$$

Here we had a tie for θ and selected the first vector. The new tableau is as follows:

Basis	c	$\mathbf{P}_0$	$\mathbf{P}_1$	$\mathbf{P}_2$	$\mathbf{P}_3$	$\mathbf{P}_4$	$\mathbf{P}_5$
$\mathbf{P}_3$	1	2	−1	0	1	−1	1
$\mathbf{P}_2$	0	3	−1	1	0	−2	2
		2	−1	0	0	−2	0

Since all $x_i \geq 0$ and since $z_j - c_j \leq 0$, we have determined an optimal admissible basis for the primal and dual consisting of vectors $\mathbf{P}_2$ and $\mathbf{P}_3$. The primal optimum solution is $x_2 = 3$ and $x_3 = 2$, and that for the dual is $w_1 = -1$ and $w_2 = 0$, with a common optimum value of the objective function of 2. The reader will note that, for the dual constraints that correspond to the vectors in the basis (the second and third constraints), equality holds, and since we have an alternate optimum solution with vector $\mathbf{P}_5$, we also have equality in the fifth constraint.

Elaborate and efficient procedures have been devised which enable one to combine the original simplex algorithm and the dual method to solve the general linear-programming problem. Two of these are the *composite simplex algorithm* (Dantzig [23] and Orchard-Hays [83]) and the *primal-dual algorithm*[1] (Dantzig, Ford, and Fulkerson [28]). These techniques place no restrictions on the signs of the c_j and enable one to obtain an initial basic optimal or feasible solution. The reader is also referred to another technique for solving the general linear-programming problem called the *method of leading variables* (Beale [5]). This procedure also offers a method for initiating the computation for any problem.

All the above techniques are intimately related to the standard simplex method. Only such methods have proved effective for solving the general linear-programming problem (Hoffman, Mannos, Sokolowsky, and Wiegmann [61*a*] and Hoffman [59]). Other iterative and convergent techniques that solve the problem but are not as efficient as the simplex method include the double-description method (Raiffa, Thompson, and Thrall [86]), numerical methods for zero-sum two-person games by Brown [8] and von Neumann [79], relaxation techniques for solving linear inequalities (Agmon [1*d*] and Motzkin and Schoenberg [77]), and a projection method for solving linear constraints (Tompkins [94]).

3. INTEGER LINEAR PROGRAMMING

Since its introduction as a tool of applied mathematics, the outstanding computational problem of linear programming has been that of finding the optimum integer solution to a linear program. The need for this type of computational procedure is emphasized by the great number of problems from the realm of combinatorial analysis and the areas of scheduling and production that have been formulated as linear-programming problems.

A search through the general literature on linear programming yields a number of references to problems stated in terms of an integer linear program. Many of them are given in Dantzig [27, 27*a*]. The list includes the fixed-charge problem (Hirsch and Dantzig [56]), the traveling-salesman problem (Dantzig et al. [28*b*, 28*c*]) formulated as a multiple-trip integer problem by Tucker [96*a*], and machine scheduling (Wagner [99*a*], Manne [74*a*]). Other appropriate discussions are found in Dantzig [27*b*, 27*c*], Gross [52,52*a*], and Wagner and Whitin [99*e*]. Recent work of Gomory [51*b*, 51*c*, 51*d*, 51*e*], Beale [5*a*], Benders, Catchpole, and Kuiken [6*a*], and others has opened the way to a practical solution for a number of such problems. We shall describe only the all-integer procedure

[1] The reader is referred to Mueller and Cooper [77*a*] for a discussion comparing the computational efficiency of the standard and the primal-dual simplex algorithms.

due to Gomory. Surveys describing a wide variety of integer-programming methods and problems have been written by Balinski [3*a*] and Beale [5*b*].

We should note that the successful application of the transportation model with integer availabilities and demands is partially due to the fact that each basic solution corresponds to an extreme point that has integer or zero values for the coordinates. This is due to the triangular matrix of 0's and 1's associated with the basis. We might ask what similar condition is sufficient for the general linear-programming model to have an integer solution. For this answer we need to refer to the basic transformation of the simplex procedure: the elimination transformations.

If x_{ij} represents the element in the ith row and jth column of the simplex tableau and x_{lk} is the pivot element, the transformed elements are given by

$$\begin{aligned} x'_{ij} &= x_{ij} - \frac{x_{lj}}{x_{lk}} x_{ik} \qquad \text{for } i \neq l \\ x'_{lj} &= \frac{x_{lj}}{x_{lk}} \end{aligned} \tag{3.1}$$

For the purpose of this discussion we shall assume that the starting tableau is all-integer. We see that a very strong condition for the successive solutions to be integer is for each fraction in (3.1) to reduce to an integer. One unusual way for this to happen is for each pivot element to equal unity. This is, of course, what happens for the transportation problem. This property is equivalent to the selection of a basis whose associated matrix has a determinant equal to 1. As we shall see below, this is one factor in the computational scheme for solving the all-integer programming problem.

The integer-programming problem is to find a vector $\mathbf{X}$ which minimizes $\mathbf{cX}$ subject to $\mathbf{AX} = \mathbf{b}, \mathbf{X} \geq \mathbf{0}$, with the added (nonlinear) condition that the optimal-solution vector have integer coefficients. The geometry of the situation can be pictured by considering the solution to the following problem:

Maximize

$$3x_1 + x_2$$

subject to

$$\begin{aligned} x_1 + 2x_2 &\leq 8 \\ 3x_1 - 4x_2 &\leq 12 \\ x_1 \qquad &\geq 0 \\ x_2 &\geq 0 \end{aligned} \tag{3.2}$$

The optimum noninteger solution is given by $x_1 = 28/5$ and $x_2 = 6/5$ with the maximum value equal to 18.

This problem is pictured in Fig. 9.1. The convex set of solutions is bounded by the heavy lines, and, as indicated with heavy dots, this region also includes a number of integer points. The problem is to determine which of these points maximizes the objective function.

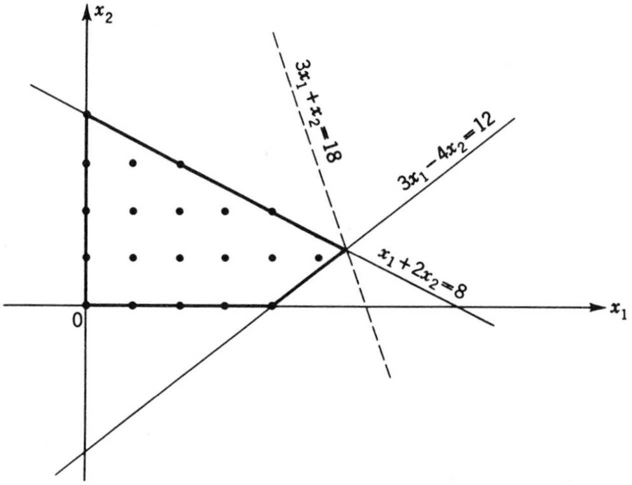

FIGURE 9.1

Since the basic computational scheme to be employed is the simplex method, we should review the geometry of this technique. The simplex method starts at an extreme point of the convex set of solutions. Each iteration, i.e., the selection of a new basis, determines a new extreme point which is a neighbor to the old point (has a boundary segment in common). This process continues and, after a finite number of steps, stops at the global maximum. We see that, in order to have an optimum integer solution, we must also have an associated extreme point that is integer-valued. For the problem in Fig. 9.1, the simplex process would probably start at the extreme point (0,0), move to (4,0), and finally to $(28/5, 6/5)$. It is impossible to indicate what integer point in the solution space optimizes the objective function.

The question that now arises considers the possibility of changing the convex set of solutions in such a manner as to make the appropriate feasible integer point an extreme point of the new convex region. For example, in Fig. 9.2 we have introduced two arbitrary constraints to the problem which do the trick. This technique of putting in cutting hyperplanes was used in the past for specific formulations. How to introduce these cuts systematically for any problem is the essence of the procedures developed by Gomory.

An early approach to the application of the cutting-plane method required solving the problem by the basic simplex algorithm, ignoring the integer constraints. If the optimum solution is not integer, then a cutting plane (i.e., a linear inequality) is introduced which preserves the optimality state of the primal problem but not the feasibility conditions. A single application of the dual algorithm restores feasibility and yields an optimum answer to the reduced problem. If this new solution is integer-valued, the process stops. If not, a new constraint is introduced and the algorithm is repeated until an integer solution is found or until an indication is given that no integer solution exists for the original problem (Gomory [51*c*]). This method has been superseded by a more efficient procedure, in which the problem, assumed to be given in integers, is transformed by a modified simplex algorithm which preserves the integer-value characteristic of the complete tableau for all iterations (Gomory [51*e*]). This algorithm is discussed below.

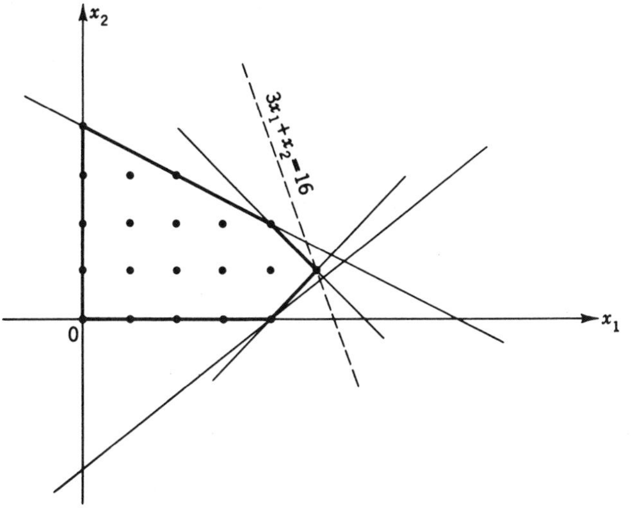

FIGURE 9.2

We assume for starting conditions that the simplex tableau is all-integer-valued and that we start with a feasible solution for the dual simplex algorithm but not for the primal. Letting $\mathbf{P}_1, \ldots, \mathbf{P}_m$ form a feasible basis for the dual problem, the initial Tableau 9.1 is given for a minimization problem.

In Tableau 9.1 x_{00} is the value of the objective function; some $x_{i0} < 0$ for $i \geq 1$ (solution is not primal feasible); $x_{0j} = z_j - c_j \leq 0$ for $j \geq 1$ (solution is dual feasible); and an additional set of $n - m$ constraints, $x_j' - x_j = 0$ for $j = m + 1, \ldots, n$, with $x_j' \geq 0$ and $c_j' = c_j$, has been added to the tableau. With this set of constraints, we are now forced to determine a basic feasible solution that contains n nonnegative variables instead of the usual m; hence we are given the means by which an interior point of the original convex set can be expressed. However, we still lack the means to force us into the interior and toward the optimum integer-valued point. This is accomplished by a proper selection of cutting planes.

We now wish to introduce new variables into the solution in a manner that will enable us to preserve the integer characteristic of the tableau and to move toward the optimum integral solution that is contained in the convex set defined by the equations of Tableau 9.1. This will be

Tableau 9.1

i	Basis	c	P_0	c_1	c_2	···	c_l	···	c_m	c_{m+1}	···	c_n	c_{m+1}	···	c_n	
				P_1	P_2	···	P_l	···	P_m	P_{m+1}	···	P_n	P'_{m+1}	···	P'_n	P_{n+s}
0			x_{00}	0	0		···		0	$x_{0,m+1}$	···	$x_{0,n}$	0	···	0	
1	P_1	c_1	x_{10}	1	0	···	0	···	0	$x_{1,m+1}$	···	x_{1n}	0	···	0	
⋮	⋮	⋮	⋮	⋮	⋮		⋮		⋮	⋮		⋮	⋮		⋮	
l	P_l	c_l	x_{l0}	0	0	···	1	···	0	$x_{l,m+1}$	···	x_{ln}	0	···	0	
⋮	⋮	⋮	⋮	⋮	⋮		⋮		⋮	⋮		⋮	⋮		⋮	
m	P_m	c_m	x_{m0}	0	0		0	···	1	$x_{m,m+1}$	···	x_{mn}	0	···	0	
$m+1$	P'_{m+1}	c_{m+1}	0	0	0	···	0	···	0	-1	···	0	1	···	0	
⋮	⋮	⋮	⋮	⋮					⋮	⋮		⋮	⋮		⋮	
n	P'_n	c_n	0	0	0	···	0	···	0	0	···	-1	0	···	1	
$n+1$																

accomplished by determining cutting hyperplanes defined as inequalities in terms of the nonbasic variables and a new slack vector. Introducing the new slack vector into the basis with a pivot element[1] of -1 will preserve the integer tableau and move the solution point closer to the optimum integer point. A finite number of applications of these new constraints leads us to the optimum integer solution.

[1] We select -1 as a pivot element instead of $+1$ as we are using the dual algorithm to change solutions.

To determine the form of the cutting constraint, let us consider any equation from Tableau 9.1 that has a corresponding $x_{i0} < 0$, for example, the lth equation:

$$x_{l0} = x_l + x_{l,\ m+1}x_{m+1} + x_{l,\ m+2}x_{m+2} + \cdots + x_{ln}x_n \tag{3.3}$$

where $x_{m+1}, \ldots, x_n$ are nonbasic variables.

We next rewrite each coefficient of (3.3) as a multiple of an integer and a remainder, i.e., in the form $b_{lj}\lambda + r_{lj}$, where b_{lj} is an integer, r_{lj} is a remainder, and λ is an unspecified positive number to be determined. The coefficients can then be expressed as

$$\begin{aligned} x_{lj} &= b_{lj}\lambda + r_{lj} = \left[\frac{x_{lj}}{\lambda}\right]\lambda + r_{lj} \qquad \text{for all } j \\ 1 &= \left[\frac{1}{\lambda}\right]\lambda + r \\ 0 &\le r_{lj} < \lambda \qquad 0 \le r < \lambda \qquad 0 < \lambda \end{aligned} \tag{3.4}$$

where the brackets indicate "integer part of." If $x_{lj}/\lambda < 0$, then $[x_{lj}/\lambda] = b_{lj} < 0$ such that $b_{lj}\lambda + r_{lj} = x_{lj}$. Note that, for $\lambda > 1$, we have $[1/\lambda] = 0$.

Substituting (3.4) into (3.3) and gathering appropriate terms, we have

$$\begin{aligned} \left[\frac{x_{l0}}{\lambda}\right]\lambda + r_{l0} &= \left(\left[\frac{1}{\lambda}\right]\lambda + r\right)x_l + \left(\left[\frac{x_{l,\ m+1}}{\lambda}\right]\lambda + r_{l,\ m+1}\right)x_{m+1} \\ &\quad + \left(\left[\frac{x_{l,\ m+2}}{\lambda}\right]\lambda + r_{l,\ m+2}\right)x_{m+2} + \cdots + \left(\left[\frac{x_{ln}}{\lambda}\right]\lambda + r_{ln}\right)x_n \end{aligned}$$

or

$$\begin{aligned} r_{l0} + \lambda \Bigg\{\left[\frac{x_{l0}}{\lambda}\right] - \left[\frac{x_{l,\ m+1}}{\lambda}\right]x_{m+1} - \left[\frac{x_{l,\ m+2}}{\lambda}\right]x_{m+2} - \cdots \\ - \left[\frac{x_{ln}}{\lambda}\right]x_n - \left[\frac{1}{\lambda}\right]x_l\Bigg\} = r_{l,\ m+1}x_{m+1} + r_{l,\ m+2}x_{m+2} + \cdots \\ + r_{ln}x_n + rx_l \end{aligned} \tag{3.5}$$

We note that any nonnegative integer values of the variables that satisfy Eq. (3.3) will also satisfy (3.5). Such a substitution will make the right-hand side of (3.5) a nonnegative number.

We denote the quantity in braces by x_{n+s} and rewrite that expression to obtain

$$x_{n+s} = \left[\frac{x_{l0}}{\lambda}\right] - \sum_{j \ne \text{basis}} \left[\frac{x_{lj}}{\lambda}\right]x_j - \left[\frac{1}{\lambda}\right]x_l \tag{3.6}$$

For (3.6) we note that x_{n+s} not only must be an integer but also must be nonnegative. The first part is clear since all variables and the bracketed quantities are integers. The second part is established by noting that

$r_{l0} < \lambda$ and, if x_{n+s} were a negative integer, the left-hand side of (3.5) would be negative, which contradicts the statement that the right-hand side is nonnegative. This nonnegative integer x_{n+s} is introduced as a new variable of the problem.

Restricting $\lambda > 1$, the expression for x_{n+s} becomes

$$x_{n+s} = \left[\frac{x_{l0}}{\lambda}\right] - \left[\frac{x_{l,\,m+1}}{\lambda}\right] x_{m+1} - \left[\frac{x_{l,\,m+2}}{\lambda}\right] x_{m+2} - \cdots - \left[\frac{x_{ln}}{\lambda}\right] x_n$$

or

$$x_{n+s} = b_{l0} - b_{l,\,m+1}x_{m+1} - b_{l,\,m+2}x_{m+2} - \cdots - b_{ln}x_n$$

or

$$b_{l0} = b_{l,\,m+1}x_{m+1} + b_{l,\,m+2}x_{m+2} + \cdots + b_{ln}x_n + x_{n+s} \tag{3.7}$$

For any $\lambda > 1$, Eq. (3.7)[1] is a constraint that must be satisfied by any integer solution to the original linear-programming problem. After a suitable selection of λ, Eq. (3.7) can be used as a cutting constraint.

We note that $b_{l0} < 0$ since it was assumed that $x_{l0} < 0$. Also, since we are applying the dual simplex method, some x_{lj} for j not in the basis must be negative or the problem is not feasible. By selecting λ large enough, all $[x_{lj}/\lambda]$ for $x_{lj} < 0$ yields a $b_{lj} = -1$; hence, a pivot element of -1 is available and the integer tableau is preserved. However, on the basis of the dual simplex transformations [Eqs. (2.2) to (2.4)] we note that a small λ will cause a larger improvement in the objective function than a larger λ since the change is a direct function of $b_{l0} = [x_{l0}/\lambda] < 0$; that is, the new value of the objective function is given by

$$x_{00} - \frac{x_{0k}}{-1} b_{l0} = x_{00} + x_{0k}b_{l0} = x_{00} + (z_k - c_k)\left[\frac{x_{l0}}{\lambda}\right]$$

[1] For $\lambda = 1$, and by substituting the value of x_l given by (3.3) into (3.6), a cutting constraint can be obtained of the form

$$x_{n+s} = -f_0 + \sum_{j \neq \text{basis}} f_j x_j$$

where the f are the fractional parts of the corresponding x_{ij}. This constraint is used in a similar fashion to (3.7) to solve the general integer-programming problem, i.e., when not all the coefficients are integers but all the variables are restricted to integers (Gomory [51c]). This algorithm calls for the solution of the problem using the regular simplex algorithm; if the optimal solution is not all-integer, then a row with a fractional value of a variable is selected to generate the cutting plane. It can be added to the set of original constraints and the new variable x_{n+s} put into the solution by an application of the dual simplex method (Sec. 2). However, λ must be specified such that it yields a pivot element of -1 in Eq. (3.7) and at the same time causes the value of the objective function to be reduced as much as possible.

The reader will note that since $b_{l0} < 0$ and $x_{n+s} \geq 0$, (3.7) "cuts" away the current basic feasible solution with $x_{m+1} = \cdots = x_n = 0$, that is, forces some x_j not in the current basis into the new basis.

where $\mathbf{P}_k$, the pivot column, must be the one such that

$$\theta = \min_{x_{lj}<0} \frac{x_{0j}}{-1} = \frac{z_k - c_k}{-1}$$

(A $z_k - c_k = 0$ indicates degeneracy in the dual problem.)

In order to select λ in a precise fashion, we note the following: Since λ must be chosen such that $b_{lk} = [x_{lk}/\lambda] = -1$, we have from Eqs. (2.2)

$$\frac{z_k - c_k}{[x_{lk}/\lambda]} = \frac{z_k - c_k}{-1} \leq \frac{z_j - c_j}{[x_{lj}/\lambda]} = \frac{z_j - c_j}{b_{lj}} \qquad \text{for } x_{lj} < 0$$

or

$$\frac{z_k - c_k}{-1} \leq \frac{z_j - c_j}{b_{lj}}$$

Let m_j be the largest integer for which

$$\frac{z_k - c_k}{-1} \leq \frac{z_j - c_j}{-m_j}$$

Then

$$-b_{lj} = -\left[\frac{x_{lj}}{\lambda}\right] \leq m_j \qquad \text{for } x_{lj} < 0 \tag{3.8}$$

Hence, for each j, the smallest λ that satisfies (3.8) and allows $\mathbf{P}_k$ to be the pivot column, i.e., yields a pivot element $b_{lk} = -1$, is given by $\lambda_j = -(x_{lj}/m_j)$. As the minimum permissible λ must be at least as great as the largest λ_j, we have $\lambda_{\min} = \max\limits_{x_{lj}<0} \lambda_j$. [If $\lambda_{\min} = \lambda_k = -x_{lk} = 1$, that is, the pivot element happens to be a -1, then we construct a suitable cutting constraint whose variables include only the nonbasic variables of the equation, as this implies a selection of $\lambda > 1$; see Eq. (3.6).]

λ is not restricted to be an integer, and for $\mathbf{P}_k$ we have $m_k = 1$, $\lambda_k = -x_{lk}$, $\lambda_{\min} \geq \lambda_k$; hence $b_{lk} = [x_{lk}/\lambda_{\min}] = -1$.

The above procedure can be summarized as follows:

1. Assume that a dual feasible solution with an all-integer tableau has been determined.
2. From the rows containing a negative constant term, select a row to be used to generate the cutting constraint (the lth row).[1] If no such rows exist, an optimum all-integer solution has been determined.

[1] The selection of the row that generates the cutting constraint is tied intimately to the proof that the all-integer algorithm converges. Gomory [51*e*] cites the following rules which are consistent with the finiteness proof for selection of a row to generate the new constraint:

1. Always select the first row from the top having a negative element.
2. Select the rows by a cyclic process; i.e., on the first step look at the first row and

3. The pivot column corresponds to the $\mathbf{P}_k$ whose

$$-(z_k - c_k) = \min_{x_{lj}<0} -(z_j - c_j) \qquad \text{for } j \geq 1$$

If no $x_{lj} < 0$ exists, then the problem is not feasible.
4. For each j having a $x_{lj} < 0$, determine the largest integer

$$-(z_k - c_k) \leq \frac{-(z_j - c_j)}{m_j}$$

5. Define $\lambda_j = -(x_{lj}/m_j)$ for $x_{lj} < 0$.
6. Determine $\lambda_{\min} = \max \lambda_j$. If $\max \lambda_j = 1$, set $\lambda_{\min} > 1$.
7. Develop the corresponding cutting constraint (3.7) and add it to Tableau 9.1, with the new equation being the $(n + 1)$st and the new variable x_{n+s} corresponding to column $\mathbf{P}_{n+s}$ (where $s = 1, 2, \ldots,$ that is, the index of the iteration).
8. Apply the simplex transformation to the new tableau using $b_{lk} = -1$ as the pivot element. This introduces x_{n+s} as a new nonbasic variable, and because the pivot element was -1, the tableau remains in integers. The new row is then dropped from further consideration and the process repeated.

We shall illustrate the procedure with an example due to Gomory [51*e*].

Minimize

$$10x_1 + 14x_2 + 21x_3$$

subject to

$$\begin{aligned} 8x_1 + 11x_2 + 9x_3 - x_4 \qquad\qquad &= 12 \\ 2x_1 + 2x_2 + 7x_3 \qquad - x_5 \qquad &= 14 \\ 9x_1 + 6x_2 + 3x_3 \qquad\qquad - x_6 &= 10 \\ x_j &\geq 0 \end{aligned}$$

---

if it does not have a negative constant, look at its successors; on the second step look at the second row and then its successors, etc.
3. If the rows are chosen at random, a finite process will result with probability 1.

The usual simplex rule of selecting the one with the largest negative term is not covered by the finiteness rule. In developing suitable computer codes for the all-integer algorithm, there has been much experimentation in the quest for a rule that is efficient for large classes of problems. Combinations of the simplex rule and other rules have worked rather well. The reader interested in developing a code would be well advised to consult the latest literature in the field.

The set of equations $x_1' = x_1$, $x_2' = x_2$, $x_3' = x_3$ and the conditions $c_j' = c_j$ have been added in Tableau 9.2, and row 7 has been added for the cutting constraint and column $\mathbf{P}_7$ for the new variable.

Tableau 9.2

i	Basis	c	$\mathbf{P}_0$	10	14	21	0	0	0	10	14	21	
				$\mathbf{P}_1$	$\mathbf{P}_2$	$\mathbf{P}_3$	$\mathbf{P}_4$	$\mathbf{P}_5$	$\mathbf{P}_6$	$\mathbf{P}_1'$	$\mathbf{P}_2'$	$\mathbf{P}_3'$	$\mathbf{P}_7$
0			0	−10	−14	−21	0	0	0	0	0	0	0
1	$\mathbf{P}_4$	0	−12	−8	−11	−9	1	0	0	0	0	0	0
2	$\mathbf{P}_5$	0	−14	−2	−2	−7	0	1	0	0	0	0	0
3	$\mathbf{P}_6$	0	−10	−9	−6	−3	0	0	1	0	0	0	0
4	$\mathbf{P}_1'$	10	0	−1	0	0	0	0	0	1	0	0	0
5	$\mathbf{P}_2'$	14	0	0	−1	0	0	0	0	0	1	0	0
6	$\mathbf{P}_3'$	21	0	0	0	−1	0	0	0	0	0	1	0
7	$\mathbf{P}_7$		−4	(−1)	−1	−2	0	0	0	0	0	0	1

The second equation of the system is used to generate the cutting constraint. We have

$$\frac{z_k - c_k}{-1} = \min_{x_{2j}<0} \frac{z_j - c_j}{-1} = \frac{z_1 - c_1}{-1} = 10$$

Hence column $\mathbf{P}_1$ is to be the pivot column. To determine the m_j, we have

$$10 \le \frac{10}{m_1} \qquad 10 \le \frac{14}{m_2} \qquad 10 \le \frac{21}{m_3}$$

or $m_1 = 1$, $m_2 = 1$, $m_3 = 2$. The $\lambda_j = -x_{lj}/m_j$ are then

$$\lambda_1 = 2/1 \qquad \lambda_2 = 2/1 \qquad \lambda_3 = 7/2$$

and

$$\lambda = \max \lambda_j = 7/2$$

The new constraint is then given by

$$\left[\frac{-14}{7/2}\right] = \left[\frac{-2}{7/2}\right] x_1 + \left[\frac{-2}{7/2}\right] x_2 + \left[\frac{-7}{7/2}\right] x_3 + x_7$$

$$-4 = -x_1 \qquad -x_2 \qquad -2x_3 + x_7$$

This equation is shown added to Tableau 9.2. By applying the elimination transformation with the pivot element as shown, Tableau 9.3 is obtained. Since all the elements of column $\mathbf{P}_1$ have been reduced to zeros, it can be eliminated from the tableau and column $\mathbf{P}_7$ substituted

Tableau 9.3

i	Basis	c	$\mathbf{P}_0$	$\mathbf{P}_1$	$\mathbf{P}_2$	$\mathbf{P}_3$	$\mathbf{P}_4$	$\mathbf{P}_5$	$\mathbf{P}_6$	$\mathbf{P}'_1$	$\mathbf{P}'_2$	$\mathbf{P}'_3$	$\mathbf{P}_7$
0			40	0	−4	−1	0	0	0	0	0	0	−10
1	$\mathbf{P}_4$	0	20	0	−3	7	1	0	0	0	0	0	−8
2	$\mathbf{P}_5$	0	−6	0	0	−3	0	1	0	0	0	0	−2
3	$\mathbf{P}_6$	0	26	0	3	15	0	0	1	0	0	0	−9
4	$\mathbf{P}'_1$	10	4	0	1	2	0	0	0	1	0	0	−1
5	$\mathbf{P}'_2$	14	0	0	−1	0	0	0	0	0	1	0	0
6	$\mathbf{P}'_3$	21	0	0	0	−1	0	0	0	0	0	1	0
7													

in its place. In fact, since vectors $\mathbf{P}_4$, $\mathbf{P}_5$, $\mathbf{P}_6$, $\mathbf{P}'_1$, $\mathbf{P}'_2$, $\mathbf{P}'_3$ will not change under the all-integer algorithm transformation, the tableau can be made quite concise as Tableau 9.4. Note that $x'_1 = x_1$ is still in the basis and now has a value of 4. Here there is only one row with a negative constant term. The pivot column is $\mathbf{P}_3$ since

$$\frac{-1}{-1} = \min\left(\frac{-10}{-1}, \frac{-1}{-1}\right) \qquad 1 \leq \frac{10}{m_7} \qquad 1 \leq \frac{1}{m_3}$$

or $m_7 \leq 10$, $m_3 \leq 1$. Hence $\lambda_7 = 2/10$, $\lambda_3 = 3/1$, $\lambda = 3$. The new constraint is given by

$$-2 = -x_7 - x_3 + x_8$$

as shown in Tableau 9.4. The new solution is shown in Tableau 9.5.

Tableau 9.4

i	Basis	$\mathbf{P}_0$	$\mathbf{P}_7$	$\mathbf{P}_2$	$\mathbf{P}_3$	$\mathbf{P}_8$
0		40	−10	−4	−1	
1	$\mathbf{P}_4$	20	−8	−3	7	
2	$\mathbf{P}_5$	−6	−2	0	−3	
3	$\mathbf{P}_6$	26	−9	3	15	
4	$\mathbf{P}'_1$	4	−1	1	2	
5	$\mathbf{P}'_2$	0	0	−1	0	
6	$\mathbf{P}'_3$	0	0	0	−1	
7	$\mathbf{P}_8$	−2	−1	0	(−1)	1

Selecting row 3 as the row to generate the cutting constraint, we have $\mathbf{P}_7$ as the pivot column and $\lambda = 24$. The new constraint is $-1 = -x_7 + x_9$. The next solution is then given by Tableau 9.6. With row 2 as the gener-

Tableau 9.5

i	Basis	$\mathbf{P}_0$	$\mathbf{P}_7$	$\mathbf{P}_2$	$\mathbf{P}_8$	$\mathbf{P}_9$
0		42	−9	−4	−1	
1	$\mathbf{P}_4$	6	−15	−3	7	
2	$\mathbf{P}_5$	0	1	0	−3	
3	$\mathbf{P}_6$	−4	−24	3	15	
4	$\mathbf{P}_1'$	0	−3	1	2	
5	$\mathbf{P}_2'$	0	0	−1	0	
6	$\mathbf{P}_3'$	2	1	0	−1	
7	$\mathbf{P}_9$	−1	(−1)	0	0	1

Tableau 9.6

i	Basis	$\mathbf{P}_0$	$\mathbf{P}_9$	$\mathbf{P}_2$	$\mathbf{P}_8$	$\mathbf{P}_{10}$
0		51	−9	−4	−1	
1	$\mathbf{P}_4$	21	−15	−3	7	
2	$\mathbf{P}_5$	−1	1	0	−3	
3	$\mathbf{P}_6$	20	−24	3	15	
4	$\mathbf{P}_1'$	3	−3	1	2	
5	$\mathbf{P}_2'$	0	0	−1	0	
6	$\mathbf{P}_3'$	1	1	0	−1	
7	$\mathbf{P}_{10}$	−1	0	0	(−1)	1

Tableau 9.7

i	Basis	$\mathbf{P}_0$	$\mathbf{P}_9$	$\mathbf{P}_2$	$\mathbf{P}_{10}$
0		52	−9	−4	−1
1	$\mathbf{P}_4$	14	−15	−3	7
2	$\mathbf{P}_5$	2	1	0	−3
3	$\mathbf{P}_6$	5	−24	3	15
4	$\mathbf{P}_1'$	1	−3	1	2
5	$\mathbf{P}_2'$	0	0	−1	0
6	$\mathbf{P}_3'$	2	1	0	−1

ating row, we have $\mathbf{P}_8$ as the pivot column and $\lambda = 3$. The new constraint is $-1 = -x_8 + x_{10}$. This yields the final Tableau 9.7 and the optimum solution. The final solution is given by

$$x_1 = x_1' = 1 \quad x_2 = x_2' = 0 \quad x_3 = x_3' = 2$$
$$x_4 = 14 \quad x_5 = 2 \quad x_6 = 5 \quad x_{00} = 52$$

An example of the type of problem that can be solved by the integer-programming algorithm is the traveling-salesman problem as developed by Miller, Tucker, and Zemlin [74*f*].

A salesman is required to visit each of n cities, indexed by $1, \ldots, n$. He leaves from a "base city" indexed by 0, visits each of the n other cities exactly once, and returns to city 0. During his travels he must return to 0 exactly t times, including his final return (here t may be allowed to vary), and he must visit no more than p cities in one tour. (By a tour we mean a succession of visits to cities without stopping at city 0.) It is required to find such an itinerary which minimizes the total distance traveled by the salesman.

Note that if t is fixed, then for the problem to have a solution we must have $tp \geqq n$. For $t = 1$, $p \geqq n$, we have the standard traveling-salesman problem.

Let d_{ij} $(i \neq j = 0, 1, \ldots, n)$ be the distance covered in traveling from city i to city j. The following integer-programming problem is shown to be a model of the problem:

Minimize the linear form

$$\sum_{0 \leq i \neq j \leq n} \sum d_{ij} x_{ij}$$

over the set determined by the relations

$$\sum_{\substack{i=0 \\ i \neq j}}^{n} x_{ij} = 1 \qquad j = 1, \ldots, n$$

$$\sum_{\substack{j=0 \\ j \neq i}}^{n} x_{ij} = 1 \qquad i = 1, \ldots, n$$

$$u_i - u_j + p x_{ij} \leqq p - 1 \qquad 1 \leqq i \neq j \leqq n$$

where the x_{ij} are nonnegative integers and the u_i $(i = 1, \ldots, n)$ are arbitrary real numbers and can be restricted to nonnegative integers. If t is fixed, we must add the restriction $\sum_{i=1}^{n} x_{i0} = t$.

From a theoretical point of view, the techniques of Gomory and others should enable us to solve any integer-programming problem. The digital-computer codes based on these algorithms have, in most instances, proved unpredictable in their ability to guarantee convergence to the optimum. The success of such algorithms varies according to the problem and the rules used to develop the cutting planes. This is in sharp contrast to the highly successful use of the simplex method to solve almost any standard (continuous) linear problem. The reader is referred to Balinski [3*a*] and Trauth and Woolsey [94*b*] for discussions comparing the efficacy of different algorithms and related computational experience.

4. THE DECOMPOSITION OF LARGE-SCALE SYSTEMS

Although it is theoretically possible to solve any given linear-programming model, the analyst is quickly made aware of certain limitations which restrict his endeavors. Chief among these limitations is the problem of dimensionality. Almost all difficulties that arise in the development of a programming problem can be related to its size. This is certainly true for such restrictive items as the cost of data gathering, matrix preparation, computing costs, and validity of the linear model.

The development of procedures for the solution of large-scale systems is reviewed in Dantzig [27*d*] and Gomory [51*f*], and a number of procedures are given in Graves and Wolfe [51*g*]. Dantzig discusses techniques to reduce the computational requirements of large systems from two points of view: (1) by decreasing the number of iterations, and (2) by finding a compact form for the inverse and/or by taking advantage of any special structure of the system of equations. To cut down the number of iterations, variants of the simplex method have been proposed to replace the usual Phase I of the simplex method: *method of leading variables*, Beale [5]; *composite simplex algorithm*, Orchard-Hays [83] and Wolfe [105*e*]; and *primal-dual algorithm*, Dantzig, Ford, and Fulkerson [28].

Proposals for finding a compact form for the inverse or for taking advantage of special structures include the sparse-basis technique, Markowitz [74*c*]; block-triangular basis, Dantzig [20, 27*f*]; the solution of a dynamic Leontief model with substitution, Dantzig [27*e*]; large-scale economic systems, Rech [86*a*]; and for the important class of block-angular systems and multistage systems of the staircase type we have the *decomposition algorithm*, Dantzig and Wolfe [32*b*]; *partition programming*, Benders [6*b*], Rosen [87*c*], and Ritter [87*a*]; *pseudo-basic variable procedure*, Beale [5*c*]; and the *dualplex method*, Gass [46*b*]. In this section we shall describe only the decomposition algorithm of Dantzig and Wolfe which, when applied to block-angular systems, has an interesting and important economic interpretation in terms of decentralized planning; see Dantzig [16*c*] and Baumol and Fabian [3*g*].

For many problems the constraints consist of rather large independent subsets of equations which refer to the same time period or same production facility. These subsets are usually tied together by a small set of equations. These "tie-in equations" might represent restrictions on the availability of basic resources, total demand of a product, or total budgetary constraint. In problems of this sort we have, in a sense, a number of separate linear-programming problems whose joint solution must satisfy a set of additional restrictions. If we boxed in the sets of constraints and corresponding part of the objective function, Fig. 9.3 would result.

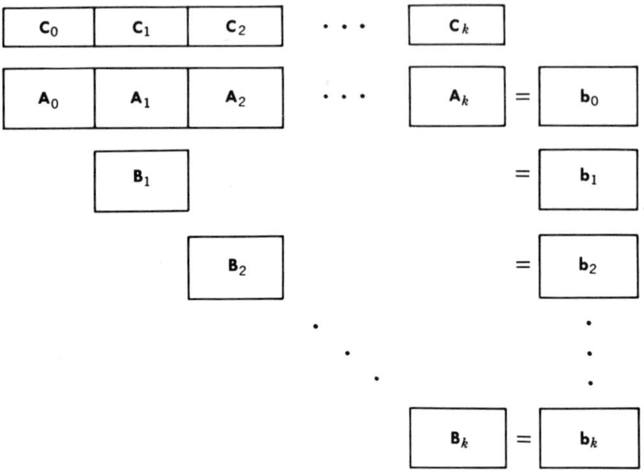

FIGURE 9.3

In Fig. 9.3 we have partitioned the original problem of $\mathbf{AX} = \mathbf{b}$, $\mathbf{X} \geq \mathbf{O}$, $\mathbf{cX}$ a minimum, into the *decomposed program*[1] of finding the vectors $\mathbf{X}_p \geq \mathbf{O}$ $(p = 0, 1, \ldots, k)$ which minimize

$$\sum_{p=0}^{k} \mathbf{C}_p \mathbf{X}_p \tag{4.1}$$

subject to

$$\sum_{p=0}^{k} \mathbf{A}_p \mathbf{X}_p = \mathbf{b}_0 \tag{4.2}$$

$$\mathbf{B}_p \mathbf{X}_p = \mathbf{b}_p \qquad p = 1, 2, \ldots, k \tag{4.3}$$

where $\mathbf{A}_p$ is an $m_0 \times n_p$ matrix, $\mathbf{B}_p$ is an $m_p \times n_p$ matrix, $\mathbf{C}_p$ is an n_p-component row vector, $\mathbf{b}_p$ is an m_p-component column vector, and $\mathbf{X}_p$ is an n_p-variable column vector. Hence, the above problem has $m = \sum_{p=0}^{k} m_p$ constraints and $n = \sum_{p=0}^{k} n_p$ variables. In explicit matrix notation, (4.2) and (4.3) are given by

$$\begin{pmatrix} \mathbf{A}_0 & \mathbf{A}_1 & \mathbf{A}_2 & \cdots & \mathbf{A}_k \\ & \mathbf{B}_1 & & & \\ & & \mathbf{B}_2 & & \\ & & & \ddots & \\ & & & & \mathbf{B}_k \end{pmatrix} \begin{pmatrix} \mathbf{X}_0 \\ \mathbf{X}_1 \\ \mathbf{X}_2 \\ \vdots \\ \mathbf{X}_k \end{pmatrix} = \begin{pmatrix} \mathbf{b}_0 \\ \mathbf{b}_1 \\ \mathbf{b}_2 \\ \vdots \\ \mathbf{b}_k \end{pmatrix}$$

[1] Here we are adhering to the notation of Harvey [53*a*].

To better understand what follows, let us look at the geometry of a very simple decomposed problem: minimize $c_{11}x_{11} + c_{12}x_{12}$ subject to $\mathbf{X}_1 = (x_{11}, x_{12}) \geq 0$ and

$$\left.\begin{array}{r} x_{11} + 4x_{12} \leq 12 \end{array}\right\} \mathbf{A}_1\mathbf{X}_1 \leq \mathbf{b}_0$$
$$\left.\begin{array}{r} -x_{11} + 2x_{12} \leq 2 \\ 3x_{11} + 4x_{12} \leq 24 \end{array}\right\} \mathbf{B}_1\mathbf{X}_1 \leq \mathbf{b}_1 \qquad (4.4)$$

The nonnegative solution set to (4.4) is shown in Fig. 9.4. The hatched area is the solution set to $\mathbf{B}_1\mathbf{X}_1 \leq \mathbf{b}_1$ and the shaded area is the solution space to the complete problem (4.4). The extreme points of $\mathbf{B}_1\mathbf{X}_1 \leq \mathbf{b}_1$ are indicated by a square, while the extreme points of the complete problem are indicated by a circle. Any solution to the complete problem must, of course, satisfy $\mathbf{B}_1\mathbf{X}_1 \leq \mathbf{b}_1$. But since the solution set of linear constraints is a convex polyhedron, any solution to $\mathbf{B}_1\mathbf{X}_1 \leq \mathbf{b}_1$ can be expressed as a convex combination of the extreme points of its solution set (we assume this convex set is bounded). Hence, if we knew all the extreme points of $\mathbf{B}_1\mathbf{X}_1 \leq \mathbf{b}_1$ or could find the right ones, all we would need to do is to find the convex combination of these points which satisfies $\mathbf{A}_1\mathbf{X}_1 \leq \mathbf{b}_0$ and optimizes the objective function. This is the essence of the Dantzig-Wolfe decomposition algorithm. In Fig. 9.4 we note that any circled point is also a convex combination of the squared points and that any point in the shaded area can be represented as a convex combination of the squared points.

Extending this idea to the general problem of (4.1) to (4.3), we assume that for each p there is available the corresponding solution set $\mathbf{S}_p$ to the constraints $\mathbf{B}_p\mathbf{X}_p = \mathbf{b}_p$, $\mathbf{X}_p \geq \mathbf{O}$. Then the solution of the original problem could be thought of as the selection of a convex combination of solu-

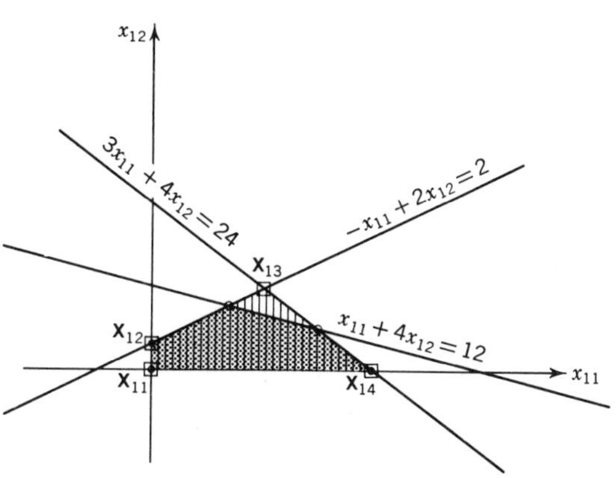

FIGURE 9.4

tion points (i.e., extreme points) from each $\mathbf{S}_p$ so as to satisfy the tie-in restrictions $\sum_{p=0}^{k} \mathbf{A}_p\mathbf{X}_p = \mathbf{b}_0$ and make $\sum_{p=0}^{k} \mathbf{C}_p\mathbf{X}_p$ a minimum. If our original problem is looked at in the above manner, we shall see that we would then be required to optimize a new problem with $m_0 + k$ constraints subject to the solution of k individual $m_p \times n_p$ subproblems instead of one large problem with $\sum_{p=0}^{k} m_p$ constraints.[1]

The new problem to be considered is called the *extremal problem* or *master problem* and arises in the following fashion: Let the extreme points of each $\mathbf{S}_p$ be denoted by $\mathbf{X}_{pj}$ for $j = 1, 2, \ldots, N_p$, where $\mathbf{X}_{pj}$ are solution vectors for the pth subproblem and N_p is the total number of extreme points for $\mathbf{S}_p$. Any feasible solution $\mathbf{X}_p$ to the pth subproblem can be expressed as a convex combination of extreme points of $\mathbf{S}_p$, that is,

$$\mathbf{X}_p = \sum_{j=1}^{N_p} \lambda_{pj}\mathbf{X}_{pj} \tag{4.4a}$$

where

$$\sum_{j=1}^{N_p} \lambda_{pj} = 1 \qquad \text{and} \qquad \lambda_{pj} \geq 0 \tag{4.4b}$$

We are temporarily assuming that each $\mathbf{S}_p$ is a bounded polyhedron.

From (4.4a), substituting the expressions for $\mathbf{X}_p$ into (4.1) and (4.2) we have, respectively,

$$\mathbf{C}_0\mathbf{X}_0 + \sum_{p=1}^{k} \mathbf{C}_p \left(\sum_{j=1}^{N_p} \lambda_{pj}\mathbf{X}_{pj}\right)$$

and

$$\mathbf{A}_0\mathbf{X}_0 + \sum_{p=1}^{k} \mathbf{A}_p \left(\sum_{j=1}^{N_p} \lambda_{pj}\mathbf{X}_{pj}\right) = \mathbf{b}_0$$

or

$$\mathbf{C}_0\mathbf{X}_0 + \sum_{p=1}^{k} \sum_{j=1}^{N_p} \mathbf{C}_p\mathbf{X}_{pj}\lambda_{pj} \tag{4.4c}$$

and

$$\mathbf{A}_0\mathbf{X}_0 + \sum_{p=1}^{k} \sum_{j=1}^{N_p} \mathbf{A}_p\mathbf{X}_{pj}\lambda_{pj} \tag{4.4d}$$

Define the transformations

$$\mathbf{P}_{pj} = \mathbf{A}_p\mathbf{X}_{pj} \qquad \text{and} \qquad f_{pj} = \mathbf{C}_p\mathbf{X}_{pj} \qquad p = 1, 2, \ldots, k \tag{4.5}$$

and substituting in (4.4c) and (4.4d) we have that problem (4.1) to (4.3) can now be written in the equivalent form of finding $\lambda_{pj} \geq 0$, $\mathbf{X}_0 \geq \mathbf{O}$,

[1] From the design of present computer procedures for solving linear-programming problems, we note that the number of constraints is usually the restrictive element.

which minimizes

$$\mathbf{C}_0\mathbf{X}_0 + \sum_{p=1}^{k} \sum_{j=1}^{N_p} f_{pj}\lambda_{pj} \tag{4.6}$$

subject to

$$\mathbf{A}_0\mathbf{X}_0 + \sum_{p=1}^{k} \sum_{j=1}^{N_p} \mathbf{P}_{pj}\lambda_{pj} = \mathbf{b}_0 \tag{4.7}$$

$$\sum_{j=1}^{N_p} \lambda_{pj} = 1 \qquad p = 1, 2, \ldots, k \tag{4.8}$$

This problem is the extremal problem. The first m_0 constraint rows are termed the *transfer rows* and the last k rows are the *convexity rows*. In (4.7) and (4.8) the columns $\begin{pmatrix}\mathbf{A}_{0j}\\ \mathbf{O}\end{pmatrix}$, that is, the columns of $\mathbf{A}_0$ with k zeros attached, are *natural columns*; and the columns $\begin{pmatrix}\mathbf{P}_{pj}\\ \mathbf{e}_p\end{pmatrix}$ are *extremal columns*, where $\mathbf{e}_p$ is the pth column of a unit matrix of order k. We should emphasize that (4.6) to (4.8) are just the representation of (4.1) to (4.3) in terms of all the extreme points of all the k subproblems. The transformations $\mathbf{P}_{pj}$ and f_{pj} are just the representation of these extreme points in terms of their contribution to the tie-in equations and objective function, respectively. Given the optimum λ_{pj} to the extremal problem (4.6) to (4.8), then the corresponding optimum solution in terms of each $\mathbf{X}_p$ is given by $\mathbf{X}_p = \sum_j \lambda_{pj}\mathbf{X}_{pj}$.

To illustrate the above let us develop the extremal problem for problem (4.4). The original problem in equation form, after the addition of slack variables x_{01}, x_{13} and x_{14}, is to minimize

$$c_{11}x_{11} + c_{12}x_{12}$$

subject to

x_{01}	$+x_{11} + 4x_{12}$	$= 12$
	$-x_{11} + 2x_{12} + x_{13}$	$= 2$
	$3x_{11} + 4x_{12} \qquad + x_{14}$	$= 24$

with all $x_{ij} \geq 0$. Here $\mathbf{A}_0 = (1)$, $\mathbf{A}_1 = (1,4,0,0)$. $\mathbf{B}_1 = \begin{pmatrix} -1 & 2 & 1 & 0\\ 3 & 4 & 0 & 1\end{pmatrix}$, $\mathbf{b}_0 = (12)$, $\mathbf{b}_1 = \begin{pmatrix}2\\ 24\end{pmatrix}$, $\mathbf{C}_0 = (0)$, $\mathbf{C}_1 = (c_{11},c_{12},0,0)$. In Fig. 9.4 the extreme points of the solution set to $\mathbf{B}_1\mathbf{X}_1 \leq \mathbf{b}_1$ are denoted by $\mathbf{X}_{11},\mathbf{X}_{12}, \mathbf{X}_{13},\mathbf{X}_{14}$. We note that we are now dealing with equations and the $\mathbf{X}_{ij}$'s have four components corresponding to x_{11}, x_{12}, and the slack variables x_{13}, x_{14}.† Following Eq. (4.5), we have

$$\begin{aligned}
\mathbf{P}_{11} &= \mathbf{A}_1\mathbf{X}_{11} = (1,4,0,0)\mathbf{X}_{11} \\
\mathbf{P}_{12} &= \mathbf{A}_1\mathbf{X}_{12} = (1,4,0,0)\mathbf{X}_{12} \\
\mathbf{P}_{13} &= \mathbf{A}_1\mathbf{X}_{13} = (1,4,0,0)\mathbf{X}_{13} \\
\mathbf{P}_{14} &= \mathbf{A}_1\mathbf{X}_{14} = (1,4,0,0)\mathbf{X}_{14}
\end{aligned}$$

and

$$\begin{aligned}
f_{11} &= \mathbf{C}_1\mathbf{X}_{11} = (c_{11},c_{12},0,0)\mathbf{X}_{11} \\
f_{12} &= \mathbf{C}_1\mathbf{X}_{12} = (c_{11},c_{12},0,0)\mathbf{X}_{12} \\
f_{13} &= \mathbf{C}_1\mathbf{X}_{13} = (c_{11},c_{12},0,0)\mathbf{X}_{13} \\
f_{14} &= \mathbf{C}_1\mathbf{X}_{14} = (c_{11},c_{12},0,0)\mathbf{X}_{14}
\end{aligned}$$

The extremal problem, that is, the original problem expressed in terms of all the extreme points of the subproblem $\mathbf{B}_1\mathbf{X}_1 \leq \mathbf{b}_1$, is to find $\lambda_{1j} \geq 0$ which minimizes

$$x_{00} + f_{11}\lambda_{11} + f_{12}\lambda_{12} + f_{13}\lambda_{13} + f_{14}\lambda_{14} \qquad (4.6a)$$

subject to

$$x_{00} + \mathbf{P}_{11}\lambda_{11} + \mathbf{P}_{12}\lambda_{12} + \mathbf{P}_{13}\lambda_{13} + \mathbf{P}_{14}\lambda_{14} = 12 \qquad (4.7a)$$

$$\lambda_{11} + \lambda_{12} + \lambda_{13} + \lambda_{14} = 1 \qquad (4.8a)$$

As mentioned above, this transformation yields a problem with only $m_0 + k$ constraints, the m_0 transfer rows (4.7), and the set of k convexity constraints (4.8). However, the number of variables has been increased to the total of all extreme points of the convex polyhedra $\mathbf{S}_p$. The saving element of the decomposition principle is that we need consider only a small number of this usually rather large total, and we need only the explicit representation of those to be considered. We shall see that the decomposition algorithm embodies the basic elements of the pricing vector and basis transformation of the revised simplex method. We next outline this algorithm and give a numerical example.

We treat the extremal problem (4.6) to (4.8) as we would any other linear-programming problem except that we have to allow for the fact that we are unable to write out the matrix of the system corresponding to (4.7). However, let us assume that we know enough about the system so that we have an initial basic feasible solution to the extremal problem. This first solution could have been obtained using an artificial basis or, based on the structure of the subproblems, we were able to find extreme-point solutions to the subproblems which combined into a feasible solution to

† To avoid notational confusion, the reader should bear in mind that $\mathbf{X}_1$ is the variable vector with variables (components) x_{11}, x_{12}, x_{13}, x_{14}, and the extreme points $\mathbf{X}_{1p}$ are particular solution vectors with specified values of the variables.

the extremal problem (see example). This basis is of order $m_0 + k$ and must contain at least one extremal column from each subproblem solution space $\mathbf{S}_p$, that is, contain in total at least k extremal columns and m_0 additional extremal or natural columns or both. Denote the associated simplex multiplier vector[1] by $\mathbf{\Pi} = (\boldsymbol{\pi},\bar{\boldsymbol{\pi}})$, where the m_0-component row vector $\boldsymbol{\pi}$ is associated with the m_0 transfer rows of (4.7) and the k-component row vector $\bar{\boldsymbol{\pi}}$ is associated with the convexity rows (4.8). The components of $\boldsymbol{\pi}$ are called the *transfer prices* and will be denoted by π_i and those of $\bar{\boldsymbol{\pi}}$ by $\bar{\pi}_p$. For the vectors in the basis we must have $f_{pj} = \boldsymbol{\pi}\mathbf{P}_{pj} + \bar{\pi}_p$; that is, the corresponding $c_j - z_j = 0$. As in the regular simplex method, we next have to determine whether or not the initial feasible solution can be improved by introducing some new natural or extremal column. Pricing out an extremal column from some subproblem p, we have from (4.5) relative costs (that is, $c_j - z_j$)

$$\begin{aligned} f_{pj} - \mathbf{\Pi}\begin{pmatrix}\mathbf{P}_{pj}\\ \mathbf{e}_p\end{pmatrix} &= f_{pj} - \boldsymbol{\pi}\mathbf{P}_{pj} - \bar{\boldsymbol{\pi}}\mathbf{e}_p = \mathbf{C}_p\mathbf{X}_{pj} - \boldsymbol{\pi}\mathbf{A}_p\mathbf{X}_{pj} - \bar{\pi}_p \\ &= (\mathbf{C}_p - \boldsymbol{\pi}\mathbf{A}_p)\mathbf{X}_{pj} - \bar{\pi}_p \qquad p = 1, 2, \ldots, k \end{aligned} \tag{4.9}$$

Pricing out the natural columns, we have, for each j, the relative costs

$$c_{0j} - \boldsymbol{\pi}\mathbf{A}_{0j} \tag{4.10}$$

where c_{0j} are the elements of $\mathbf{C}_0$. Since the extremal problem is a minimization problem, if all the relative costs (4.9) and (4.10) are nonnegative for all p and j, then the simplex criterion tells us that we have an optimum, since (4.9) and (4.10) correspond to the set of $c_j - z_j$ of the simplex algorithm, and for a minimization problem we are optimal if all $z_j - c_j \leq 0$ or if all $c_j - z_j \geq 0$. The costs (4.10) are readily calculated, as we know all the $\mathbf{A}_{0j}$. However, in order to determine if any of the relative costs (4.9) are negative, we need to solve p optimization problems which arise as follows. For any p we need to know whether or not for the set of extreme points $\mathbf{X}_{pj}$ we have

$$\underset{\substack{\mathbf{B}_p\mathbf{X}_{pj}=\mathbf{b}_p\\ \mathbf{X}_{pj}\geq 0}}{\text{minimum}}\,[(\mathbf{C}_p - \boldsymbol{\pi}\mathbf{A}_p)\mathbf{X}_{pj} - \bar{\pi}_p] < 0 \qquad p = 1, 2, \ldots, k \tag{4.11}$$

But (4.11) states that for each p the quantity in brackets is minimized over those extreme points $\mathbf{X}_{pj}$ which satisfy the constraints of the pth subproblem. If the minimum is nonnegative, then all extreme points of the pth subproblem price out optimally.

Since the condition (4.11) is independent of the scalar $\bar{\pi}_p$, (4.11) reduces to solving for each p the linear-programming problem, which is to minimize

$$(\mathbf{C}_p - \boldsymbol{\pi}\mathbf{A}_p)\mathbf{X}_p \tag{4.12a}$$

[1] We assume we are employing the revised simplex method to perform the optimization of the extremal problem; see Chap. 6.

subject to

$$\mathbf{B}_p\mathbf{X}_p = \mathbf{b}_p \qquad (4.12b)$$
$$\mathbf{X}_p \geq \mathbf{O}$$

We note that since the solution to (4.12*b*) by the simplex procedure will generate only extreme-point solutions, then our solution to subproblem (4.12*a* and *b*) will also satisfy (4.11). We also note that each subproblem (4.12*a* and *b*) is equivalent to the given subproblem except that the corresponding cost coefficients have been reduced by $\pi\mathbf{A}_p$; the $(\mathbf{C}_p - \pi\mathbf{A}_p)$ are called *adjusted costs*. We see that $\pi\mathbf{A}_p$ is found by taking n_p inner products and can be readily calculated. We let $\mathbf{X}_{pq}$ be an optimum solution to (4.12*a* and *b*) for a given p ($\mathbf{X}_{pq}$ will be an extreme point if $\mathbf{S}_p$ is bounded) and let the corresponding optimum value of the objective function be denoted by

$$Z_{pq} = (\mathbf{C}_p - \pi\mathbf{A}_p)\mathbf{X}_{pq}$$

If $Z_{pq} - \bar{\pi}_p < 0$, then the new extremal column formed by the transformation (4.5), that is, $\mathbf{P}_{pq} = \mathbf{A}_p\mathbf{X}_{pq}$ with its associated cost $f_{pq} = \mathbf{C}_p\mathbf{X}_{pq}$, is a candidate to enter the basis. The new extreme point $\mathbf{X}_{pq}$ is called a *proposal vector*, the transformation vector $\mathbf{P}_{pq}$ its *transfer vector*, and f_{pq} its *transfer cost.*

Given the set of minimum costs (4.9) and the costs (4.10), the simplex method tells us to take as the next column to enter the extremal-problem basis the one corresponding to the min $(c_j - z_j)$, that is,

$$\min[\min_p(Z_{pq} - \bar{\pi}_p), \min_j(c_{0j} - \pi\mathbf{A}_{0j})] \qquad (4.13)$$

where p ranges over the set of subproblems and j ranges over the nonbasic natural columns. If (4.13) is nonnegative, then the current basic feasible solution of the extremal problem yields an optimum to the original problem. If (4.13) is negative, we select the corresponding extremal or natural column to enter the basis. For this latter case, we determine the pivot row as usual, determine the new $\mathbf{\Pi}$, and repeat the procedure until an optimum is reached. The optimal solution to the original problem is given by

$$[\mathbf{X}_0,\ \mathbf{X}_p = \sum_j \lambda_{pj}\mathbf{X}_{pj} \qquad (p = 1, 2, \ldots, k)] \qquad (4.14)$$

where the summation is taken over those extreme points of the pth subproblem which are in the optimal solution to the extremal problem.

Up to this point we have assumed that the polyhedra $\mathbf{S}_p$ are bounded. For the general situation let us assume that this is not the case. Then in optimizing one of the subproblems (4.12*a* and *b*), we could obtain an optimum solution with an unbounded value Z_{pq} of the objective function;

see Sec. 1, Chap. 4. For this case the optimum solution to the subproblem will yield a corresponding transfer vector $\mathbf{P}_{pq}$ as a candidate for the extremal-problem basis, but we cannot restrict the associated extremal-problem variable λ_{pq} to be bounded by the convexity equation; that is, λ_{pq} is not bounded above. The reason for this is as follows: In the simplex method an unbounded solution is represented by a basic feasible solution plus a nonnegative sum of vectors (see Exercise 19). For the standard linear-programming problem this latter sum is a solution to the homogeneous equations $\mathbf{AX} = \mathbf{O}$ and the constraints $\mathbf{X} \geq \mathbf{O}$. For the decomposition problem, we would have a nonnegative solution to the homogeneous equations $\mathbf{B}_p\mathbf{X}_p = \mathbf{O}$. We let $\mathbf{X}_{pq}$ denote this solution, $\mathbf{B}_p\mathbf{X}_{pq} = \mathbf{O}$, and treat it as any other proposal vector (note that $\mathbf{X}_{pq}$ is not an extreme-point solution to the subproblem). As $\mathbf{B}_p(\lambda_{pq}\mathbf{X}_{pq}) = \mathbf{O}$ for all $\lambda_{pq} \geq 0$, we have that a nonnegative sum of $\mathbf{X}_{pq}$ with a convex combination of extreme points of $\mathbf{S}_p$ can be found which yields a vector satisfying the constraints $\mathbf{B}_p\mathbf{X}_p = \mathbf{b}_p$. Whenever we encounter an unbounded solution $\mathbf{X}_{pq}$ in any subproblem p, the corresponding convexity row is rewritten so as not to include the appropriate λ_{pq}, that is,

$$\sum_{\substack{j=1 \\ j \neq q}}^{N_p} \lambda_{pj} = 1$$

From a computational point of view we should note the following. Each time it is necessary to reoptimize a subproblem for a new objective function, the previous optimal feasible solution should be used as the first feasible solution to the new problem. In the selection of the new nonbasic variable to enter the basis, we need not obtain the true minimum (4.13), but can select any vector which would yield an improvement. Computational experience has shown that it appears to be best if a number of vectors which would individually yield an improvement, possibly some from each nonoptimal subproblem, are simultaneously considered as transfer vectors in the optimization of the extremal problem. That is, we do not restrict an iteration of the extremal problem to just a preselected basis change, but allow the old basic vectors and a number of new transfer vectors to be candidates in the formation of the new basis. Large systems of the order of 5,900 rows and 8,000 columns, with 175 tie-in restrictions and 10 subproblems, have been solved in 18 hours on an IBM 7094 computer (Hellerman [54*a*]). We should note that the application of the decomposition algorithm is not always successful in that convergence to the optimum might require an inordinate number of iterations. Even successful applications, as noted above, can use a great deal of costly computer time. The field of large-scale linear-programming systems will always be a challenging one and still requires much experimentation and new developments. For a general discussion on approaches to solving large-scale systems, see Gomory [51*f*].

It should be stressed that although the decomposition algorithm was developed for structured problems, any linear-programming problem can be decomposed. For example, if a computer code is limited to the handling of a problem with 500 constraints, an appropriately designed decomposition code could subdivide a larger problem into two parts which could be readily solved. Also problems which involve transportation or network problems as the subproblems, e.g., the multicommodity problem, can use the special algorithms to solve the subproblems and the revised simplex method to solve the extremal problem. The fact that any linear-programming problem can be decomposed in a manner that best suits its special form is the basis of an application of the decomposition principle to the transportation problem (Williams [100*d*]).

Extensions of the decomposition principle include (1) the generalized-programming problem of Wolfe (Dantzig [16*c*]), where each column vector of a linear-programming problem is allowed to be selected from a convex set (see Gilmore and Gomory [49*a*] for an application of this technique), and (2) the dual solution of the extremal problem and parametric decomposition procedures of Abadie and Williams [1*c*]. As the solution to the extremal problem can yield a nonbasic optimal solution to the original problem, Abadie [1*a*] modifies the computational procedure so that the current solution to the extremal problem is always a basic one for the original problem.

Example. Minimize

$$-x_{11} + x_{12} - 3x_{21} - 2x_{22}$$

subject to

$$\begin{array}{|r|r|l|}
\hline
x_{11} - 3x_{12} & -2x_{21} + x_{22} & \le 6 \\
x_{12} & +3x_{21} - x_{22} & \le 4 \\
\hline
3x_{11} + 2x_{12} & & \le 12 \\
-x_{11} + 3x_{12} & & \le 6 \\
\hline
 & 3x_{21} + x_{22} & \le 12 \\
 & x_{21} + 2x_{22} & \le 8 \\
\hline
\end{array} \qquad (4.15)$$

and

$$x_{ij} \ge 0$$

Rewriting (4.15) in terms of equations with slack variables x_{01},x_{02}, $x_{13},x_{14},x_{23},x_{24}$, we have

$$
\begin{array}{|ll|llll|lllll|l|}
\hline
x_{01} & & +x_{11} & -3x_{12} & & & -2x_{21} & +x_{22} & & & & = 6 \\
 & x_{02} & & +x_{12} & & & +3x_{21} & -x_{22} & & & & = 4 \\
\hline
 & & 3x_{11} & +2x_{12} & +x_{13} & & & & & & & = 12 \\
 & & -x_{11} & +3x_{12} & & +x_{14} & & & & & & = 6 \\
\hline
 & & & & & & 3x_{21} & +x_{22} & +x_{23} & & & = 12 \\
 & & & & & & x_{21} & +2x_{22} & & +x_{24} & & = 8 \\
\hline
\end{array}
\qquad (4.16)
$$

$$
\mathbf{A}_0 = \begin{pmatrix} 1 & 0 \\ 0 & 1 \end{pmatrix} \qquad \mathbf{A}_1 = \begin{pmatrix} 1 & -3 & 0 & 0 \\ 0 & 1 & 0 & 0 \end{pmatrix} \qquad \mathbf{A}_2 = \begin{pmatrix} -2 & 1 & 0 & 0 \\ 3 & -1 & 0 & 0 \end{pmatrix}
$$

$$
\mathbf{B}_1 = \begin{pmatrix} 3 & 2 & 1 & 0 \\ -1 & 3 & 0 & 1 \end{pmatrix} \qquad \mathbf{B}_2 = \begin{pmatrix} 3 & 1 & 1 & 0 \\ 1 & 2 & 0 & 1 \end{pmatrix}
$$

$$
\mathbf{b}_0 = \begin{pmatrix} 6 \\ 4 \end{pmatrix} \qquad \mathbf{b}_1 = \begin{pmatrix} 12 \\ 6 \end{pmatrix} \qquad \mathbf{b}_2 = \begin{pmatrix} 12 \\ 8 \end{pmatrix}
$$

$$
\mathbf{C}_0 = (0,0) \qquad \mathbf{C}_1 = (-1,1,0,0) \qquad \mathbf{C}_2 = (-3,-2,0,0)
$$

Treating the subproblems as inequalities with nonnegative slack vectors, the solution spaces $\mathbf{S}_1$ and $\mathbf{S}_2$ to the subproblems are given in Fig. 9.5*a* and *b*. We have indicated all the feasible extreme-point solutions.

To find a first feasible solution to the extremal problem,[1] we consider the extreme-point solution to the first subproblem $\mathbf{X}_{11} = (0,0,12,6)$ and the extreme-point solution of the second subproblem $\mathbf{X}_{21} = (0,0,12,8)$, where the last two components of $\mathbf{X}_{11}$ are the slack variables x_{13} and x_{14}, and the last two components of $\mathbf{X}_{21}$ are the slack variables x_{23} and x_{24}. Then $\mathbf{P}_{11} = \begin{pmatrix} 0 \\ 0 \end{pmatrix}$, $\mathbf{P}_{21} = \begin{pmatrix} 0 \\ 0 \end{pmatrix}$, $f_{11} = 0$, $f_{21} = 0$, and the extremal problem is given by

$$
\begin{pmatrix} 1 \\ 0 \\ 0 \\ 0 \end{pmatrix} x_{01} + \begin{pmatrix} 0 \\ 1 \\ 0 \\ 0 \end{pmatrix} x_{02} + \begin{pmatrix} 0 \\ 0 \\ 1 \\ 0 \end{pmatrix} \lambda_{11} + \begin{pmatrix} 0 \\ 0 \\ 0 \\ 1 \end{pmatrix} \lambda_{21} = \begin{pmatrix} 6 \\ 4 \\ 1 \\ 1 \end{pmatrix}
$$

with the first feasible solution of $x_{01} = 6$, $x_{02} = 4$, $\lambda_{11} = 1$, $\lambda_{21} = 1$, and as all c_{0j} and f_{pj} are zero, the value of the objective function is zero. Note that $\mathbf{\Pi} = (0,0,0,0)$. For this pricing vector $\mathbf{\Pi}$, we need to see if any other extreme points from the subproblems can improve the value of the objective function. Since both $\boldsymbol{\pi} = 0$ and $\bar{\boldsymbol{\pi}} = 0$, the optimization of the subproblems just involves the original data of the subproblems; i.e., the objective functions of the subproblems

[1] See Exercise 24 for a general procedure for finding a first feasible solution.

For $\mathbf{B}_1\mathbf{X}_1 = \mathbf{b}_1$

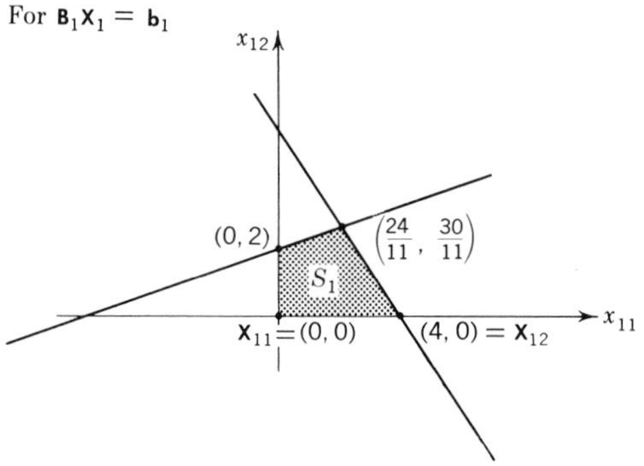

FIGURE 9.5*a*

For $\mathbf{B}_2\mathbf{X}_2 = \mathbf{b}_2$

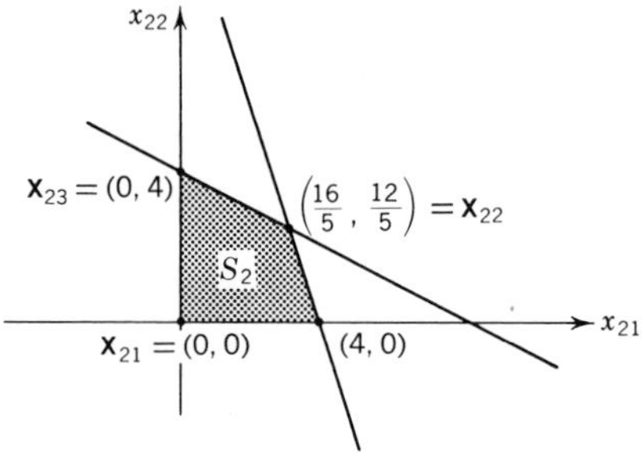

FIGURE 9.5*b*

are not changed. We then have to find the extreme-point solutions of the following problems:

Minimize

$$-x_{11} + x_{12}$$

subject to

$$\begin{aligned} 3x_{11} + 2x_{12} + x_{13} \qquad\qquad &= 12 \\ -x_{11} + 3x_{12} \qquad\qquad + x_{14} &= 6 \\ x_{ij} &\geq 0 \end{aligned} \tag{4.17}$$

and minimize

$$-3x_{21} - 2x_{22}$$

subject to

$$\begin{aligned} 3x_{21} + x_{22} + x_{23} \qquad\qquad &= 12 \\ x_{21} + 2x_{22} \qquad\qquad + x_{24} &= 8 \\ x_{ij} &\geq 0 \end{aligned} \tag{4.18}$$

We would ordinarily use the simplex method to solve (4.17) and (4.18), but since we have the complete set of extreme-point solutions in Fig. 9.5*a* and *b*, respectively, we shall just evaluate each extreme point in the corresponding objective function and select the minimum. For the first subproblem we have $\mathbf{X}_{12} = (4,0,0,10)$, $Z_{12} = -4$, and for the second subproblem we have $\mathbf{X}_{22} = (16/5, 12/5, 0, 0)$, $Z_{22} = -72/5$. We note that all the vectors of $\mathbf{A}_0$ are in the current extremal basis and that $\min[Z_{12}, Z_{22}] = -72/5$. Therefore, by (4.13) we select $\mathbf{X}_{22}$ to enter the basis of the extremal problem. Its transfer vector $\mathbf{P}_{12} = \begin{pmatrix} -4 \\ 36/5 \end{pmatrix}$ and its transfer cost $f_{12} = -72/5$. The extremal problem in tableau form is now

b	x_{01}	x_{02}	λ_{11}	λ_{21}	λ_{22}
6	1	0	0	0	-4
4	0	1	0	0	(36/5)
1	0	0	1	0	0
1	0	0	0	1	1
0	0	0	0	0	$-72/5$

To improve the solution, we introduce λ_{22} into the basis, and to preserve feasibility, we eliminate x_{02}. This is shown in the next tableau.

	x_{01}		λ_{11}	λ_{21}	λ_{22}
74/9	1		0	0	0
5/9	0		0	0	1
1	0		1	0	0
4/9	0		0	1	0
8	0		0	0	0

We drop the column associated with x_{02}; if it is needed for a future basis, it will be brought back to the extremal problem at that time.[1] Our extremal solution is $x_{01} = 74/9$, $\lambda_{11} = 1$, $\lambda_{21} = 4/9$, $\lambda_{22} = 5/9$, $Z = -8$. The corresponding feasible solution to the original problem (4.16) is $\mathbf{X}_0 = (74/9, 0)$, $\mathbf{X}_1 = \lambda_{11}\mathbf{X}_{11} = (0,0,12,6)$, and

$$\begin{aligned}\mathbf{X}_2 = \lambda_{21}\mathbf{X}_{21} + \lambda_{22}\mathbf{X}_{22} &= 4/9(0,0,12,8) + 5/9(16/5, 12/5, 0, 0) \\ &= 1/9(16,12,48,32)\end{aligned}$$

[1] For more involved problems, it appears as if it pays to keep the past extreme points as candidates to reenter the extremal basis as long as it is computationally convenient.

with $Z = -8$. The corresponding $\mathbf{\Pi} = (0,-2,0,0)$; that is, $\pi = (0,-2)$, $\bar{\pi} = (0,0)$. The reader should recall that $\mathbf{\Pi} = \mathbf{f}\mathbf{B}^{-1}$, where $\mathbf{f}$ is the vector of cost coefficients of the vectors in the current basis $\mathbf{B}$. We leave it to the reader to calculate $\mathbf{B}^{-1}$. Using this pricing vector, we repeat the process to see if a new extreme-point solution from the subproblems can improve the solution. From (4.12a) the subproblems now have the respective objective functions of

$$\text{minimize } -x_{11} + 3x_{12} \qquad \text{and} \qquad \text{minimize } 3x_{21} - 4x_{22}$$

For subproblem 1, $\mathbf{X}_{12}$ yields the optimum with $Z_{12} = -4$; for subproblem 2, $\mathbf{X}_{23}$ yields the optimum with $Z_{23} = -16$. Since $\min_j (c_{0j} - \pi\mathbf{A}_{0j}) = 0$, we then select $\mathbf{X}_{23}$ to be the proposal vector with the transfer vector of $\mathbf{P}_{23} = \begin{pmatrix} 4 \\ -4 \end{pmatrix}$ and transfer cost of $f_{23} = -8$. The *complete* new tableau of the extremal problem is now

b	x_{01}	x_{02}	λ_{11}	λ_{21}	λ_{22}	λ_{23}
6	1	0	0	0	-4	4
4	0	1	0	0	(36/5)	-4
1	0	0	1	0	0	0
1	0	0	0	1	1	1
0	0	0	0	0	$-72/5$	-8

Introducing λ_{22} as before, we have

74/9	1		0	0	0	16/9
5/9	0		0	0	1	$-5/9$
1	0		1	0	0	0
4/9	0		0	1	0	(14/9)
8	0		0	0	0	-16

The reader will note that the extremal column associated with $\mathbf{P}_{23}$ has now been transformed, as would be the case in the revised simplex method; i.e., it has been multiplied by the inverse of the current basis of the extremal problem. In practice, we would not set up the first complete tableau, but just form the transformed extremal column and introduce the new variable. We next introduce λ_{23} and eliminate λ_{21} to obtain

	x_{01}	λ_{11}	λ_{22}	λ_{23}
$54/7$	1	0	0	0
$5/7$	0	0	1	0
1	0	1	0	0
$2/7$	0	0	0	1
$88/7$	0	0	0	0

For this solution $\mathbf{\Pi} = (0, -4/7, 0, -72/7)$; that is, $\pi = (0, -4/7)$, $\bar{\pi} = (0, -72/7)$, and the new objective functions for the subproblems are

$$\text{minimize } -x_{11} + 11/7\, x_{12} \qquad \text{and} \qquad \text{minimize } -9/7\, x_{21} - 18/7\, x_{22}$$

For subproblem 1 the minimum is assumed for $\mathbf{X}_{12} = (4,0,0,10)$ with $Z_{12} = -4$ and $f_{12} = -4$; for subproblem 2 the minimum is assumed for both $\mathbf{X}_{22} = (16/5, 12/5, 0, 0)$ and $\mathbf{X}_{23} = (0,4,0,0)$ with $Z_{22} = Z_{23} = -72/7$. Also $c_{02} - \pi\mathbf{A}_{02} = 4/7 > 0$. We then select $\mathbf{X}_{12}$ as the proposal vector since $Z_{12} - \bar{\pi}_1 = -4 - 0 = -4 < 0$ and $Z_{22} - \bar{\pi}_2 = Z_{23} - \bar{\pi}_2 = -72/7 + 72/7 = 0$. (Note that the extremal column corresponding to $\mathbf{X}_{22}$ is a current basis vector of the extremal problem; hence $Z_{22} - \bar{\pi}_2 = 0$.) $\mathbf{P}_{12} = \begin{pmatrix} 4 \\ 0 \end{pmatrix}$ and $f_{12} = -4$.

	x_{01}	λ_{11}	λ_{22}	λ_{23}	λ_{12}
$54/7$	1	0	0	0	4
$5/7$	0	0	1	0	0
1	0	1	0	0	(1)
$2/7$	0	0	0	1	0
$88/7$	0	0	0	0	−4

We introduce λ_{12} and eliminate λ_{11} to obtain

	x_{01}	λ_{22}	λ_{23}	λ_{12}
$26/7$	1	0	0	0
$5/7$	0	1	0	0
1	0	0	0	1
$2/7$	0	0	1	0
$116/7$	0	0	0	0

The new $\mathbf{\Pi} = (0, -\frac{4}{7}, -4, -\frac{72}{7})$, and the reader can verify that for this pricing vector no improvement can be obtained. The optimum solution to the original problem is

$$\mathbf{X}_0 = (\tfrac{26}{7}, 0) \qquad \mathbf{X}_1 = (4,0,0,10)$$

$$\mathbf{X}_2 = \tfrac{5}{7}(\tfrac{16}{5}, \tfrac{12}{5}, 0, 0) + \tfrac{2}{7}(0,4,8,0) = \tfrac{1}{7}(16,20,16,0)$$

with $-\frac{116}{7}$ the minimum value of the objective function.

5. BOUNDED-VARIABLE PROBLEMS

An important class of linear-programming problems is one in which all or some of the variables are bounded from above. Mathematically we have the following problem:

Minimize

$$z = \mathbf{cX} \tag{5.1}$$

subject to

$$\mathbf{AX} = \mathbf{b} \tag{5.2}$$

$$\mathbf{X} \geq \mathbf{0} \tag{5.3}$$

$$x_j \leq u_j \text{ for all or some } j \tag{5.4}$$

We have, of course, each $u_j > 0$. To generalize the discussion, we assume that all x_j are bounded above as we can always bound a variable by a suitably large number. The given upper-bound restrictions (or secondary constraints) can represent capacity limitations, production requirements, or resource limitations, and arise naturally in many applications.

One way of solving this type of problem would be to replace the set of $x_j \leq u_j$ by an equivalent set of equalities in nonnegative variables (using slack variables). This, however, greatly increases the number of constraints to be handled. A more efficient technique has been developed (Charnes and Lemke [14] and Dantzig [24]) which entails only a few modifications of the standard simplex procedure and enables one to solve the problem without explicit representation of the upper-bound constraints. (The reader will note that lower-bound conditions offer no difficulty; see footnote on page 248.)

The justification of the special procedure for solving the bounded-variable problem can be seen from the following (Dantzig [16*c*] and Simonnard [90*b*]). In the general theory of the simplex method we deal with basic feasible solutions which are obtained by arbitrarily setting $n - m$ nonbasic variables equal to zero and solving the resulting $m \times m$ square system, assuming we have an admissible basis. For the bounded-variable problem (5.1) to (5.4), in order to employ a similar technique, we must allow the nonbasic variables to take on values other than zero. However, the situation is simplified, as we can restrict the value of a nonbasic variable to be either zero or the corresponding upper bound. For this problem, we then define an *extended basic feasible solution* to be one for which $n - m$ variables have been set equal to either their lower or upper bounds (zero or u_j), the matrix $\mathbf{B}$ of the associated square system is a basis, and the solution to the square system yields values of the basic variables which satisfy the upper- and lower-bound constraints (5.3) and (5.4). We then have the following theorem.

Theorem 1. *An extended basic feasible solution is optimal if the $z_j - c_j$ of the nonbasic variables satisfy the conditions*

$$z_j - c_j \leq 0 \qquad \textit{if } x_j = 0 \tag{5.5}$$

and

$$z_j - c_j \geq 0 \qquad \textit{if } x_j = u_j \tag{5.6}$$

Proof. Theorem 1 can be proved by noting the following. For $\mathbf{B}$, the basis of the extended basic feasible solution, let $\boldsymbol{\pi} = (\pi_1, \ldots, \pi_m)$ be the associated simplex multipliers (see discussion in Chap. 6). Then multiplying (5.2) by $\boldsymbol{\pi}$ and (5.1) by -1 and adding, we have

$$\boldsymbol{\pi}\mathbf{AX} - \mathbf{cX} = \boldsymbol{\pi}\mathbf{b} - z$$

or

$$z = \boldsymbol{\pi}\mathbf{b} - (\boldsymbol{\pi}\mathbf{A} - \mathbf{c})\mathbf{X}$$

For $\mathbf{B}$, we have $z = \bar{z} - \sum_{j \notin \mathbf{B}} (z_j - c_j)x_j$, where $\bar{z} = \boldsymbol{\pi}\mathbf{b}$ is the value of the objective function for the basic variables.[1] Thus

$$\min z = \min \left[\bar{z} - \sum_{j \notin \mathbf{B}} (z_j - c_j)x_j\right] \tag{5.7}$$

If conditions (5.5) and (5.6) hold, then any changes in the nonbasic variables will increase the total value of the objective function; thus (5.7) is at its minimum.

If the conditions of Theorem 1 do not hold, we can change the current solution and determine a new, improved one. An analysis of the computational considerations enables us to apply the simplex algorithm to the primary constraints (5.2) and (5.3), but modified to ensure that the secondary constraints (5.4) are not violated. Thus, the size of the basis will be only of dimensions $m \times m$. Compared with solving the complete

[1] The notation $\sum_{j \notin \mathbf{B}}$ means that the sum is taken for the set of indices j of those vectors not in the basis $\mathbf{B}$.

bounded-variable problem by the simplex method, the modified procedure represents quite a computational saving. By using the bounded-variable algorithm, we do not reduce the number of iterations, but decrease the time per iteration and are able to solve rather large bounded-variable problems. A discussion of the bounded-variable algorithm follows (Garvin [45*a*] and Simonnard [90*b*]).

Assume that we have an extended basic feasible solution with basis **B** and that conditions (5.5) or (5.6) do not hold for some nonbasic variable x_k. If we want to increase x_k, then either some basic variable will reach its upper or lower bound or x_k will attain its upper bound. Similarly, if we want to decrease x_k, then either some basic variable will reach its upper or lower bound or x_k will attain its lower bound. If x_k reaches one of its bounds first, then we will not have to change the current basis **B**. If some basic variable, say x_l, attains one of its bounds first, then x_l will be replaced by x_k in the basis. We next have to determine the appropriate value of x_k for which all the constraints of the problem will remain satisfied. An improved solution will be obtained as we selected an x_k for which (5.5) or (5.6) is not satisfied.

Case 1. Determination of a new basis if $x_k = 0$.

For the given solution we have $x_k = 0$ and $z_k - c_k > 0$. In terms of the simplex transformations for the current basis (see Tableau 4.1, Chap. 4), each variable x_i in the basis is given by

$$x_i = x_{i0} - \sum_{j \in \mathbf{U}} x_{ij}u_j - x_{ik}x_k \tag{5.8}$$

where **U** is the set of indices for those nonbasic variables which equal their upper bounds.[1] Letting $\bar{b}_i = x_{i0} - \sum_{j \in \mathbf{U}} x_{ij}u_j$, (5.8) becomes

$$x_i = \bar{b}_i - x_{ik}x_k \qquad \text{for all } i \in \mathbf{B}$$

For those i for which $x_{ik} > 0$, we see that an increase in x_k from its current value of zero will decrease the corresponding x_i. Thus we must restrict any increase to x_k to a value which does not allow any of the current basic variables to become negative. We must have for $x_{ik} > 0$,

$$\bar{b}_i - x_{ik}x_k \geq 0$$

or

$$x_k \leq \frac{\bar{b}_i}{x_{ik}}$$

or

$$x_k \leq \frac{\bar{b}_p}{x_{pk}} = \min_{x_{ik}>0} \frac{\bar{b}_i}{x_{ik}} \tag{5.9}$$

[1] The notation $\sum_{j \in \mathbf{U}}$ means that the sum is taken for all i in the set of indices **U**; $i \in \mathbf{B}$ represents the set of indices of the basis vectors.

Similarly, for $x_{ik} < 0$, an increase in x_k will also increase the corresponding x_i. But each x_i is bounded above. Thus we must have for $x_{ik} < 0$,

$$\bar{b}_i - x_{ik}x_k \leq u_i$$

$$x_k \leq \frac{u_i - \bar{b}_i}{-x_{ik}}$$

$$x_k \leq \frac{u_q - \bar{b}_q}{-x_{qk}} = \min_{x_{ik}<0} \frac{u_i - \bar{b}_i}{-x_{ik}} \tag{5.10}$$

Finally, we also must have that

$$x_k \leq u_k \tag{5.11}$$

Taking (5.9) to (5.11) together, the maximum increase in x_k is given by

$$\max x_k = \min_{i \in \mathbf{B}} \left[\frac{\bar{b}_i}{x_{ik}} \text{ for } x_{ik} > 0, \frac{u_i - \bar{b}_i}{-x_{ik}} \text{ for } x_{ik} < 0, u_k \right]$$

$$\max x_k = \min \left[\frac{\bar{b}_p}{x_{pk}}, \frac{u_q - \bar{b}_q}{-x_{qk}}, u_k \right]$$

If max x_k is limited by (5.9), then x_k replaces x_p in the basis; if it is limited by (5.10), then x_k replaces x_q in the basis; if it is limited by (5.11), then x_k stays nonbasic but its value is now u_k. In the first two situations the basis is changed using x_{pk} or x_{qk} as the corresponding pivot. We note that the pivot could be negative.

Case 2. Determination of a new basis if $x_k = u_k$ and $z_k - c_k < 0$. For this case we have

$$x_i = x_{i0} - \sum_{\substack{j \in \mathbf{U} \\ j \neq k}} x_{ij}u_j - x_{ik}x_k \tag{5.12}$$

Letting $\bar{b}'_i = x_{i0} - \sum_{\substack{j \in \mathbf{U} \\ j \neq k}} x_{ij}u_j$, (5.12) becomes $x_i = \bar{b}'_i - x_{ik}x_k$ for all $i \in \mathbf{B}$.

For $x_{ik} > 0$, a decrease in x_k will increase x_i, while for $x_{ik} < 0$, a decrease in x_k will decrease x_i. We must ensure that a decrease in x_k will not allow an x_i in the basis to go below zero or above its upper bound. Thus we must have

$$x_i = \bar{b}'_i - x_{ik}x_k \leq u_i \qquad \text{for } x_{ik} > 0$$

or

$$x_k \geq \frac{\bar{b}'_p - u_p}{x_{pk}} = \max_{x_{ik}>0} \frac{\bar{b}'_i - u_i}{x_{ik}} \tag{5.13}$$

and

$$x_i = \bar{b}'_i - x_{ik}x_k \geq 0 \qquad \text{for } x_{ik} < 0$$

or

$$x_k \geq \frac{\bar{b}'_q}{x_{qk}} = \max_{x_{ik}<0} \frac{\bar{b}'_i}{x_{ik}} \tag{5.14}$$

and finally,

$$x_k \geq 0 \tag{5.15}$$

Combining (5.13) to (5.15), we have that the minimum value of x_k is given by

$$\min x_k = \max \left[\frac{\bar{b}'_p - u_p}{x_{pk}}, \frac{\bar{b}'_q}{x_{qk}}, 0 \right]$$

If min x_k is limited by (5.13), then x_k replaces x_p in the basis; if it is limited by (5.14), then x_k replaces x_q in the basis; if it is limited by (5.15), the old basis stays the same and x_k is a nonbasic variable equal to zero.

In either case, we obtain the largest decrease in the value of the objective function by making the variable x_k equal to the value which constrains it; that is, we increase x_k as much as possible in Case 1 and decrease it as much as possible in Case 2. After an appropriate change in the variable x_k, the above procedure is repeated until the conditions of Theorem 1 hold. Since all $x_j \leq u_j$, the given problem has a finite minimum. The development of a computational tableau and related rule for the selection of the variable x_k to enter the basis is left to the reader in Exercise 26.

Other approaches to the bounded-variable problem employing dual relationships are given in Eisemann [41*c*], Gass [46*b*], and Wagner [99*d*]. Special algorithms exist for solving the capacitated-transportation problem in which each shipment $x_{ij} \leq u_{ij}$ (Ford and Fulkerson [44*a*] and Garvin [45*a*]). A more general approach to upper bounds is given by the algorithms of Dantzig and Van Slyke [32*a*]. They show that a linear-programming problem with $m + l$ rows, such that each variable has at most one nonzero coefficient in the last l equations, can be solved using a basis of order m or $2m$, depending on the algorithm. The last l equations are general sums of variables, with each variable appearing in at most one of these equations. Problems of this sort include the transportation problem, bounded-variable problems, and distribution problems.

All the above procedures can be looked at as special approaches which enable us to solve large-scale linear-programming problems. These algorithms, along with the decomposition and related techniques for handling the distinctive structures which usually arise in large-scale programming models, have added immeasurably to the utility of such models. However, as we now see the need to solve extremely large problems (of the order of many hundred thousands of constraints and millions of variables) and as this need is now coupled with the incredible computing power of today's electronic computers, the development of further advances in this area should be a prime field of research. It should be noted that the mechanics of the bounded-variable algorithm were discovered by the application of the standard simplex method to the enlarged problem.

This fact, plus similar experiences, leads us to suggest that, whenever one is faced with a new linear-programming model, one should construct numerical examples, arrange them in the simplex tableau, and solve the problems. An investigation of the computational tableaus might reveal hidden characteristics of the equations which could possibly lead to a special and more efficient computational method.

6. DIGITAL-COMPUTER CODES

The first successful solution of a linear-programming problem on a high-speed electronic digital computer occurred in January, 1952, on the National Bureau of Standards computer, the SEAC. The computational method used was the original simplex procedure, and the application was an Air Force programming problem dealing with the deployment and support of aircraft to meet stipulated requirements. Since that time, the simplex algorithm, or variations of this procedure, has been coded for practically all general-purpose electronic computers.

For past editions of this book, we gathered from the computer manufacturers information concerning linear-programming codes for their computers. Because of the great variety of computers available today, it is rather difficult to maintain such a listing. For the current status of codes for any particular computer, the reader should inquire of the manufacturer.

The reader interested in computational characteristics of the simplex method is referred to Cutler and Wolfe [16*b*]. This study summarizes the results of a set of computations involving the solution of nine linear-programming problems using 30 variations of the simplex method. The statistics collected allow a comparison of most of the variations proposed in recent years and indicate the important features in the efficiency of linear-programming codes.

EXERCISES

1. Selecting vectors $\mathbf{P}_3$, $\mathbf{P}_4$, $\mathbf{P}_5$ as an initial basis to the following equations, determine the corresponding solution and, if necessary, transform the system to a form suitable for Phase I computation:

$$\begin{aligned} x_1 - 3x_2 + 4x_3 + x_4 + x_5 &= 9 \\ -x_1 + 2x_2 - x_3 - 7x_4 + x_5 &= 2 \\ -x_1 + x_2 - 3x_3 - 2x_4 + 6x_5 &= 5 \\ x_j &\geq 0 \end{aligned}$$

2. Solve the following problem by means of the dual simplex method: Maximize

$$7x_1 + 7x_2 + 2x_3 + x_4 + 6x_5$$

subject to

$$
\begin{aligned}
3x_1 - x_2 + x_3 - 2x_4 \qquad\qquad &= -3 \\
2x_1 + x_2 \qquad + x_4 + x_5 \qquad &= 4 \\
-x_1 + 3x_2 \qquad - 3x_4 \qquad + x_6 &= 12 \\
x_j &\geq 0
\end{aligned}
$$

3. Solve the following problems in terms of integers, using the all-integer algorithm.

a. Minimize

$$x_1 + 8x_2$$

subject to

$$
\begin{aligned}
2x_1 + 3x_2 &\geq 6 \\
-3x_1 + x_2 &\leq 1 \\
x_1 + x_2 &\leq 7 \\
x_1 - 3x_2 &\leq 1 \\
x_j &\geq 0
\end{aligned}
$$

b. Minimize

$$x_1$$

subject to

$$
\begin{aligned}
3x_1 + 2x_2 &\geq 12 \\
-2x_1 + 6x_2 &\geq 12 \\
x_2 &\leq 5 \\
x_j &\geq 0
\end{aligned}
$$

c. Minimize

$$x_1 + 3x_2 + x_3$$

subject to

$$
\begin{aligned}
x_1 + 2x_2 + 3x_3 &\geq 7 \\
-2x_1 + x_2 + x_3 &\leq 13 \\
x_1 - 2x_2 + 2x_3 &\leq 5 \\
3x_1 + x_2 + 7x_3 &\geq 14 \\
x_j &\geq 0
\end{aligned}
$$

4. Solve *a* and *b* of Exercise 3 by graphing the constraints.

5. Determine the noninteger solution to the example in Sec. 3.

6. Solve the following problem using the decomposition algorithm: Minimize

$$2x_1 - 3x_2 + x_3 - 2x_4 + 6x_5 - 4x_6$$

subject to

$$\begin{aligned} x_1 + 2x_2 + 3x_3 + 4x_4 - 5x_5 - 2x_6 &= 13 \\ x_1 + 2x_2 + x_3 &= 8 \\ 3x_1 - 2x_2 + 2x_3 &= 7 \\ x_4 + x_5 - x_6 &= 3 \\ 2x_4 - x_5 + 3x_6 &= 21 \\ x_j &\geq 0 \end{aligned}$$

7. Phase I procedure for the dual-simplex algorithm (Vajda [97*a*] and Beale [5]). For the dual simplex algorithm let **B** be a basis such that $\mathbf{B}^{-1}\mathbf{b}$ has some negative components (not a feasible solution) with some associated $z_j - c_j > 0$ (the optimality conditions are not all satisfied). Let $\{\mathbf{J}\}$ be the set of indices for the nonbasic vectors for which $z_j - c_j > 0$. Add the equation $\sum_{j \epsilon \mathbf{J}} x_j + x_0 = M$, where M is a very large positive number and $x_0 \geq 0$. Let $z_k - c_k = \max_{j \epsilon \mathbf{J}} z_j - c_j$ and in the new equation, pivot x_k into the basis, which is now an $(m + 1) \times (m + 1)$ matrix. This procedure makes all the transformed $z_j - c_j \leq 0$, and the values of the basic variables are functions of M. From this point on, show how the problem can be solved in a manner similar to the Phase I procedure of the standard simplex method. What happens when x_0 goes into the basis? What if x_0 is not in the optimal feasible basis of this new problem? This procedure is the analogous dual-problem approach to the one artificial-vector technique of Sec. 1, Chap. 9.

8. Describe geometrically the phenomenon of cycling in the dual simplex method.

9. Solve the dual of Example 2 of Chap. 7.

10. Transformation of problems into integer problems (Dantzig [16*c*]).

a. The *fixed-charge problem* has as constraints $\mathbf{AX} = \mathbf{b}$, $\mathbf{X} \geq \mathbf{O}$ and the objective function $\mathbf{cX}$ to be minimized, where $c_j = d_jx_j + k_j$ if $x_j > 0$, and $c_j = 0$ if $x_j = 0$; that is, the k_j are fixed, one-time charges applied to the cost if the activity x_j is used at a positive level. Transform this problem into a mixed-integer problem, where the x_j are not restricted to be integers, but special integer variables are needed to express the fixed-charge requirements. HINT: Assume each x_j has a known, finite upper bound.

b. Show how any integer problem can be transformed to one in which all variables are restricted to be either 0 or 1.

c. Show how to express the condition that x_j can equal b_1, b_2, or b_3, using integer variables.

d. Using a variable restricted to be either 0 or 1, express the nonlinear condition $x_1x_2 = 0$ by two constraints which involve the nonnegative variables x_1 and x_2 by themselves. HINT: Assume that x_1 and x_2 have known, finite upper bounds.

11. Formulate the following *delivery problem* as an integer-programming problem (Balinski and Quandt [3*c*]). A delivery firm has three large boxes to deliver and six different trucks which can carry loads of varying sizes. Truck 1 can carry only order 1, truck 2 must carry orders 1 and 3 together, truck 3 must carry orders 1 and 2 together, truck 4 must carry all the orders at the same time, truck 5 can carry only order 2 and truck 6 can carry only order 3. Each truck can make one trip a day and all deliveries must be made on the same day. The cost of each truckload combination is c_j. Are there feasible solutions to this problem?

12. Solve the following continuous linear-programming problem geometrically (Trauth and Woolsey [94*b*]). Is the optimum solution rounded to the nearest integer a feasible solution to the problem? What is the integer optimum solution?

Maximize

$$x_1 + x_2$$

subject to

$$\begin{aligned} -9x_1 + 10x_2 &\leq 0 \\ x_1 - 3x_2 &\leq 0 \\ 23x_1 - 22x_2 &\leq 24 \\ x_j &\geq 0 \end{aligned}$$

13. *The knapsack problem* (Dantzig [16c]). Formulate the following problem as an integer linear-programming problem. A hiker has a choice of carrying a combination of n objects subject to a total weight restriction. Each object has a certain value c_j to the hiker and a weight a_j. If b is the total weight he can carry, how should he determine the combination of objects which maximizes the total value? How would you solve the continuous-variable version of this problem?

14. What assumptions are required to prove the finiteness of the decomposition algorithm?

15. Lower bound for the objective function in the decomposition algorithm (Dantzig [16c]). Denote the objective function (4.6) of the extremal problem by Z and let Z_0 be the current value of the objective function. Show that for $\mathbf{A}_0 = (0)$ a lower bound for Z is given by

$$\min Z \geq Z_0 + \sum_{p=1}^{k} (Z_{pq} - \bar{\pi}_p)$$

HINT: See Exercise 4 in Chap. 4. Multiply Eqs. (4.7) and (4.8) by π and $\bar{\pi}$, respectively, and subtract from Z.

16. Develop a decomposition algorithm for the transportation problem.

17. Discuss the decomposition of a bounded-variable problem, where the full set of bounding constraints represents one subproblem (Hadley [52d]).

18. Formulate the multicommodity problem as a block-angular system and discuss how it should be decomposed.

19. Prove that if an unbounded feasible solution exists for

$$\begin{aligned} \mathbf{BX} &= \mathbf{b} \\ \mathbf{X} &\geq \mathbf{O} \end{aligned}$$

then we can find a family of feasible solutions which satisfies the homogeneous equations

$$\begin{aligned} \mathbf{BX} &= \mathbf{O} \\ \mathbf{X} &\geq \mathbf{O} \end{aligned}$$

HINT: See Sec. 1, Chap. 4.

20. Solve the following problem using the decomposition algorithm (Harvey [53a]). Minimize

$$-6x_{01} - 4x_{02} \quad -5x_{11} - 5x_{12} \qquad\qquad -7x_{21} - 7x_{22}$$

subject to

$$
\begin{array}{|cc|cccc|cccc|r|}
\hline
x_{01} & & x_{11} & & & & +\ x_{21} & & & & = 10 \\
& x_{02} & & +\ x_{12} & & & & +\ x_{22} & & & = 20 \\
\hline
& & x_{11} & +\ x_{12} & -\ x_{13} & & & & & & = 6 \\
& & -2x_{11} & +\ x_{12} & & +\ x_{14} & & & & & = 0 \\
\hline
& & & & & & x_{21} & +\ x_{22} & +\ x_{23} & & = 25 \\
& & & & & & -2x_{21} & +\ x_{22} & & +\ x_{24} & = 0 \\
\hline
\end{array}
$$

$$x_{ij} \geq 0$$

Answer: Minimum $= -198$,

$$\mathbf{X}_0 = (0,0) \qquad \mathbf{X}_1 = 0\begin{pmatrix} 6 \\ 0 \\ 0 \\ 12 \end{pmatrix} + 1\begin{pmatrix} 2 \\ 4 \\ 0 \\ 0 \end{pmatrix} = \begin{pmatrix} 2 \\ 4 \\ 0 \\ 0 \end{pmatrix}$$

$$\mathbf{X}_2 = \tfrac{1}{25}\begin{pmatrix} 0 \\ 0 \\ 25 \\ 0 \end{pmatrix} + \tfrac{24}{25}\begin{pmatrix} 25/3 \\ 50/3 \\ 0 \\ 0 \end{pmatrix} = \begin{pmatrix} 8 \\ 16 \\ 1 \\ 0 \end{pmatrix}$$

21. Solve the following problem by the direct simplex method and the decomposition algorithm (Müller-Merbach [77*b*]).

Minimize

$$-8x_{11} - 3x_{12} - 8x_{21} - 6x_{22}$$

subject to

$$
\begin{aligned}
4x_{11} + 3x_{12} + x_{21} + 3x_{22} &\leq 16 \\
4x_{11} - x_{12} + 3x_{21} \qquad &\leq 12 \\
x_{11} + 2x_{12} \qquad &\leq 8 \\
3x_{11} + x_{12} \qquad &\leq 10 \\
2x_{21} + 3x_{22} &\leq 9 \\
4x_{21} + x_{22} &\leq 12 \\
x_{ij} &\geq 0
\end{aligned}
$$

Answer: Minimum $= -{}^{877}/_{20}$,

$$x_{11} = {}^{107}/_{80} \qquad x_{12} = {}^{29}/_{20} \qquad x_{21} = {}^{27}/_{10} \qquad x_{22} = {}^{6}/_{5}$$

22. For problem (4.4), compute the numerical values of the $\mathbf{X}_{ij}$. Determine the values of the $\mathbf{P}_{ij}$ and f_{ij} and solve the related extremal problem (4.6*a*) to (*c*) by the standard simplex method; then solve problem (4.4) by the decomposition algorithm. Let $c_{11} = -2$, $c_{12} = 4$.

23. Solve the example (4.15) by the regular simplex method and the revised simplex method.

24. Develop an artificial-basis procedure for determining a first feasible solution to the extremal problem. HINT: Use positive and negative unit vectors which correspond, respectively, to the positive and negative elements of $\mathbf{b}_0$. Each vector has a nonnegative artificial variable. Employ an additional constraint which forces the sum of the artificial variables to zero if a feasible solution exists; i.e., Phase I corresponds to minimizing a variable which represents the sum of the artificial variables (Dantzig [16c]).

25. Solve the following problem by the decomposition algorithm using the first constraint for the extremal problem and the last two constraints as a single subproblem.

Maximize

$$x_1 + 2x_2$$

subject to

$$\begin{aligned} x_1 + x_2 &\geq 3 \\ -2x_1 + x_2 &\leq 2 \\ x_1 - x_2 &\leq 1 \\ x_j &\geq 0 \end{aligned}$$

26. For the bounded-variable problem, develop a rule for selection of the variable x_k to enter the basis and a computational tableau.

27. Solve the following problems by the bounded-variable algorithm.

a. Maximize

$$x_1 + x_2$$

subject to

$$\begin{aligned} 2x_1 + x_2 &\leq 18 \\ x_1 - 2x_2 &\leq 6 \\ -3x_1 + 6x_2 &\leq 9 \\ x_1 &\leq 8 \\ x_2 &\leq 3 \\ x_j &\geq 0 \end{aligned}$$

b. Maximize $x_1 - x_2$ subject to the constraints in *a*.

28. Make the necessary changes to the bounded-variable algorithm if some x_j are not bounded above.

29. Modify the bounded-variable algorithm and the transportation-problem algorithm to solve the capacitated-transportation problem.

PART THREE / **APPLICATIONS**

CHAPTER 10 / THE TRANSPORTATION PROBLEM

One of the earliest and most fruitful applications of linear-programming techniques has been the formulation and solution of the transportation problem as a linear-programming problem. The basic transportation problem was originally stated by Hitchcock [57] and later discussed in detail by Koopmans [66]. The linear-programming formulation and the associated systematic method of solution were first given by Dantzig [18]. The computational procedure is an adaptation of the simplex method applied to the system of equations of the associated linear-programming problem. We shall discuss but not develop this special procedure for solving the transportation problem. The reader is referred to Dantzig [18] for the complete details of the method. An alternate procedure for solving this problem is given in Ford and Fulkerson [43].

As the transportation problem is of great interest, we have included a number of basic theorems concerning the equations of its corresponding mathematical system. So as not to digress from the main task of formulating and solving the problem, the proof of a few theorems has been left to the reader. As preliminary reading it is suggested that the reader refer to the applications described in Sec. 2 of Chap. 1.

1. THE GENERAL TRANSPORTATION PROBLEM

A homogeneous product is to be shipped in the amounts $a_1, a_2, \ldots, a_m$, respectively, from each of m *shipping origins* and received in amounts $b_1, b_2, \ldots, b_n$, respectively, by each of n *shipping destinations*. The cost of shipping a unit amount from the ith origin to the jth destination is c_{ij} and is known for all combinations (i,j). The problem is to determine the amounts x_{ij} to be shipped over all routes (i,j) so as to minimize the total cost of transportation.

To develop the constraints of the problem, we set up Tableau 10.1. The amount shipped from origin i to destination j is x_{ij}; the total shipped

Tableau 10.1

Origins \ Destinations (i) \ (j)	(1)	(2)	...	(j)	...	(n)	
(1)	x_{11}	x_{12}	...	x_{1j}	...	x_{1n}	a_1
(2)	x_{21}	x_{22}	...	x_{2j}	...	x_{2n}	a_2
.	.	.	...	.	...	.	.
.	.	.	...	.	...	.	.
(i)	x_{i1}	x_{i2}	...	x_{ij}	...	x_{in}	a_i
.	.	.	...	.	...	.	.
.	.	.	...	.	...	.	.
(m)	x_{m1}	x_{m2}	...	x_{mj}	...	x_{mn}	a_m
	b_1	b_2	...	b_j	...	b_n	

from origin i is $a_i \geq 0$, and the total received by destination j is $b_j \geq 0$. Here we temporarily impose the restriction that the total amount shipped is equal to the total amount received; that is, $\sum_i a_i = \sum_j b_j = A$. The total cost of shipping x_{ij} units is $c_{ij}x_{ij}$. Since a negative shipment has no valid interpretation for the problem as stated, we restrict each $x_{ij} \geq 0$. From the tableau we have the mathematical statement of the transportation problem: Find values for the variables x_{ij} which minimize the total cost

$$\sum_{i=1}^{m} \sum_{j=1}^{n} c_{ij}x_{ij} \tag{1.1}$$

subject to the constraints

$$\sum_{j=1}^{n} x_{ij} = a_i \qquad i = 1, 2, \ldots, m \tag{1.2}$$

$$\sum_{i=1}^{m} x_{ij} = b_j \qquad j = 1, 2, \ldots, n \tag{1.3}$$

and

$$x_{ij} \geq 0 \tag{1.4}$$

Equations (1.2) represent the row sums of Tableau 10.1 and (1.3) the column sums. In order for Eqs. (1.2) and (1.3) to be consistent, we must have the sum of Eqs. (1.2) equal to the sum of Eqs. (1.3); that is,

$$\sum_{i=1}^{m} \sum_{j=1}^{n} x_{ij} = \sum_{j=1}^{n} \sum_{i=1}^{m} x_{ij} = \sum_i a_i = \sum_j b_j = A$$

We note that the system of Eqs. (1.1) to (1.4) is a linear-programming problem with $m + n$ equations in mn variables.

Theorem 1. *The transportation problem has a feasible solution.*

Proof. Since $\sum_{i=1}^{m} a_i = \sum_{j=1}^{n} b_j = A$, we have the feasible solution $x_{ij} = a_i b_j / A$ for all (i,j). Each $x_{ij} \geq 0$, and (1.2) is satisfied, since

$$\sum_{j=1}^{n} x_{ij} = \sum_{j=1}^{n} \frac{a_i b_j}{A} = \frac{a_i \sum_{j=1}^{n} b_j}{A} = a_i$$

and (1.3) is satisfied, since

$$\sum_{i=1}^{m} x_{ij} = \sum_{i=1}^{m} \frac{a_i b_j}{A} = \frac{b_j \sum_{i=1}^{m} a_i}{A} = b_j$$

For $m = 3$ and $n = 5$, let us write out the equations that correspond to (1.2) and (1.3). We then obtain the following 8 (that is, $m + n$) equations in 15 (that is, mn) unknowns.

$$\begin{array}{ll}
(a) & x_{11} + x_{12} + x_{13} + x_{14} + x_{15} = a_1 \\
(b) & x_{21} + x_{22} + x_{23} + x_{24} + x_{25} = a_2 \\
(c) & x_{31} + x_{32} + x_{33} + x_{34} + x_{35} = a_3 \\
(d) & x_{11} + x_{21} + x_{31} = b_1 \\
(e) & x_{12} + x_{22} + x_{32} = b_2 \\
(f) & x_{13} + x_{23} + x_{33} = b_3 \\
(g) & x_{14} + x_{24} + x_{34} = b_4 \\
(h) & x_{15} + x_{25} + x_{35} = b_5
\end{array}$$

If we sum Eqs. (d) to (h) and subtract Eqs. (b) and (c) from this sum, the result is Eq. (a). Hence Eq. (a) is redundant and does not need to be included in the system. Generalizing, we note that one equation from the system (1.2) or (1.3) can be eliminated, and the transportation problem reduces to $m + n - 1$ independent equations in mn variables. For the preceding equations we have the following matrix for the reduced system [here we have eliminated the first equations of (1.2)]:

$$\mathbf{A} = \begin{array}{c} \begin{array}{ccccccccccccccc} \mathbf{P}_{11} & \mathbf{P}_{12} & \mathbf{P}_{13} & \mathbf{P}_{14} & \mathbf{P}_{15} & \mathbf{P}_{21} & \mathbf{P}_{22} & \mathbf{P}_{23} & \mathbf{P}_{24} & \mathbf{P}_{25} & \mathbf{P}_{31} & \mathbf{P}_{32} & \mathbf{P}_{33} & \mathbf{P}_{34} & \mathbf{P}_{35} \end{array} \\ \begin{pmatrix} 0 & 0 & 0 & 0 & 0 & 1 & 1 & 1 & 1 & 1 & 0 & 0 & 0 & 0 & 0 \\ 0 & 0 & 0 & 0 & 0 & 0 & 0 & 0 & 0 & 0 & 1 & 1 & 1 & 1 & 1 \\ 1 & 0 & 0 & 0 & 0 & 1 & 0 & 0 & 0 & 0 & 1 & 0 & 0 & 0 & 0 \\ 0 & 1 & 0 & 0 & 0 & 0 & 1 & 0 & 0 & 0 & 0 & 1 & 0 & 0 & 0 \\ 0 & 0 & 1 & 0 & 0 & 0 & 0 & 1 & 0 & 0 & 0 & 0 & 1 & 0 & 0 \\ 0 & 0 & 0 & 1 & 0 & 0 & 0 & 0 & 1 & 0 & 0 & 0 & 0 & 1 & 0 \\ 0 & 0 & 0 & 0 & 1 & 0 & 0 & 0 & 0 & 1 & 0 & 0 & 0 & 0 & 1 \end{pmatrix} \end{array} \qquad (1.5)$$

and

$$\mathbf{P}_0 = \begin{pmatrix} a_2 \\ a_3 \\ b_1 \\ b_2 \\ b_3 \\ b_4 \\ b_5 \end{pmatrix}$$

The vector $\mathbf{P}_{ij}$ corresponds to the variable x_{ij}. Letting

$$\mathbf{X} = (x_{11}, \ldots, x_{ij}, \ldots, x_{mn})$$

be a column vector with mn variables and $\mathbf{c} = (c_{11}, \ldots, c_{ij}, \ldots, c_{mn})$ be a row vector of cost coefficients, the reduced transportation problem can be represented by

$$\begin{aligned} \mathbf{AX} &= \mathbf{P}_0 \\ \mathbf{X} &\geq 0 \\ \mathbf{cX} &\quad \text{a minimum} \end{aligned} \tag{1.6}$$

From the discussion in Chaps. 3 and 4, we have that the minimum feasible solution to (1.6) requires at most $m + n - 1$ positive x_{ij}'s.

Theorem 2. *Construction of a basic feasible solution.*[1] *A solution of at most $m + n - 1$ positive x_{ij}'s exists.*

To demonstrate this theorem, we shall use the procedure for obtaining a first basic feasible solution proposed by Dantzig [18] and termed the "northwest-corner rule" by Charnes and Cooper [10]. We shall apply this scheme to the following 3×4 tableau:

x_{11}	x_{12}	x_{13}	x_{14}	a_1
x_{21}	x_{22}	x_{23}	x_{24}	a_2
x_{31}	x_{32}	x_{33}	x_{34}	a_3
b_1	b_2	b_3	b_4	

[1] In the paper by Demuth [32*d*] (translated by Doig [37*a*]), it is shown that if N is the minimum number of basic feasible solutions to a transportation problem with m origins and n destinations ($m \leq n$), then N is bounded as follows:

$$n^{m-1} \leq N \leq n^{m-1}m^{n-1} \qquad \text{(Nondegenerate problem)}$$

$$\frac{n!}{(n - m + 1)!} \leq N \leq n^{m-1}m^{n-1} \qquad \text{(Degenerate problem)}$$

The upper bounds are due to Simonnard and Hadley [90*c*].

We first determine a value for the northwest-corner variable x_{11}. We let $x_{11} = \min (a_1,b_1)$; and if $a_1 \leq b_1$, then $x_{11} = a_1$, and all $x_{1j} = 0$ for $j = 2$, 3, 4. If $a_1 \geq b_1$, then $x_{11} = b_1$, and all $x_{i1} = 0$ for $i = 2, 3$. For discussion purposes, assume that the former is true; then this initial step transforms the tableau as shown in Step 1 below. Here the total left to be shipped from origin 1 has been reduced to 0, and the total left to be shipped to destination 1 is $b_1 - a_1$.

Step 1. Assume $b_1 > a_1$.

$x_{11} = a_1$	0	0	0	0
x_{21}	x_{22}	x_{23}	x_{24}	a_2
x_{31}	x_{32}	x_{33}	x_{34}	a_3
$b_1 - a_1$	b_2	b_3	b_4	

We next determine a value for the first variable in row 2. We let $x_{21} = \min (a_2, b_1 - a_1)$. If we assume $a_2 > b_1 - a_1$, we have $x_{21} = b_1 - a_1$ and $x_{31} = 0$. This is shown in Step 2. The total left to be shipped from origin 2 is now $a_2 - b_1 + a_1$, and the total to be shipped to destination 1 is 0.

Step 2. Assume $a_2 > b_1 - a_1$.

$x_{11} = a_1$	0	0	0	0
$x_{21} = b_1 - a_1$	x_{22}	x_{23}	x_{24}	$a_2 - b_1 + a_1$
0	x_{32}	x_{33}	x_{34}	a_3
0	b_2	b_3	b_4	

Similarly, for the following steps, we determine a value of a variable x_{ij} and reduce to zero either the amount to be shipped from i or the amount to be shipped to j, or both.

Step 3. Assume $a_2 - b_1 + a_1 > b_2$.

$x_{11} = a_1$	0	0	0	0
$x_{21} = b_1 - a_1$	$x_{22} = b_2$	x_{23}	x_{24}	$a_2 - b_1 + a_1 - b_2$
0	0	x_{33}	x_{34}	a_3
0	0	b_3	b_4	

Step 4. Assume $a_2 - b_1 + a_1 - b_2 < b_3$.

$x_{11} = a_1$	0	0	0	0
$x_{21} = b_1 - a_1$	$x_{22} = b_2$	$x_{23} = a_2 - b_1 + a_1 - b_2$	0	0
0	0	x_{33}	x_{34}	a_3
0	0	$b_3 - a_2 + b_1 - a_1 + b_2$	b_4	

From Step 4, we see that $x_{33} = b_3 - a_2 + b_1 - a_1 + b_2$ and $x_{34} = b_4$. It should be noted that the values for the x_{ij} were all obtained by adding and subtracting various combinations of the a_i and b_j. Hence, if the a_i and b_j were originally nonnegative integers, then the feasible solution obtained by the above procedure would also consist of nonnegative integers. Since the northwest-corner rule of determining each x_{ij} eliminates either a row or a column from further consideration, while the last allocation eliminates both a row and a column, this feasible solution can have at most $m + n - 1$ positive x_{ij}'s. Depending on the values of the a_i and b_j and on the assumptions made in obtaining the feasible solution to the above example, we have as the six variables with possible positive values

$$x_{11} = a_1 \qquad x_{21} = b_1 - a_1$$
$$x_{22} = b_2 \qquad x_{23} = a_2 - b_1 + a_1 - b_2$$
$$x_{33} = b_3 - a_2 + b_1 - a_1 + b_2 \qquad x_{34} = b_4$$

We next exhibit two numerical examples and their northwest-corner-rule basic feasible solutions.

Example 1

2	0	0	0	0	2
1	2	1	0	0	4
0	0	3	2	2	7
3	2	4	2	2	

Example 2

1	2	0	0	0	3
0	1	3	0	0	4
0	0	0	2	5	7
1	3	3	2	5	

In Example 2, only $m + n - 2 = 6$ of the x_{ij}'s are positive, and we have determined a degenerate basic feasible solution. This will happen whenever the amount left to be shipped from origin i is exactly equal to the amount to be shipped to destination j, except when both $i = m$ and $j = n$. In Example 2 this is true for $x_{23} = \min (a_2 - x_{22}, b_3) = (3,3) = 3$.

Associated with each such solution is a set of $m + n - 1$ linearly independent vectors. For Example 1, the vectors associated with the positive x_{ij} are $\mathbf{P}_{11}$, $\mathbf{P}_{21}$, $\mathbf{P}_{22}$, $\mathbf{P}_{23}$, $\mathbf{P}_{33}$, $\mathbf{P}_{34}$, $\mathbf{P}_{35}$. Let the basis formed by these vectors be denoted by $\mathbf{B}$. Then the solution to the system $\mathbf{BX} = \mathbf{P}_0$ must yield the solution obtained by the northwest-corner rule. Here $\mathbf{X} = (x_{11},x_{21},x_{22},x_{23},x_{33},x_{34},x_{35})$ is a column vector. We can similarly select an associated set of $m + n - 1$ linearly independent vectors for the degenerate solution of Example 2. This will be discussed below. The solutions obtained by the northwest-corner rule (and similar schemes) are extreme-point solutions, and only such solutions need be considered as candidates for the minimum feasible solution.

Since virtually all applications of the transportation problem require the shipping of only whole units of the item being considered, it is useful to establish the following important property of the transportation problem:

Theorem 3. *Assuming the a_i and b_i are nonnegative integers, then every basic feasible solution (i.e., extreme-point solution) has integral values.*

We prove Theorem 3 with the aid of the following statement:

Lemma 1. *Every set of $m + n - 1$ linearly independent vectors of the reduced-transportation-problem system of equations can be arranged into a triangular matrix.*

To illustrate this point, let us rearrange the basis of Example 1 into a triangular matrix. The original system given by $\mathbf{BX} = \mathbf{P}_0$ is

$$\begin{array}{c} \\ (a)\\(b)\\(c)\\(d)\\(e)\\(f)\\(g)\end{array}
\begin{array}{c}\begin{array}{ccccccc}\mathbf{P}_{11}&\mathbf{P}_{21}&\mathbf{P}_{22}&\mathbf{P}_{23}&\mathbf{P}_{33}&\mathbf{P}_{34}&\mathbf{P}_{35}\end{array}\\
\begin{pmatrix}0&1&1&1&0&0&0\\0&0&0&0&1&1&1\\1&1&0&0&0&0&0\\0&0&1&0&0&0&0\\0&0&0&1&1&0&0\\0&0&0&0&0&1&0\\0&0&0&0&0&0&1\end{pmatrix}\end{array}
\begin{array}{c}\mathbf{X}\\\begin{pmatrix}x_{11}\\x_{21}\\x_{22}\\x_{23}\\x_{33}\\x_{34}\\x_{35}\end{pmatrix}\end{array}
=
\begin{array}{c}\mathbf{P}_0\\\begin{pmatrix}4\\7\\3\\2\\4\\2\\2\end{pmatrix}\end{array}$$

To rearrange this $\mathbf{B}$, we need only interchange the rows of $\mathbf{B}$ and corresponding elements of $\mathbf{P}_0$, as shown below. Since the rearrangement of $\mathbf{B}$ does not require any column interchanges, the elements of vector $\mathbf{X}$ are not permuted. The transformed system is

$$\begin{array}{c} \\ (c)\\(a)\\(d)\\(e)\\(b)\\(f)\\(g)\end{array}
\begin{array}{c}\begin{array}{ccccccc}\mathbf{P}_{11}&\mathbf{P}_{21}&\mathbf{P}_{22}&\mathbf{P}_{23}&\mathbf{P}_{33}&\mathbf{P}_{34}&\mathbf{P}_{35}\end{array}\\
\begin{pmatrix}1&1&0&0&0&0&0\\0&1&1&1&0&0&0\\0&0&1&0&0&0&0\\0&0&0&1&1&0&0\\0&0&0&0&1&1&1\\0&0&0&0&0&1&0\\0&0&0&0&0&0&1\end{pmatrix}\end{array}
\begin{array}{c}\mathbf{X}\\\begin{pmatrix}x_{11}\\x_{21}\\x_{22}\\x_{23}\\x_{33}\\x_{34}\\x_{35}\end{pmatrix}\end{array}
=
\begin{array}{c}\mathbf{P}_0\\\begin{pmatrix}3\\4\\2\\4\\7\\2\\2\end{pmatrix}\end{array}$$

From the system of equations determined by the above triangular system, we see immediately that $x_{35} = 2$ and $x_{34} = 2$. Then by substituting these values in Eq. (*b*), we have $x_{33} = 3$. In a similar manner we solve in order Eqs. (*e*), (*d*), (*a*), and (*c*) to obtain $x_{23} = 1$, $x_{22} = 2$, $x_{21} = 1$, and $x_{11} = 2$. Since the matrix associated with a basic feasible solution is triangular and consists of elements that are either 1 or 0 and since the solution of a set of equations determined by this type of triangular matrix requires only additions and subtractions, the values of the variables will be nonnegative integers.

Theorem 4. *A finite minimum feasible solution always exists.*

Proof. By Theorem 1 the problem is feasible. Since the coefficients of Eqs. (1.2) and (1.3) are nonnegative and since we assume that all a_i and b_j are nonnegative and finite, no x_{ij} can be made arbitrarily large. In fact, the x_{ij} cannot be greater than the corresponding a_i or b_j. (We may note that, for any basic feasible solution, at least one x_{ij} equals its corresponding a_i or b_j.) If the a_i and b_j are integral, then a basic minimum feasible solution has finite integral values.

The reduced system of $m + n - 1$ equations in mn variables can, of course, be solved by the general simplex procedure. However, for even small values of m and n, the resulting system of equations becomes unwieldy for manual computation. Such systems could possibly be too large to be solved efficiently on computing machinery. This dilemma is resolved by the special adaptation of the simplex algorithm to the transportation problem. This iterative procedure requires the knowledge of exactly $m + n - 1$ nonnegative variables that are associated with $m + n - 1$ linearly independent vectors. We know that the vectors that are associated with the variables of a basic feasible solution (i.e., an extreme-point solution) are linearly independent. If the basic solution is nondegenerate, it satisfies the requirement of the procedure. This is not true if the solution is degenerate, as is the solution for Example 2. If a degenerate solution has $k < m + n - 1$ positive variables, then we have to select $m + n - 1 - k$ zero variables to be in the solution. The $m + n - 1 - k$ vectors associated with these zero variables plus the k vectors associated with the positive variables must be linearly independent. This selection is accomplished by the following "ϵ perturbation" method for the transportation problem (Dantzig [18]).

Let us refer to the 3×4 tableaus used to obtain a first feasible solution by the northwest-corner rule. In Step 1, if $x_{21} = b_1 - a_1 = a_2$, then neither x_{31} nor x_{22} could be positive. Or in Step 2, if

$$x_{22} = b_2 = a_2 - b_1 + a_1$$

then neither x_{32} nor x_{23} could be positive. Whenever situations of this sort arise, the number of positive variables in the basic solution is reduced by one. These degenerate cases arise when, in evaluating some x_{ij}, its two possible values are equal. We have then encountered a partial sum of the a_i and b_j which is equal to some a_i or b_j. In Example 2 we have a degenerate solution occurring because $x_{23} = \min (b_3, a_2 - b_2 + a_1 - b_1)$ and $b_3 = a_2 - b_2 + a_1 - b_1 = 3$.

To avoid these degenerate situations, we have to make sure that no partial sums of the a_i and b_j can be equal to some a_i or b_j. This is done by "perturbing" (i.e., modifying) the values of the a_i and b_j. We set up a new problem where

$$\bar{a}_i = a_i + \epsilon \qquad i = 1, 2, \ldots, m$$

and

$$\bar{b}_j = b_j \qquad j = 1, 2, \ldots, n - 1$$

$$\bar{b}_n = b_n + m\epsilon$$

for $\epsilon > 0$.

We select an ϵ small enough that the final solution rounded to the same number of significant digits as the original a_i and b_j will yield the correct solution. This is possible since the computational procedure involves only additions and subtractions. Orden [84] proves that any $\epsilon < \delta/2m$ will do, where δ is a 1 in the last significant place in the quantities a_i and b_j. In practice, ϵ is taken to be 1×10^{-d}, where d is determined by $\delta/2m$. (For machine computation, it is easier always to add a 1 to each a_i and to add m to b_n in the last significant place that can be carried in the computer.)

Applying the ϵ procedure to Example 2, we have the solution

1	$2 + \epsilon$	0	0	0	$3 + \epsilon$
0	$1 - \epsilon$	3	2ϵ	0	$4 + \epsilon$
0	0	0	$2 - 2\epsilon$	$5 + 3\epsilon$	$7 + \epsilon$
1	3	3	2	$5 + 3\epsilon$	

The basic solution now contains $x_{24} = 2\epsilon > 0$. We see that the ϵ determines which of two variables with real values of zero should be brought into the solution. Here we had a choice of x_{24} or x_{33}. The ϵ procedure does away with any such choice and enables the computation to procede without any degenerate solutions. Theoretically, the selection of either x_{24} or x_{33} would have yielded a basis of $m + n - 1$ vectors. As will be seen, the computational procedure requires that we need only note which of these zero variables is selected to be in the basic solution. We can then dispense with the ϵ procedure and select either one of the two zero variables. It seems best to select the one with the smallest c_{ij}. No transportation problem has ever been known to cycle.

2. COMPUTATIONAL PROCEDURE FOR SOLVING THE TRANSPORTATION PROBLEM

In [18], Dantzig shows that, for any basic feasible solution, numbers u_i and v_j can be found such that, for those x_{ij} in the basic solution, we have $u_i + v_j = c_{ij}$. Dantzig also shows that, if $u_i + v_j = \bar{c}_{ij}$ for those variables not in the basic solution and if all $\bar{c}_{ij} - c_{ij} \leq 0$, then the basic feasible solution is also a minimum solution. If this condition of optimality is not satisfied, then we can very readily obtain a new basic feasible solution whose corresponding value of the objective function is less than (nondegeneracy assumed) the preceding value. Dantzig's ingenious computational procedure enables us to obtain basic feasible solutions without setting up the usual simplex tableau and to test for optimality, i.e., to compute the $z_{ij} = \bar{c}_{ij}$, without the explicit representation of the vectors not in the basis in terms of the basis vectors. (See Henderson and Schlaifer [55] and Charnes and Cooper [10] for nontechnical discussions of the computational procedure.) We shall illustrate this procedure with a sample 3×4 problem.

In the problem we have three origins, with availabilities of $a_1 = 6$, $a_2 = 8$, and $a_3 = 10$, and four destinations, with requirements of $b_1 = 4$, $b_2 = 6$, $b_3 = 8$, and $b_4 = 6$. We note that $\sum_i a_i = \sum_j b_j = 24$. The costs between each origin and destination are given in the following cost matrix:

Destinations

	1	2	3	4
Origins 1	1	2	3	4
Origins 2	4	3	2	0
Origins 3	0	2	2	1

$= (c_{ij})$

For example, the cost for shipping one unit between origin 3 and destination 2 is $c_{32} = 2$. In general, the c_{ij} can be any positive or negative numbers. Using the northwest-corner rule, we obtain the first feasible solution:

$(x_{ij}) =$

4	2			6
	4	4		8
		4	6	10
4	6	8	6	

where $x_{11} = 4$, $x_{12} = 2$, $x_{22} = 4$, $x_{23} = 4$, $x_{33} = 4$, $x_{34} = 6$, and all the other $x_{ij} = 0$. The value of the objective function is 42. Since this is a nondegenerate basic feasible solution, the first solution does not require our perturbing the problem. As we are doing a simple hand computation, it is easier to resolve degenerate situations by not employing the technique at all. Instead, when a degenerate solution occurs, we shall select, from the two variables that can be in the solution with a value of zero, the one having the smaller c_{ij}. The example will illustrate this point. We shall always keep track of a solution with exactly $m + n - 1$ nonnegative variables. If the problem were to be solved on an electronic computer, it would be more efficient to start the computation with the perturbed problem.

We are next required to determine, for those variables in the basic solution, m numbers u_i and n numbers v_j such that

$$\begin{aligned} u_1 + v_1 &= c_{11} = 1 \\ u_1 + v_2 &= c_{12} = 2 \\ u_2 + v_2 &= c_{22} = 3 \\ u_2 + v_3 &= c_{23} = 2 \\ u_3 + v_3 &= c_{33} = 2 \\ u_3 + v_4 &= c_{34} = 1 \end{aligned} \tag{2.1}$$

Here we have seven variables in six (that is, $m + n - 1$) equations. Since (2.1) is an underdetermined set of linear equations (i.e., the number of unknowns exceeds the number of equations), this system has an infinite number of solutions. We determine a solution by arbitrarily letting any one of the variables equal its corresponding c_{ij}. This reduces the number of unknowns by one and forces a unique solution of the $m + n - 1$ equations in the remaining $m + n - 1$ variables.

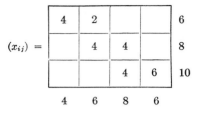

In (2.1) let $u_1 = 1$; then it is an easy matter to solve for the remaining u_i and v_j. We see that, with $u_1 = 1$, then $v_1 = 0$ and $v_2 = 1$; with $v_2 = 1$ then $u_2 = 2$; with $u_2 = 2$ then $v_3 = 0$; with $v_3 = 0$ then $u_3 = 2$; and with $u_3 = 2$ then $v_4 = -1$. This computation can be readily accomplished by setting up the following table containing the cost coefficients (in bold type) of the variables in the basic solution:

u \ v				
	1	**2**		
		3	**2**	
			2	**1**

By letting $u_1 = 1$ and $v_1 = 0$, we can compute, as was done above, the resulting u_i and v_j and enter them in the corresponding positions as follows:

u \ v	0	1	0	−1
1	**1**	**2**		
2		**3**	**2**	
2			**2**	**1**

Since all the equations of (2.1) are satisfied, we have $u_i + v_j = c_{ij}$ for those x_{ij} in the basic feasible solution. We then compute $\bar{c}_{ij} = u_i + v_j$ for all combinations (i,j) and place these figures in their corresponding cells of the *indirect-cost* table. ($\bar{c}_{ij} = c_{ij}$ for all x_{ij} in the solution.) The indirect-cost table ($\bar{c}_{ij}$) for the first solution is as follows:[1]

[1] A concise way of combining the allocation table and the indirect-cost table is to form the following tableau (shown for a 3 × 4 problem):

u_i \ v_j	v_1	v_2	v_3	v_4	
u_1	c_{11} (x_{11})	c_{12} (x_{12})	c_{13} $\bar{c}_{13} - c_{13}$	c_{14} $\bar{c}_{14} - c_{14}$	a_1
u_2	c_{21} $\bar{c}_{21} - c_{21}$	c_{22} (x_{22})	c_{23} (x_{23})	c_{24} $\bar{c}_{24} - c_{24}$	a_2
u_3	c_{31} $\bar{c}_{31} - c_{31}$	c_{32} $\bar{c}_{32} - c_{32}$	c_{33} (x_{33})	c_{34} (x_{34})	a_3
	b_1	b_2	b_3	b_4	A

$(\bar{c}_{ij}) =$

1	**2**	1	0
2	**3**	**2**	1
2	3	**2**	**1**

For example, $\bar{c}_{14} = u_1 + v_4 = 0$. As will be shown below, the above three boxes actually can be combined into one efficient computational step. We next compute the differences $\bar{c}_{ij} - c_{ij}$. If all $\bar{c}_{ij} - c_{ij} \leq 0$, then the solution that yielded the indirect-cost table is a minimum feasible solution. If at least one $\bar{c}_{ij} - c_{ij} > 0$, we have not found a minimum solution. We can, as will be described below, readily obtain a new basic feasible solution which contains a variable associated with a $\bar{c}_{ij} - c_{ij} > 0$. As in the general simplex procedure, we select for entry into the new basis the variable corresponding to the max $(\bar{c}_{ij} - c_{ij} > 0)$. This scheme will yield a new basic solution whose value of the objective function will be less than the value for the preceding solution (or possibly equal to it, if the preceding solution is degenerate).

For our first solution, we find that max $(\bar{c}_{ij} - c_{ij}) = \bar{c}_{31} - c_{31} = 2$. Hence we select x_{31} to be introduced into the solution. If there were ties, we would select the one with the smaller c_{ij}. Referring to the first solution matrix (x_{ij}), we first introduce x_{31} into the solution at an unknown nonnegative level θ_1. As the row and column sums of the variables must equal the corresponding values of the a_i and b_j, we must add or subtract θ_1 from some of the other x_{ij} in the first solution, as follows:

$4 - \theta_1$	$2 + \theta_1$			6
	$4 - \theta_1$	$4 + \theta_1$		8
θ_1		$4 - \theta_1$	6	10
4	6	8	6	

Since we put $\theta_1 \geq 0$ in cell (3,1), we must subtract θ_1 from x_{11}, x_{22}, and x_{33} and add θ_1 to x_{12} and x_{23}, in order to keep the row and column sums correct. We see that the size of θ_1 is restricted by those x_{ij} from which it is subtracted. θ_1 cannot be larger than the smallest x_{ij} from which it is subtracted. Here θ_1 must be less than or equal to 4 and must be greater than zero in order to preserve feasibility. Since we wish to eliminate one of the variables from the old solution and to introduce x_{31}, we let $\theta_1 = 4$. However, since $x_{11} - \theta_1 = x_{22} - \theta_1 = x_{33} - \theta_1 = 4 - \theta_1$, we shall eliminate three variables and obtain a degenerate solution with four positive variables. To keep a solution with exactly $m + n - 1$ nonnegative variables, we retain two of these three variables with values of zero. We select x_{11} and x_{33}, because they correspond to the smaller c_{ij}. The new solution is shown below:

$(x_{ij}) =$

0	6			6
		8		8
4		0	6	10
4	6	8	6	

The objective function for this solution is equal to

$$42 - [\max (\bar{c}_{ij} - c_{ij} > 0)]\theta_1 = 42 - (2)(4) = 34$$

The corresponding combined (u_i,v_j) table and $\bar{c}_{ij}$ matrix for this solution is given in the following table:

$(\bar{c}_{ij}) =$

$u \backslash v$	0	1	2	1
1	**1**	**2**	3	2
0	0	1	**2**	1
0	**0**	1	**2**	**1**

where the bold numbers correspond to the c_{ij} for the basis vectors. We see that $\max (\bar{c}_{ij} - c_{ij} > 0) = \bar{c}_{24} - c_{24} = 1$. Introducing $\theta_2 > 0$ in cell (2,4), we have to add and subtract θ_2 from the x_{ij} as follows:

0	6			6
		$8 - \theta_2$	θ_2	8
4		$0 + \theta_2$	$6 - \theta_2$	10
4	6	8	6	

We have $\theta_2 = 6$; x_{34} is eliminated, and x_{24} is introduced with a value of $\theta_2 = 6$. The new value of the objective function is

$$34 - (\bar{c}_{24} - c_{24})\theta_2 = 34 - (1)(6) = 28$$

The new solution is

$$(x_{ij}) = \begin{array}{|c|c|c|c|}\hline 0 & 6 & & \\\hline & & 2 & 6 \\\hline 4 & & 6 & \\\hline\end{array} \begin{array}{c} 6 \\ 8 \\ 10 \end{array}$$
$$\qquad\quad 4\quad 6\quad 8\quad 6$$

The corresponding (u_i, v_j) table and $(\bar{c}_{ij})$ matrix is given by

$$(\bar{c}_{ij}) = \begin{array}{c|c|c|c|c}\diagdown v & 0 & 1 & 2 & 0 \\ u \diagdown & & & & \\\hline \begin{array}{c}1\\0\\0\end{array} & \begin{array}{c}\mathbf{1}\\0\\0\end{array} & \begin{array}{c}\mathbf{2}\\1\\1\end{array} & \begin{array}{c}3\\\mathbf{2}\\\mathbf{2}\end{array} & \begin{array}{c}1\\0\\0\end{array}\end{array}$$

Here, all $\bar{c}_{ij} - c_{ij} \le 0$, and this last solution ($x_{11} = 0$, $x_{12} = 6$, $x_{23} = 2$, $x_{24} = 6$, $x_{31} = 4$, $x_{33} = 6$, and all other $x_{ij} = 0$) is a degenerate minimum feasible solution. The value of the objective function is 28. We note that $\bar{c}_{13} - c_{13} = 0$ and that x_{13} is not in the solution. We can then introduce x_{13} into this last solution and obtain an alternate minimum solution. We put $\theta_3 \ge 0$ into cell (1,3) and obtain

$$\begin{array}{|c|c|c|c|}\hline 0 - \theta_3 & 6 & \theta_3 & \\\hline & & 2 & 6 \\\hline 4 + \theta_3 & & 6 - \theta_3 & \\\hline\end{array} \begin{array}{c} 6 \\ 8 \\ 10 \end{array}$$
$$\qquad\quad 4\quad 6\quad 8\quad 6$$

We then have $x_{13} = \theta_3 = 0$ and a new degenerate basic minimum solution as follows:

$$(x_{ij}) = \begin{array}{|c|c|c|c|}\hline & 6 & 0 & \\\hline & & 2 & 6 \\\hline 4 & & 6 & \\\hline\end{array} \begin{array}{c} 6 \\ 8 \\ 10 \end{array}$$
$$\qquad\quad 4\quad 6\quad 8\quad 6$$

Of course, the value of the objective function is 28. For this multiple solution the value of the objective function did not change because

$\bar{c}_{13} - c_{13} = 0$, while the values of the variables did not change because $\theta_3 = 0$.

As a constant can be added or subtracted from each element of a row or column in the cost matrix without changing the optimum allocation (see Exercise 5), the final set of u_i and v_j elements can be subtracted from the corresponding rows or columns of the original cost table. This results in a cost table whose elements are either zero or positive, that is,

$$c_{ij} - u_i - v_j = c_{ij} - \bar{c}_{ij} \geq 0$$

When this transformed cost table is used for our transportation problem, an optimum solution would, of course, be one that allocated shipments only to those routes with zero costs. This is just the situation encountered in the final allocation determined by the above variation of the simplex method; hence the solution is optimum.

Some transportation problems might have the situation where the total of availabilities, $\sum_i a_i$, is less than the total of requirements, $\sum_j b_j$. Even though we cannot satisfy all the requirements, we can still allocate the items at the origins to the destinations in a manner that minimizes the total shipping cost. For this case, we assume that we have a "fictitious" origin that has on hand a total of $\sum_j b_j - \sum_i a_i > 0$ units. The costs of shipping a unit between this fictitious or $(m + 1)$st origin and the destinations are assumed to be zero. If our problem was originally a 3×4 problem, we would set up the following 4×4 problem and solve this problem with the fictitious origin like any other transportation problem:

Destinations

Origins	1	2	3	4	
1	c_{11}	c_{12}	c_{13}	c_{14}	a_1
2	c_{21}	c_{22}	c_{23}	c_{24}	a_2
3	c_{31}	c_{32}	c_{33}	c_{34}	a_3
4	0	0	0	0	$\sum_j b_j - \sum_i a_i$
	b_1	b_2	b_3	b_4	

By letting all the $c_{m+1,j} = c_{4j} = 0$, we make the minimum value of the objective function for the fictitious problem equal to the minimum value of the original problem.

If the problem is given with $\sum_i a_i > \sum_j b_j$, we set up a similar fictitious problem. Here we assume a fictitious destination that requires $\sum_i a_i - \sum_j b_j > 0$ units. The fictitious shipping costs are again zero. For a 3×4 problem we would then solve the following 3×5 problem:

		Destinations				
	1	2	3	4	5	
1	c_{11}	c_{12}	c_{13}	c_{14}	0	a_1
Origins 2	c_{21}	c_{22}	c_{23}	c_{24}	0	a_2
3	c_{31}	c_{32}	c_{33}	c_{34}	0	a_3
	b_1	b_2	b_3	b_4	$\sum_i a_i - \sum_j b_j$	

In solving any linear-programming problem, we should, in general, expect the total number of iterations required to depend on how close the value of the objective function for the first feasible solution is to the actual minimum. Since the northwest-corner rule does not consider the size of the c_{ij}, we cannot expect the corresponding value of the objective function to be close to the minimum. A number of alternative methods have been suggested that can be adapted to machine computation. We shall illustrate three of them for the following example taken from Dantzig [18]:

3	2	1	2	3	1
5	4	3	−1	1	5
0	2	3	4	5	7
3	3	3	2	2	

The northwest-corner solution yields a value of the objective function of 52.

a. Row minimum. Let the minimum element in the first row be c_{1k}. (If there is more than one minimum element, select the one with the smallest index j.) We let $x_{1k} = a_1$ if $a_1 \leq b_k$ or $x_{1k} = b_k$ if $a_1 > b_k$. In the first case, we have shipped all the a_1 units and go on to the second row after changing b_k to $b_k - a_1$. We next find the minimum element in the second row and repeat the process. In the second case, we have allocated only b_k units of the a_1; hence we change a_1 to $a_1 - b_k$ and b_k to zero and find the next smallest c_{ij} in the first row and repeat the process. Using this scheme, we have a first feasible solution as shown below:

$(x_{ij}) =$

		1			1
		1	2	2	5
3	3	1			7
3	3	3	2	2	

This is a minimum feasible solution with an objective-function value of 13.

b. Column minimum. We do a calculation similar to that using the row minimum, except that we start with the first column and proceed to the last column. Using this scheme, the first feasible solution is

$(x_{ij}) =$

		1			1
		2	2	1	5
3	3			1	7
3	3	3	2	2	

with an objective-function value of 17.

c. Matrix minimum. Here we search the whole matrix for the smallest element and allocate accordingly. We repeat this procedure until all the units are shipped. Using this scheme on the preceding example, we obtain the minimum solution obtained by the row-minimum procedure. Even though these schemes produce good results for the example in question, they are not foolproof. Examples have been constructed whose northwest-corner solution is much better than solutions obtained by any of the above procedures. However, experience has shown that, in general, the extra computational time required to obtain a first feasible solution by one of the above methods will be more than made up because the total iterations will be fewer. This is especially true for large problems.

For high-speed digital computers, where searching for elements can be time-consuming, a number of procedures have been employed for determining which variable should enter the basis, i.e., a rule that does not select the x_{ij} associated with the max $(\bar{c}_{ij} - c_{ij} > 0)$ for all (i,j). Dennis [32*e*] describes rules for searching the cost data which appear to be very effective. In particular, a complete row of the indirect-cost matrix is examined and the x_{ij} in this row with the greatest $\bar{c}_{ij} - c_{ij} > 0$ is chosen. For the next iteration, the following row is examined in the same way, and the procedure continues in this cyclic fashion.

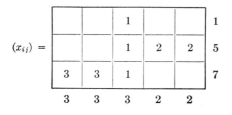

3. VARIATIONS OF THE TRANSPORTATION PROBLEM

a. The personnel-assignment problem.[1] We shall illustrate the basic problem by the following example: A business concern needs to fill three jobs, which demand different abilities and training. Three applicants, who can be hired for identical salaries, are available. However, because of their different abilities, training, and experience, the value of each applicant to the company depends upon the job in which he is placed. The estimate of the value of each applicant to the company each year if he were to be assigned to any one of the three jobs is given below.

		Jobs 1	2	3
	1	5	4	7
Applicants	2	6	7	3
	3	8	11	2

It is desired to assign the applicants to the jobs in such a way that the total value to the company is a maximum. In this problem, there are $3! = 6$ possible assignments. Here we wish to determine the *value of the assignment* x_{ij} of the ith applicant to the jth job. We can then restrict x_{ij} to be either 0 or 1. An assignment value of 0 means that the ith man does not get the jth job, and a value of 1 means that the ith man does get the jth job. Since each man can be assigned only one job and since, conversely, each job requires only one man, the total assignment value of the ith man, $\sum_j x_{ij}$, must equal 1. Similarly, the total assignment value associated with each job, $\sum_i x_{ij}$, must also equal 1. If we let the contribution of the ith man to the common effort if he is assigned to job j be c_{ij}, then the above example has the following linear-programming interpretation:

$$\sum_{j=1}^{3} x_{ij} = 1 \qquad i = 1, 2, 3$$

$$\sum_{i=1}^{3} x_{ij} = 1 \qquad j = 1, 2, 3$$

$$x_{ij} \geq 0$$

and

$$\sum_{i=1}^{3} \sum_{j=1}^{3} c_{ij}x_{ij} \qquad \text{to be a maximum}$$

[1] The material in this section is from Dwyer [41].

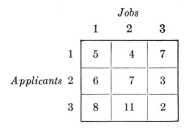

The form of this problem is that of the transportation problem if we minimize the negative of the objective function, and hence it can be solved by means of the same computational procedure.

Frequently there are many identical jobs which demand the same basic qualifications. Such jobs can be combined into a job category. We shall assume there are n such categories and let b_j denote the number of such jobs grouped in the jth job category. If different individuals have identical or approximately the same c_{ij} values, then these individuals can be grouped into personnel categories. We shall assume there are m such categories and let a_i denote the number of men in the ith personnel category. The linear-programming formulation of this general personnel-assignment problem is then

$$\sum_{j=1}^{n} x_{ij} = a_i \qquad i = 1, 2, \ldots, m$$

$$\sum_{i=1}^{m} x_{ij} = b_j \qquad j = 1, 2, \ldots, n$$

$$x_{ij} \geq 0$$

and

$$\sum_{i=1}^{m} \sum_{j=1}^{n} c_{ij}x_{ij} \qquad \text{to be a maximum}$$

Here, x_{ij} is either zero or a positive integer indicating the number of persons in personnel category i assigned to job category j. We imply that the total number of jobs available is equal to the total number of men to be assigned, that is, that $\sum_i a_i = \sum_j b_j$. If $\sum_i a_i < \sum_j b_j$, then a fictitious personnel category containing $\sum_j b_j - \sum_i a_i$ men is added to the problem. When $\sum_j b_j < \sum_i a_i$, then a fictitious job category containing $\sum_i a_i - \sum_j b_j$ jobs is used.

Both Dwyer [41] and Votaw and Orden [97*f*] discuss methods for obtaining "near-optimal" solutions to the personnel-assignment problem and, of course, to the transportation problem. The special problem of assigning n individuals to n jobs, i.e., all $a_i = 1$ and all $b_j = 1$, has led to the development of a number of computationally efficient algorithms (Balinski and Gomory [3*b*], Kuhn [67*a*], and Ford and Fulkerson [44*a*]). Another problem which takes on this formulation is the *machine-assignment problem* in which n tasks are to be assigned n different machines with the cost or time of setting up the jth machine for the ith job being c_{ij}. The objective is to assign the jobs so as to minimize the total set-up cost or time. For additional techniques for handling the assignment problem, the reader is referred to von Neumann [78] and Egerváry [42].

b. The contract-award problem.[1] Although the contract-award problem is peculiar to government operations, its formulation and solution as a linear-programming problem represent an outstanding application of linear-programming techniques. Whenever a government agency wishes to procure items from civilian sources, producers of the items must be invited to participate in the bidding for contracts. The individual manufacturer submits bids in which he states:

1. The price per unit of article or articles
2. The maximum and minimum quantity of each item that can be produced at the stated price
3. *Any other conditions he wishes to impose*

The bid reflects the manufacturer's desire for profit, his guess about the other fellow's bid, and his own peculiar limitations. The agency must award contracts in such a way that the total dollar cost to the government is at a minimum.

In evaluating the bids, the procurement office must add shipping and other related costs to each of the bidders' quoted prices. Similarly, any savings that could be effected by agreeing to certain conditions are subtracted; e.g., a discount may be allowed for payments made within a certain time. Once the contracts have been awarded, the procurement office must be ready to demonstrate that the total cost to the government was the least possible. The application of linear-programming techniques to this problem guarantees that the solution is minimum.

Let us first consider the *simple* contract-award problem. Here there are m separate bidders willing to furnish the needs of n supply depots. The ith bidder wishes to produce an amount not exceeding a_i, for $i = 1, 2, \ldots, m$, and the jth depot requires the amount b_j, for $j = 1, 2, \ldots, n$. It costs c_{ij} dollars to purchase and deliver a unit from the ith bidder to the jth depot. If x_{ij} denotes the quantity purchased from the ith manufacturer for shipment to the jth depot, then our problem is to find a set of nonnegative x_{ij} such that

$$\sum_{j=1}^{n} x_{ij} \leq a_i \qquad i = 1, 2, \ldots, m$$

$$\sum_{i=1}^{m} x_{ij} = b_j \qquad j = 1, 2, \ldots, n$$

and minimizing

$$\sum_{i=1}^{m} \sum_{j=1}^{n} c_{ij}x_{ij}$$

[1] For a more detailed discussion, see Stanley, Honig, and Gainen [91].

This is a problem of the same type as the transportation problem. Note that $\sum_i a_i \neq \sum_j b_j$. In general, we would expect more to be offered for sale than is required in the total award; that is, $\sum_i a_i \geq \sum_j b_j$. If we establish a fictitious depot, the $(n + 1)$st depot, which absorbs the amount $\sum_i a_i - \sum_j b_j = b_{n+1}$, assigning zero costs $c_{i,\,n+1}$ for all i, we have then formulated the simple contract-award problem as a general transportation problem. The problem is called simple because it does not consider the many conditions that can be stipulated by the manufacturers. For example, bidder one may wish to make a_1 units but is willing to accept an award for any part of a_1. Bidder two bids for a_2 units and will not accept a contract for less than a_2. Bidder three bids for a_3 units but will accept as little as $a_3' < a_3$ units. Bidder four is bidding for a total of a_4 units but has different costs on the sublots that total to his final award. By using various computational techniques, these and other such conditions can be handled while the problem still retains the basic form of the transportation problem. Some of these devices are discussed in Stanley, Honig, and Gainen [91]. A typical contract-award problem and its solution is given in Directorate of Management Analysis [36].

EXERCISES

1. For the transportation problem of Exercise 1, Chap. 1, obtain a first feasible solution by means of the northwest-corner rule and the column-minimum and matrix-minimum procedures.

2. Using the northwest-corner solution as the first feasible solution, determine a minimum feasible solution to Exercise 1, Chap. 1. Are there multiple minimum feasible solutions?

3. Write the dual problem to the transportation problem of Exercise 2 and show that the u_i and v_j associated with any minimum feasible solution are also a solution to the dual.

4. Solve the following problem (Joseph [63]). Consider four bases of operation B_i and three targets T_j. Because of differences in aircraft, range to target, and flying altitude, the tons of bombs per aircraft from any base that can be delivered to any target differ according to the following table:

Target (T_j)

Base of Operations (B_i)	1	2	3
1	8	6	5
2	6	6	6
3	10	8	4
4	8	6	4

$= (c_{ij})$

where c_{ij} = tons of bombs per aircraft. The daily sortie capability of each of the four bases is 150 sorties per day. The daily requirement in sorties over each individual target is 200. Find the allocation of sorties from each base to each target which maximizes the total tonnage over all three targets. Determine all multiple optimum solutions.

5. Show that the optimum set of x_{ij} to any transportation problem

$$\sum_{j=1}^{n} x_{ij} = a_i \qquad i = 1, 2, \ldots, m$$

$$\sum_{i=1}^{m} x_{ij} = b_j \qquad j = 1, 2, \ldots, n$$

$$x_{ij} \geq 0$$

$$\sum_{j=1}^{n} \sum_{i=1}^{m} c_{ij}x_{ij} \qquad \text{a minimum}$$

is equal to the optimum set of x_{ij} of the corresponding transportation problem for which a constant k has been added to or subtracted from all c_{ij} elements belonging to either the same row or the same column of the cost-coefficient matrix.

6. An assignment problem that cycles (due to B. J. Gassner and L. R. Johnson): Given a 4×4 assignment problem (i.e., all $a_i = b_j = 1$) with the cost matrix

$(c_{ij}) =$

3	5	5	11
9	7	9	15
7	7	11	13
13	13	13	17

show that the northwest-corner solution of

$(x_{ij}) =$

1	0		
	1	0	
		1	0
			1

will repeat if the following sequence of new variables is introduced into the basis:

$$x_{13},\ x_{42},\ x_{32},\ x_{41},\ x_{43},\ x_{21},\ x_{31},\ x_{24},\ x_{23},\ x_{14},\ x_{34},\ x_{21}$$

In the first iteration let x_{13} eliminate x_{12} and do not allow the variable coming into the basis in iteration k to replace the variable that came into the basis in iteration $k - 1$.

For this set of iterations, $u_i + v_j - c_{ij} = \Delta = 2$ for all basis changes. For the northwest-corner solution, the initial u_i's are given by $u = (1,3,5,9)$ and v_j's by $v = (2,4,6,8)$. Letting u' and v' be the set of u_i's and v_j's for the final iteration, note that $u' = (u_1 + 6\Delta, u_2 + 6\Delta, u_3 + 6\Delta, u_4 + 6\Delta)$ and $v' = (v_1 - 6\Delta, v_2 - 6\Delta, v_3 - 6\Delta, v_4 - 6\Delta)$.

7. Solve Exercise 6 for the maximum of the objective function by the degeneracy procedure of the transportation algorithm.

8. The transshipment problem is a transportation problem in which each origin and destination can act as an intermediate point through which goods can be temporarily received and then transshipped to other points or to the final destination. Show how this problem can be formulated and solved as a transportation problem (Orden [85*a*]). HINT: Set up an enlarged tableau with $m + n$ origins and $m + n$ destinations, assuming costs c_{ij} between all combinations of the origins and destinations are known. Make sure enough units are available to allow for transshipment.

9. Prove Lemma 1.

10. Justify the computational procedure of Sec. 2 by using the pricing-vector concept of the revised simplex algorithm. Show that for a triangular basis we can compute the components of the pricing vector, and thus the $z_j - c_j$ elements, as described in Sec. 2.

CHAPTER 11 / GENERAL LINEAR-PROGRAMMING APPLICATIONS

In the first part of this chapter we shall discuss and formulate examples of some of the more important (and classical) applications of linear-programming techniques. For purposes of exposition we have grouped linear-programming applications into four general categories: production-scheduling and inventory-control problems, interindustry problems, diet problems, and network problems.[1] In each section we shall give a verbal statement of the problem and then formulate the problem as a linear-programming problem. For some problems this is followed by a solution of a numerical example. Although these applications will give the reader some idea of the adaptability of linear-programming methods, it is not our purpose here to describe the complete range of linear-programming applications. Instead, these applications were selected mainly for their ability to illustrate the basic techniques of formulating linear-programming models. It is hoped that these discussions and the exercises will enable the reader to investigate and develop new applications.

In order to demonstrate the versatility of the linear-programming model, we have already compiled (see Sec. 3 of Chap. 1) verbal descriptions of a number of applications. This survey is designed to inform investigators in the various fields of industry, management science, and operations research of the areas that have been successfully treated by linear-programming methods.

At this point we should like to stress that there are no set procedures and techniques one can apply to determine and test the linear-programming formulation of a given problem. The casting of a new and complex problem into a linear-programming form appears to be a process of evolution. One's first impulse is to define what appear to be the variables

[1] The second and third categories are sometimes consolidated into the category of product mix, i.e., the combination of raw materials and resources to yield finished products. The important category of transportation problems is discussed in Chap. 10. See the Bibliography for a more detailed listing of applications.

of the problem and to determine the interrelationships between these variables, the resulting constraints, and appropriate objective function. Problems based on either actual or test data are then solved, using the initial model, and the solutions compared with expected results. These studies will in all probability suggest that new variables must be defined and introduced into the system and that previously defined variables can be eliminated; changes in the objective function and constraints, a reevaluation of the test data, etc., may also result. Additional problems are then solved, and this process for evolving a correct model continues until the investigator is satisfied that his resulting model approximates the real situation to an acceptable degree (or, possibly, is satisfied that his problem cannot be handled by linear-programming techniques). If, in developing the constraints of the problem, it is determined that some of the relationships are not linear, we are faced with three alternatives:

1. Approximate the troublesome expressions by appropriate linear functions
2. Redefine the problem to fit the linear-programming format
3. Use other techniques to solve the problem

In selecting the first two alternatives, the investigator should avoid a drastic transformation of his problem, for this would tend to dilute any valid interpretation of his solution. For the third alternative, we might possibly use some of the techniques of nonlinear programming (see Chap. 13).

To simplify the formulation of the problems described below, we shall discuss all but the diet problem from the ideal point of view; i.e., we shall start with a complete and valid statement of the problem being considered.

1. PRODUCTION-SCHEDULING AND INVENTORY-CONTROL PROBLEMS

Let us consider the plight of the manufacturer of a seasonal item who must determine his monthly production schedule of the item for the next 12 months. The demand for his product fluctuates, as is illustrated in his sales-forecast chart for the coming year (Fig. 11.1). The manufacturer must always meet the monthly requirements given on the chart. He can fulfill the individual demands either by producing the desired amount during the month or by producing part of the desired amount and making up the difference by using the overproduction from previous months.

In general, any such scheduling problem has many different schedules that will satisfy the requirements. For example, the manufacturer could produce each month the exact number of units required by the sales forecast. However, since a fluctuating production schedule is costly to maintain, because of overtime costs in high-production months and because of the costs associated with the release of personnel and machinery in low-production months, this type of production schedule is not an efficient one. On the other hand, the manufacturer faced with fluctuating requirements

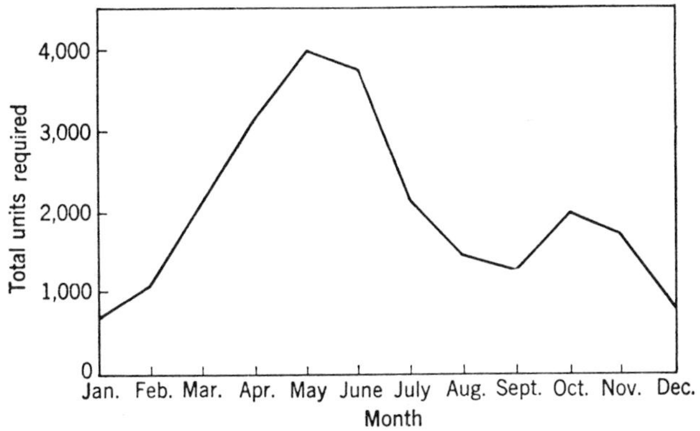

FIGURE 11.1

could overproduce in periods of low requirements, store the surplus, and use the excess in periods of high requirements. The production pattern can thus be made quite stable. However, because of the cost of keeping a manufactured item in storage, such a solution may be undesirable if it yields comparatively large monthly surpluses.

Problems of this nature illustrate the difficulties that arise whenever conflicting objectives are inherent in a problem. Here the desire is to determine a production schedule that minimizes the sum of the costs due to output fluctuations and to inventories. For these problems, efficient scheduling means the determination and acceptance of a middle-ground solution lying between the two extreme solutions—the one that minimizes surpluses and the one that minimizes fluctuations. The optimum production schedule will depend on the relative costs assigned to the conflicting objectives.

Let us next develop a mathematical model of this scheduling problem. At the beginning of the first month the manufacturer has in storage a certain amount, say s_0, of the item available from previous production. If the item to be produced is of a new design or style, then $s_0 = 0$. Let

x_t = the number of units produced in month t, that is, the production

r_t = the number of finished units that must be available in month t, that is, the requirement

s_t = the number of finished units that are not required in month t, that is, the storage

Owing to the nature of the problem, we have $x_t \geq 0$, $r_t \geq 0$, and $s_t \geq 0$ for all values of t. For the first month, the production x_1 and the previously stored items s_0 must be such that their sum is greater than or equal to the requirement r_1. This yields the relationship

$$x_1 + s_0 \geq r_1 \tag{1.1}$$

If the equality in (1.1) holds, then the storage in the first month, s_1, must equal zero. If the inequality holds, then $s_1 > 0$. In either case, we have

$$x_1 + s_0 - r_1 = s_1$$

or

$$x_1 + s_0 - s_1 = r_1$$

For the second month, the production x_2 plus the previously stored items s_1 must be greater than or equal to the second month's requirement r_2. We then have

$$x_2 + s_1 \geq r_2$$

or

$$x_2 + s_1 - s_2 = r_2$$

In general, the production x_t, the storage s_t, and the requirement r_t are related by

$$x_t + s_{t-1} - s_t = r_t \tag{1.2}$$

The manufacturer desires to minimize the fluctuations in the production schedule and obtain a smooth pattern of production. The difference between any two successive months' production, say $x_t - x_{t-1}$, will represent the corresponding increase or decrease in production. Since any number has an equivalent representation in terms of two nonnegative numbers, we set

$$x_t - x_{t-1} = y_t - z_t \tag{1.3}$$

where $y_t \geq 0$ represents an increase in production and $z_t \geq 0$ represents a decrease in production. From (1.2) and (1.3) our basic equations for this production model are

$$\begin{aligned} x_t + s_{t-1} - s_t &= r_t \\ x_t - x_{t-1} - y_t + z_t &= 0 \end{aligned} \tag{1.4}$$

where $x_t \geq 0$, $s_t \geq 0$, $y_t \geq 0$, $z_t \geq 0$, and $t = 1, 2, \ldots, n$. If at the end of the production year we desire the surplus of finished items to be zero, then we set $s_n = 0$. Depending on the conditions of the model, $x_0 \geq 0$ and $s_0 \geq 0$.

The manufacturer is, of course, interested in maximizing his profit.

As we have seen, his profit will depend on the fluctuations in his production schedule and the associated monthly storage of manufactured items. From his cost records in previous years, he knows how much it costs to increase production by one unit from month $t - 1$ to month t and also how much it costs to store one unit for 1 month. Let these costs be a dollars and b dollars, respectively, where $a > 0$ and $b > 0$. The manufacturer then wants to minimize

$$b \sum_{t=1}^{n} s_t + a \sum_{t=1}^{n} y_t$$

If we let $a/b = \lambda$, where λ measures the cost of a unit increase in output relative to that of storing a unit for 1 month, we then wish to minimize

$$\sum_{t=1}^{n} s_t + \lambda \sum_{t=1}^{n} y_t \tag{1.5}$$

If a value of λ is known, it can be substituted in (1.5), and the corresponding linear-programming problem can be solved by means of the simplex method. It should be noted that a minimum solution cannot, for the same month, t, have both a $y_t > 0$ and a $z_t > 0$. Solving the problem in this manner will yield a unique optimum production schedule or possibly additional production schedules having the same value of the objective function. In many instances a value of λ cannot be determined, or the manufacturer wishes to study various production schedules associated with a range of values for λ. For these situations an explicit value of λ need not be given, and by means of a modification of the simplex method, optimum production schedules for ranges of λ extending from $\lambda = 0$ to $\lambda < +\infty$ can be computed.[1]

If a is very small, then the corresponding optimal solution will be the schedule that requires the most fluctuation and the least storage. If b is very small, then the associated minimum solution will be the schedule that calls for the most storage and least fluctuation. In the first case we are minimizing only storage, and in the second case we are concerned with minimizing only fluctuations.

The following example from Directorate of Management Analysis [36] will illustrate the preceding discussion:

The number of monthly graduates required from an aircraft mechanics' training school (for a 12-month period) is shown in Fig. 11.2. The total production of mechanics over the 12-month period must equal the total required; i.e., there cannot be a surplus or shortage of mechanics at the end of the program.

For this problem, the possible schedules range from the one where

[1] See Chap. 8 for a discussion of parametric methods.

each man is assigned to duty on the day he graduates (minimum surplus, Fig. 11.2) to one where a fixed training load is provided (minimum fluctuation, Fig. 11.3). The former schedule, by producing each month the graduates as needed, requires no replacement pool but creates the heaviest problems in operating the training school. The second schedule elimi-

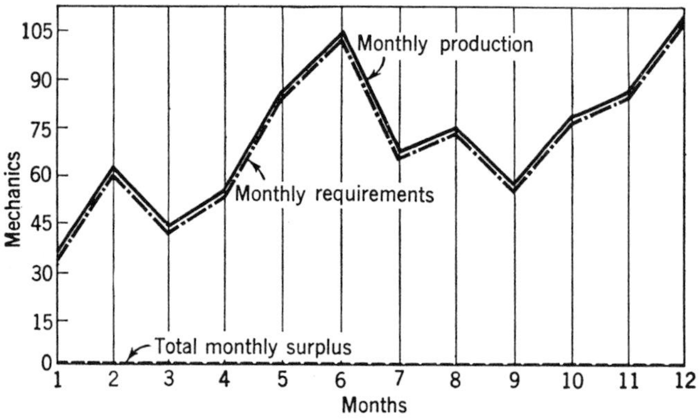

FIGURE 11.2 Minimum surplus solution. *(From Directorate of Management Analysis* [36].)

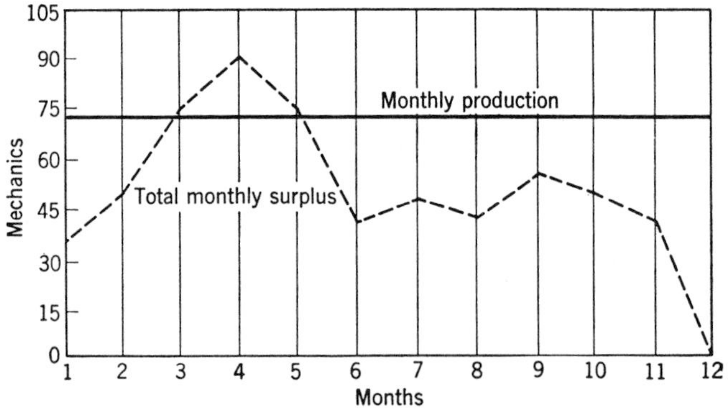

FIGURE 11.3 Minimum fluctuation solution. *(From Directorate of Management Analysis* [36].)

nates fluctuations in training output but puts the greatest burden on the replacement pool. An intermediate solution (Fig. 11.4) results in a more manageable training program, which at the same time places a lighter load on the replacement pool.

The reader will note that for only 12 time periods, that is, for $n = 12$, the linear-programming model involves 24 equations in 48 variables.

Problems of this size are difficult to solve by manual computation. It is advantageous to reduce the size of the problem whenever possible, not only for manual computations, but also for computations being done on electronic computers. In some instances, a careful analysis of the problem will reveal equations and variables that can be eliminated from the computation. The above production problem is such a problem.

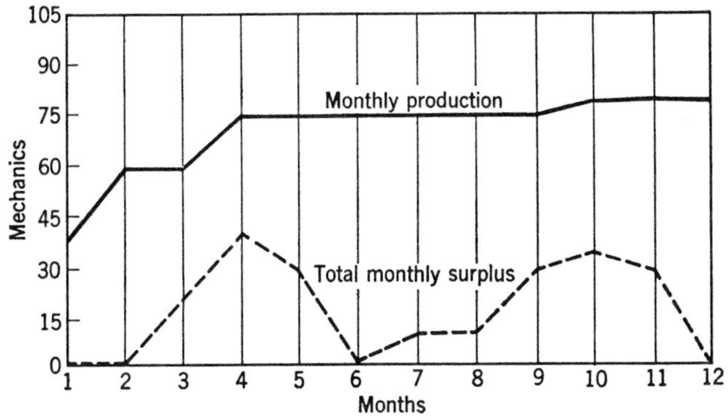

FIGURE 11.4 An intermediate solution. (*From Directorate of Management Analysis* [36].)

From Eq. (1.2) we have

$$x_t = r_t + s_t - s_{t-1}$$

and

$$x_{t-1} = r_{t-1} + s_{t-1} - s_{t-2}$$

By substituting the above expressions for x_t and x_{t-1} in (1.3), we eliminate all the x_t from the original model and obtain the following set of equations:

$$y_t - z_t + 2s_{t-1} - s_t - s_{t-2} = r_t - r_{t-1} \tag{1.6}$$

and

$$t = 1, 2, \ldots, n \qquad \text{and} \qquad s_{-1} = r_0 = 0$$

Equations (1.6) and the objective function (1.5) now represent the linear-programming model of the production problem. We have eliminated one equation from each time period and the 12 variables x_t. When we solve this reduced system, it is an easy matter to go back to the original equations and obtain the desired production schedule consisting of the x_t's. However, since the reduced set of equations does not restrict the $x_t \geq 0$, we have no assurance that the nonnegative values of y_t, z_t, and s_t associated with the optimum solution to the reduced problem will yield nonnegative values of x_t. Before any such elimination is accepted, the validity of the reduction must be demonstrated. To do this, we first assume that an optimum solution to the reduced system is given. We then need to show that the optimum values of y_t, z_t, and s_t cannot yield a negative x_t and at the same time be optimal. The proof is left to the reader.

In addition to the above reduction in the dimensions of the problem, this formulation of the production-scheduling problem has also been subjected to further interpretations which enable one to solve it using much simpler computational devices than the general simplex method. Bowman [7] gives a variation of the problem that is written as a transportation problem and solved by the corresponding algorithm. Dantzig and Johnson [28*a*] have developed a systematic and rapid graphical method for determining the optimal solutions.

For additional reading concerning scheduling problems, the reader is referred to Hoffman and Jacobs [60], Antosiewicz and Hoffman [2*a*], Jacobs [62], Magee [73], Directorate of Management Analysis [36], Charnes, Cooper, and Farr [11], and Charnes, Cooper, and Mellon [13].

2. INTERINDUSTRY PROBLEMS

The first application of linear-programming techniques in the field of economics was to the area of interindustry or input-output analysis (Koopmans [65], Leontief [71], and Morgenstern [75]). By way of introduction to the general formulation of input-output models, let us consider an economy that consists of only the three basic industries—railroad, steel, and coal—and a fourth category of all other industries (Glaser [50]). We wish to analyze the interrelationships of these industries in terms of sales to each other and to other elements of the economy, as will be described below, during a base time period, e.g., 1 year. This analysis can be readily accomplished by reference to the Input-Output Table.

Each element in the table represents a total sales activity of each industry that occurred during the base time period. For example, the first row of the table describes the sales of the railroad industry (in terms of dollars) to each of the other industries. The first element of row 1 represents the total sale of the railroad industry to the railroad industry (intra-industry sale); the second element describes the total sale of the railroad industry to the steel industry; the third element, the total sale to the coal industry; the fourth element, the total sale to the other industries; and the fifth element, the total sale of the railroad industry to what is termed the final-demand consumers. In general, the final demand consists of those elements of the economy which consume the various commodities produced but do not contribute, or feed back, to the economy a product of their own. In the final-demand category we usually include

Input-Output Table

	Sales to RR		Sales to steel		Sales to coal		Sales to other		Sales to FD		Total sales
Sales by RR	RR to RR x_{11}	+	RR to steel x_{12}	+	RR to coal x_{13}	+	RR to other x_{14}	+	RR to FD y_1	=	Total sales by RR x_1
Sales by steel	Steel to RR x_{21}	+	Steel to steel x_{22}	+	Steel to coal x_{23}	+	Steel to other x_{24}	+	Steel to FD y_2	=	Total sales by steel x_2
Sales by coal	Coal to RR x_{31}	+	Coal to steel x_{32}	+	Coal to coal x_{33}	+	Coal to other x_{34}	+	Coal to FD y_3	=	Total sales by coal x_3
Sales by other	Other to RR x_{41}	+	Other to steel x_{42}	+	Other to coal x_{43}	+	Other to other x_{44}	+	Other to FD y_4	=	Total sales by other x_4

foreign trade, government operations, and households. The total production required by the final-demand segments of the economy is called the bill of goods. The sum of the sales included in these five elements represents the total sales of the railroad industry during the base period. A similar discussion holds for the other three rows of the table. By columns, we see that each element represents the total purchase of an industry from the other industries to support its operations during the base period.

Let us define

$0 < x_i =$ the total output, in dollars, of industry i during the base period (only active industries are considered)

$0 \leq x_{ij} =$ total sales, in dollars, of industry i to industry j during the base period

$0 \leq y_i =$ amount of final demand, in dollars, for industry i

where $i, j = 1, 2, \ldots, m$.

With these definitions we can construct the following set of four linear equations, which define the interrelationships of our reduced economy, as described in the table:

$$\begin{aligned}
x_1 - x_{11} - x_{12} - x_{13} - x_{14} &= y_1 \\
x_2 - x_{21} - x_{22} - x_{23} - x_{24} &= y_2 \\
x_3 - x_{31} - x_{32} - x_{33} - x_{34} &= y_3 \\
x_4 - x_{41} - x_{42} - x_{43} - x_{44} &= y_4
\end{aligned} \tag{2.1}$$

In the form (2.1), our input-output information explicitly defines the relationship between total outputs, total sales, and final demands for a given time period.

In the 1930's Leontief was able to gather the necessary data and write sets of equations similar to (2.1) for a 45-industry classification for the years 1919 and 1929. He assumed that this linear model of the economy for a base time period could be used to analyze the structure of the economy in future time periods. Let us denote the x_i and x_{ij} that are known for the given time period by $\bar{x}_i$ and $\bar{x}_{ij}$. For the railroad industry we see that the ratios

$$\frac{\bar{x}_{11}}{\bar{x}_1} \qquad \frac{\bar{x}_{21}}{\bar{x}_1} \qquad \frac{\bar{x}_{31}}{\bar{x}_1} \qquad \frac{\bar{x}_{41}}{\bar{x}_1}$$

represent the percentages of input by each of the other industries required to produce one unit of output by the railroad industry; for example, $\bar{x}_{21}/\bar{x}_1$ is the input of the steel industry necessary to produce one unit of output of the railroad industry.

Let $0 \leq a_{ij} = \dfrac{\bar{x}_{ij}}{\bar{x}_j}$ = amount of industry i which is necessary to produce one unit of commodity j

The a_{ij} are called *input-output coefficients.*

For the base time period we have

$$\begin{aligned}
\bar{x}_1 - a_{11}\bar{x}_1 - a_{12}\bar{x}_2 - a_{13}\bar{x}_3 - a_{14}\bar{x}_4 &= \bar{y}_1 \\
\bar{x}_2 - a_{21}\bar{x}_1 - a_{22}\bar{x}_2 - a_{23}\bar{x}_3 - a_{24}\bar{x}_4 &= \bar{y}_2 \\
\bar{x}_3 - a_{31}\bar{x}_1 - a_{32}\bar{x}_2 - a_{33}\bar{x}_3 - a_{34}\bar{x}_4 &= \bar{y}_3 \\
\bar{x}_4 - a_{41}\bar{x}_1 - a_{42}\bar{x}_2 - a_{43}\bar{x}_3 - a_{44}\bar{x}_4 &= \bar{y}_4
\end{aligned} \tag{2.2}$$

We can then write (2.2) in the form

$$(\mathbf{I} - \mathbf{A})\bar{\mathbf{X}} = \bar{\mathbf{Y}}$$

where

$$\mathbf{A} = (a_{ij}) \qquad \bar{\mathbf{X}} = \begin{pmatrix} \bar{x}_1 \\ \bar{x}_2 \\ \bar{x}_3 \\ \bar{x}_4 \end{pmatrix} \qquad \bar{\mathbf{Y}} = \begin{pmatrix} \bar{y}_1 \\ \bar{y}_2 \\ \bar{y}_3 \\ \bar{y}_4 \end{pmatrix}$$

The matrix $(\mathbf{I} - \mathbf{A})$ is known as a *Leontief matrix.* By assuming that this linear structure of the economy, i.e., the input-output coefficients, describes the activity of the economy not only for the base time period but for future time periods as well, we can determine a production vector $\mathbf{X}$ which satisfies a predicted final-demand vector $\mathbf{Y}$. From this we have that the general problem for input-output economics is to find a vector $\mathbf{X}$ which satisfies the constraints

$$\begin{aligned}
\mathbf{X} &\geq 0 \\
(\mathbf{I} - \mathbf{A})\,\mathbf{X} &= \mathbf{Y}
\end{aligned} \tag{2.3}$$

where $\mathbf{Y}$ is a given nonnegative and nonzero final-demand vector and $\mathbf{A}$ is a given matrix of input-output coefficients. For the problem being considered, the one with the nonzero final demand, it can be shown not only that $a_{ij} \geq 0$ but also that $\sum_{i=1}^{m} a_{ij} < 1$ for $j = 1, 2, \ldots, m$ (Morgenstern [75]). This structure is called an open model. It is also shown in Morgenstern [75] that, if a matrix $\mathbf{A}$ satisfies the condition

$$\sum_{i=1}^{m} |a_{ij}| < 1 \qquad \text{for} \qquad j = 1, 2, \ldots, m$$

then $(\mathbf{I} - \mathbf{A})$ is nonsingular. We then have that Eqs. (2.3) have the solution

$$\mathbf{X} = (\mathbf{I} - \mathbf{A})^{-1}\mathbf{Y}$$

The initial treatment of the interindustry model as a linear-programming problem interpreted the equalities of (2.3) as a system of inequalities

$$(\mathbf{I} - \mathbf{A})\mathbf{X} \leq \mathbf{Y} \tag{2.4}$$

Here we are interested in solutions that do not have to satisfy the final-demand requirements. An obvious but impractical solution to (2.4) is $\mathbf{X} = \mathbf{O}$; hence the problem is feasible. An alternate formulation of the problem might require the system (2.4) to be a mixture of equalities and inequalities. However, for the inequalities (2.4), the corresponding system of equalities is

$$(\mathbf{I} - \mathbf{A})\mathbf{X} + \mathbf{W} = \mathbf{Y} \tag{2.5}$$

where $\mathbf{W}$ is an m-dimensional column vector whose components w_i are nonnegative slack vectors. The system (2.5) consists of m equations in $2m$ variables. The objective function can take on various interpretations. For example, if c_j is the profit per unit of commodity j produced, then an appropriate objective function would be to maximize the total profit $\mathbf{cX}$, where $\mathbf{c} = (c_1,c_2, \ldots ,c_m)$ is a row vector. The cost coefficients of the slack variables are taken to be zero. Another objective function could call for the solution that maximizes the output of a particular industry or combination of industries, e.g., to maximize $x_1 + x_3 + x_7$.

In addition to the constraints (2.3) or (2.4), the interindustry model might stipulate that the production (activity) level x_i of industry i must not exceed a known available capacity for production by industry i. Let us denote these capacity levels by l_i and let the capacity-level vector $\mathbf{L} = (l_1,l_2, \ldots ,l_m)$ be a nonnegative column vector. Then $\mathbf{X} \leq \mathbf{L}$. As we are here looking at our economy for a particular time period, e.g., a month, we can assume that for some industries a portion of the production in the previous time period was used to stockpile finished units. These units will be made available for distribution in the following time period and can be applied to satisfy the new final-demand requirements. In our model let s_{0i} represent the stock of product i available from previous production, and let the vector of stocks, $\mathbf{S}_0 = (s_{01},s_{02}, \ldots ,s_{0m})$, be a nonnegative column vector. We can then expand the system (2.5) into the following linear-programming problem: To maximize

$$\mathbf{cX}$$

subject to

$$\begin{aligned} (\mathbf{I} - \mathbf{A})\mathbf{X} + \mathbf{W} &= \mathbf{Y} - \mathbf{S}_0 \\ \mathbf{X} + \mathbf{U} &= \mathbf{L} \\ \mathbf{X} &\geq \mathbf{O} \end{aligned} \tag{2.6}$$

where $\mathbf{U} = (u_1,u_2, \ldots ,u_m)$ is a nonnegative column vector and where the u_i represent the unused capacity of industry i. If, for some i, $y_i - s_{0i} < 0$, then the final demand for industry i will be satisfied from stocks, and the effective final demand for industry i during this time period is set equal to zero. However, we assume that at least one $y_i - s_{0i} > 0$.

The system (2.6) is called a static Leontief model because it considers the economy over only a single time period. A more challenging application of the interindustry model is its use in interpreting the behavior of an economy over several time periods. We shall next discuss a particular linear-programming formulation of a similar but dynamic model, as given in Wagner [98]. To do this, we must introduce the following changes in our notation: Let n equal the total number of time periods considered, and let $t = 1, 2, \ldots , n$ be any particular time period. Then for any t we have the nonnegative column vectors

$\mathbf{X}_t = (x_{t1},x_{t2}, \ldots ,x_{tm})$ = the production vector
$\mathbf{Y}_t = (y_{t1},y_{t2}, \ldots ,y_{tm})$ = the final-demand vector
$\mathbf{S}_t = (s_{t1},s_{t2}, \ldots ,s_{tm})$ = the storage vector due to unused production up to and including month t. These stocks are available in month $t + 1$
$\mathbf{U}_t = (u_{t1},u_{t2}, \ldots ,u_{tm})$ = the unused-capacity vector

We assume that we are given the initial storage vector $\mathbf{S}_0$ and that the basic capacity-level vector $\mathbf{L}$ is the same for all time periods.

The main difference between our static and dynamic models is that in the dynamic model we shall make provision for the expansion of the capacity levels of each industry to meet the requirements of future final demands. From time period to time period we shall be adding to $\mathbf{L}$ a nonnegative column vector $\mathbf{V}_t$, which represents additional available capacities. To do this, we need, in addition to the appropriate Leontief matrix, $(\mathbf{I} - \mathbf{A})$, the knowledge of the corresponding m-dimensional square matrix $\mathbf{B}$ of capital coefficients. $\mathbf{B}$ is a matrix of nonnegative numbers in which the jth column represents the inputs from each industry needed to build an additional unit of capacity for the jth industry. Let

$$\mathbf{V}_t = (v_{t1},v_{t2}, \ldots ,v_{tm}) = \text{the nonnegative capacity-expansion vector}$$

where v_{ti} is the additional capacity for industry i in period t. Then the ith row of the product $\mathbf{BV}_t$, that is, $b_{i1}v_{t1} + b_{i2}v_{t2} + \cdots + b_{im}v_{tm}$, represents the amount of the ith industry's production that is used to build additional capacity in time period t for all the industries in our economy. This production is not available to meet the final-demand requirements.

In this model we assume that the additional production capacity is available in the next time period $(t + 1)$. We also assume that the matrices $(\mathbf{I} - \mathbf{A})$ and $\mathbf{B}$ are applicable to all the n time periods and that the final-demand requirements will be satisfied. We can, for each time period, summarize the above conditions as follows:

$$(\mathbf{I} - \mathbf{A})\mathbf{X}_t + \mathbf{S}_{t-1} = \mathbf{Y}_t + \mathbf{B}\mathbf{V}_t + \mathbf{S}_t \tag{2.7}$$

$$\mathbf{X}_t + \mathbf{U}_t = \mathbf{L} + \sum_{q=1}^{t-1} \mathbf{V}_q \tag{2.8}$$

for $t = 1, 2, \ldots, n$. Equation (2.7) states that, for any t, the total output plus the previous stocks is equal to the final-demand and capacity-expansion requirements for output plus the current period's unused stocks. Equation (2.8) equates the total used and unused production to the sum of the initial production capacity and the previous additional increases in production capacity. We can rearrange these equations to read

$$\begin{aligned} (\mathbf{I} - \mathbf{A})\mathbf{X}_t - \mathbf{B}\mathbf{V}_t - \mathbf{S}_t + \mathbf{S}_{t-1} &= \mathbf{Y}_t \\ \mathbf{X}_t - \sum_{q=1}^{t-1} \mathbf{V}_q + \mathbf{U}_t &= \mathbf{L} \end{aligned} \tag{2.9}$$

For any given set of $\mathbf{Y}_t$, $\mathbf{L}$, and $\mathbf{S}_0$ we can set up the tableau of coefficients, as shown in Tableau 11.1. Systems of this sort are termed *block-triangular*. Given an appropriate objective function, which could involve production, expansion, and storage costs, the problem could be solved by the standard simplex procedure. However, even for just a few time periods, the number of equations and variables becomes too

Tableau 11.1

	X_1	V_1	S_1	U_1	X_2	V_2	S_2	U_2	$\cdots$	X_n	V_n	S_n	U_n
$Y_1 - S_0$	$(I - A)$	$-B$	$-I$										
L	I			I									
Y_2			I		$(I - A)$	$-B$	$-I$						
L		$-I$			I			I					
$\vdots$									$\ddots$				
Y_n										$(I - A)$	$-B$	$-I$	
L		$-I$				$-I$				I			I

large to be handled efficiently by most of the electronic computers, let alone by manual methods. Some work has been done that enables one to reduce the amount of computation by taking advantage of the block-triangular configuration (with its high density of zero elements) or to reduce the number of equations by transforming the system. The former is discussed in Dantzig [20] and the latter in Wagner [98].

3. DIET PROBLEMS

In Sec. 2 of Chap. 1 we defined and formulated the basic diet problem in terms of a linear-programming problem. We next discuss a particular linear-programming model of a diet problem and review its corresponding solution to determine whether the formulation of Chap. 1 does apply. Such an analysis is typical for linear-programming problems in general, and it is recommended when one attempts to describe a complex situation by an elementary linear model. The models of such situations usually start as simple ones, and using the solutions to these modest models for a base, the investigator is able to evolve more realistic models. To illustrate this process, we shall briefly describe various adjustments to a diet-problem model that tend to make it a truer representation of the real-life situation.

Historically, the diet problem of Stigler [92] was the first rather long and complicated linear-programming problem to be solved by the simplex method. Here the problem was to determine what quantities of 77 foods should be bought in order not only to yield the minimum cost, but also to satisfy the minimum requirements of nine nutritive elements, e.g., vitamin A, niacin, and thiamine. The resulting purchases were to form the complete diet to sustain a person for a year. The basic linear-programming formulation of this problem was in terms of 9 equations and 86 variables, including 9 slack variables. The final solution obtained by the simplex method was, of course, a true minimum solution. However, since the simplex procedure deals only in terms of basic solutions, only nine of the possible foods were represented at a positive level in the minimum solution. Thus the diet obtained by linear-programming methods called for the purchase of varying amounts of wheat flour, corn meal, evaporated milk, peanut butter, lard, beef liver, cabbage, potatoes, and spinach, and cost $39.67 (for the year 1939). The solution obtained by Stigler using a systematic trial-and-error procedure required only five foods: wheat flour, evaporated milk, cabbage, spinach, and dried navy beans, and cost $39.93. Such diets, although quite inexpensive, are certainly unpalatable over any period of time, and the selection foods would do justice to the chief dietician of a slave-labor camp. As Stigler points out, "No one recommends these diets (*i.e., true minimum-cost diets*) to anyone, let alone everyone." He also cites a low-cost diet for 1939 that was constructed by a dietician and cost \$115. The difference in cost was attributed to the dietician's concern with the requirements of palatability, variety of diet, and prestige value of certain foods. Starting with our basic formulation, how can we modify our procedure or model to meet these additional dietary requirements?

Here we have a correct solution to the problem as stated, but this correct solution turns out to be unacceptable to execute. In order to overcome this defect, we could look for alternate optimum solutions and form various convex combinations of these solutions. In this manner we would be able to select a diet that included more than nine foods. Or we could look at the diets that just preceded the minimum-cost diet in the hope that one of them would be acceptable. Or, finally, we could reevaluate the linear-programming formulation of the problem to see if it really described the problem under investigation.

To correct the deficiencies inherent in the solution to the original problem a reformulation of this linear-programming problem is called for. The new model should enable more than nine foods to be in the minimum solution and also take into consideration human taste preferences for certain foods. These elements can be introduced into the model by inequalities that force some foods to be into the final solution in at least a minimum amount.[1] We could also determine preference weights and add them to the corresponding cost coefficients. Another approach to the general diet problem that would introduce more variety would be to subdivide the problem into smaller diet problems, each of which would involve only a single class of foods. In this process of suboptimizing, we might have the problem of selecting the minimum-cost diets for vegetables, or fruits, or meats, and the composite diet would be a solution to the general problem. In these attempts to introduce more realism into the problem, we suffer a corresponding increase in the cost of the resulting diet. If the problem is further constrained, the cost, in general, will increase, and it is up to the investigator to determine the relative value of dollar cost versus taste and variety.[2]

[1] It should be noted that such lower-bound inequalities do not increase the size of the linear-programming model. For example, if our lower bounds are a_j, then the set of inequalities is of the form $x_j \geq a_j$. Introducing nonnegative slack variables, we have $x_j - y_j = a_j$ or $x_j = y_j + a_j$. In our original set of constraints we need only substitute $y_j + a_j$ for the corresponding x_j, and hence the number of constraints and variables remains the same as in the original formulation.

[2] The reader is referred to work by Wolfe [102] where he solves a modified Stigler diet problem with a quadratic preference (objective) function. This treatment introduces more realism and variety into the optimum diet. Balintfy [3cc] develops a linear-programming approach to total menu planning which produces quite acceptable diets.

There are, however, many situations where the linear-programming formulation of the basic diet problem is applicable. These problems are concerned with minimum-cost feed mixtures for farm animals or with the mixing of various elements, e.g., chemicals or fertilizers, to meet minimum requirements at least cost. The application of linear-programming techniques to these problems is fairly straightforward.[1] To illustrate a typical problem and also to make available to the reader a rather long example of a typical simplex computation, we next present the computation of a minimum-cost dairy-feed diet as proposed by Waugh [100]. The formulation and computation of this problem are due to Goldstein [51*a*]. The computations are given in Tableau 11.2, pages 250 to 253.

This problem involved the selection of 10 feeds to meet the minimum requirements for digestible nutrients, digestible protein, calcium, and phosphorus. The feeds are listed in Tableau 11.2. The linear-programming model consisted of 4 equations in 14 variables, and the computation started with a full artificial basis. The costs and variables for the feeds are in terms of units of 100 lb. The excess elements are in units of 1 lb and have a zero cost. The minimum solution calls for the purchase of 18.771 lb of milo maize; 0.06142 lb of excess phosphorus; 17.020 lb of gluten feed; and 58.528 lb of flour middlings, for a total cost of $2.2798. In the final section of Tableau 11.2 we have presented only the $z_j - c_j$ elements and the $\mathbf{P}_0$ column and have left the rest of the elements to be computed by the reader. It is evident that these elements are not required for the simplex tableau corresponding to the minimum solution unless one wants to develop all the multiple minimum solutions. Here there are no such solutions.

Goldstein, in his computations, illustrates three different criteria for selecting a vector to go into the basis. The standard criterion, and the one most easily applied— that is, selecting the max $(z_j - c_j) > 0$ element— was used in going from the first to the second basis, the fourth to the fifth, and sixth to the seventh. The criterion used in going from the second to the third basis was that the vector selected should reduce the value of the objective function as much as possible. To simplify the application of this rule, the choice was limited to either x_1 or x_{10}, that is, to the two variables with the largest $z_j - c_j$ elements. In going from the third to the fourth basis, the vector was selected so as not only to reduce the value of the objective function but also to eliminate an artificial vector. This criterion enables the solution to "come closer" to a feasible solution. Finally, as there was some interest in the minimum feasible solution that involved no excess nutritive element, variable x_3 was chosen instead of x_{14} in going from the fifth to the sixth solution.

[1] For additional discussions concerning the formulation of realistic feed-blending problems, see Brigham [7*b*] and Jewell [62*a*].

Tableau 11.2. Simplex solution of Waugh's problem

	i	Basis	c		Corn	Oats	Milo maize	Bran	Flour middlings	Linseed meal	Cotton meal	Soybean meal
					2.40	2.52	2.18	2.14	2.44	3.82	3.55	3.70
				P_0	P_1	P_2	P_3	P_4	P_5	P_6	P_7	P_8
I.	1	P_{15}	w	74.2	78.6	70.1	80.1	67.2	78.9	77.0	70.6	78.5
	2	P_{16}	w	14.7	6.5	9.4	8.8	13.7	16.1	30.4	32.8	(37.1)
	3	P_{17}	w	0.14	0.02	0.09	0.03	0.14	0.09	0.41	0.20	0.26
	4	P_{18}	w	0.55	0.27	0.34	0.30	1.29	0.71	0.86	1.22	0.59
	5			0	−2.40	−2.52	−2.18	−2.14	−2.44	−3.82	−3.55	−3.70
	6			89.59	85.39	79.93	89.23	82.33	95.8	108.67	104.82	116.45
II.		P_{15}	w	43.09623	(64.84663)	50.2105	61.4800	38.2121	44.8340	12.6766	1.19838	0
		P_8	3.70	0.39623	0.17520	0.25337	0.23720	0.36927	0.43396	0.81941	0.88410	1
		P_{17}	w	0.03698	−0.025552	0.024125	−0.03167	0.04399	−0.02283	0.19696	−0.02986	0
		P_{18}	w	0.31623	0.16663	0.19051	0.16005	1.07213	0.45396	0.37655	0.69838	0
				1.46603	−1.75176	−1.58254	−1.3024	−0.77369	−0.83434	−0.7882	−0.27886	0
				43.44943	64.98771	50.4251	61.6084	39.3282	45.2651	13.2501	1.86690	0
III.		P_1	2.40	0.66459	1	0.77430	0.94808	0.58927	0.69138	0.19549	0.018480	0
		P_8	3.70	0.27979	0	0.11771	0.071090	0.26603	0.31283	0.78516	0.88086	1
		P_{17}	w	0.053964	0	0.043910	−0.007445	0.05905	−0.005162	0.20195	−0.02939	0
		P_{18}	w	0.205485	0	0.061490	0.002074	0.97394	(0.33876)	0.34398	0.69530	0
				2.63023	0	−0.22616	0.35844	0.25856	0.37680	−0.44575	−0.24647	0
				0.25945	0	0.10540	−0.00537	1.03299	0.33359	0.54593	0.66591	0

IV.	P_1	2.40	0.24520	1	0.64880	0.94385	−1.39850	0	−0.50655	−1.4006	0
	P_8	3.70	0.09003	0	0.06093	0.069175	−0.63337	0	0.46751	0.23877	1
	P_{17}	w	0.057096	0	0.04485	−0.007413	0.07389	0	0.20719	−0.01880	0
	P_5	2.44	0.60659	0	0.18152	0.006122	2.87505	1	1.01541	2.05252	0
			2.40167	0	−0.29455	0.35613	−0.82475	0	−0.82835	−1.01985	0
			0.05710	0	0.04485	−0.007413	0.07389	0	0.20719	−0.01880	0
V.	P_1	2.40	0.30028	1	0.69206	(0.93670)	−1.32721	0	−0.30666	−1.41873	0
	P_8	3.70	0.07050	0	0.04559	0.07171	−0.65865	0	0.39664	0.24520	1
	P_9	2.60	0.16353	0	0.12845	−0.02123	0.21163	0	0.59343	−0.05383	0
	P_5	2.44	0.413005	0	0.02946	0.03126	2.62452	1	0.31291	2.11624	0
			2.41444	0	−0.28452	0.35447	−0.80822	0	−0.78200	−1.02406	0
			0	0	0	0	0	0	0	0	0
VI.	P_3	2.18	0.32057	1.06758	0.73883	1	−1.4169	0	−0.32739	−1.5146	0
	P_8	3.70	0.04751	−0.076556	−0.007395	0	−0.55704	0	0.42012	0.35381	1
	P_9	2.60	0.17034	0.02267	0.14413	0	0.18155	0	0.58648	−0.08599	0
	P_5	2.44	0.40298	−0.03337	0.006371	0	2.66881	1	0.32315	2.16358	0
			2.3008	−0.37842	−0.54642	0	−0.30598	0	−0.66595	−0.48717	0
VII.	P_3	2.18	0.18771								
	P_{14}	0	0.06142								
	P_9	2.60	0.17020								
	P_5	2.44	0.58528								
			2.2798	−0.3446	−0.5432	0	−0.05964	0	−0.85174	−0.64364	−0.44224

Tableau 11.2. Simplex solution of Waugh's problem (continued)

	i	Basis	c	Gluten feed	Hominy feed	Excess nutrients	Excess protein	Excess calcium	Excess phosphorus	Artificial basis			
				2.60	2.54	0	0	0	0	w	w	w	w
				$\mathbf{P}_9$	$\mathbf{P}_{10}$	$\mathbf{P}_{11}$	$\mathbf{P}_{12}$	$\mathbf{P}_{13}$	$\mathbf{P}_{14}$	$\mathbf{P}_{15}$	$\mathbf{P}_{16}$	$\mathbf{P}_{17}$	$\mathbf{P}_{18}$
I.	1	$\mathbf{P}_{15}$	w	76.3	84.5	−1	0	0	0	1	0	0	0
	2	$\mathbf{P}_{16}$	w	21.3	8.0	0	−1	0	0	0	1	0	0
	3	$\mathbf{P}_{17}$	w	0.48	0.22	0	0	−1	0	0	0	1	0
	4	$\mathbf{P}_{18}$	w	0.82	0.71	0	0	0	−1	0	0	0	1
	5			−2.60	−2.54	0	0	0	0	0	0	0	0
	6			98.90	93.43	−1	−1	−1	−1	0	0	0	0
II.		$\mathbf{P}_{15}$	w	31.2313	67.5728	−1	2.11590	0	0	1		0	0
		$\mathbf{P}_8$	3.70	0.57412	0.21563	0	−0.02695	0	0	0		0	0
		$\mathbf{P}_{17}$	w	0.33073	0.16394	0	0.007008	−1	0	0		1	0
		$\mathbf{P}_{18}$	w	0.48127	0.58278	0	0.015903	0	−1	0		0	1
				−0.47575	−1.7422	0	−0.099730	0	0	0		0	0
				32.0432	68.3195	−1	2.13881	−1	−1	0		0	0
III.		$\mathbf{P}_1$	2.40	0.48162	1.04204	−0.015421	0.03263	0	0			0	0
		$\mathbf{P}_8$	3.70	0.48974	0.03307	0.0027018	−0.03267	0	0			0	0
		$\mathbf{P}_{17}$	w	0.34304	0.19056	−0.000394	0.007842	−1	0			1	0
		$\mathbf{P}_{18}$	w	0.40101	0.40914	0.002570	0.010466	0	−1			0	1
				0.36793	0.08324	−0.027014	−0.042572	0	0			0	0
				0.74405	0.59970	1.002176	0.018302	−1	−1			0	0

IV.	P_1	2.40	−0.33683	0.20701	−0.02067	0.011269	0	2.04095		0	
	P_8	3.70	0.11942	−0.34476	0.000329	−0.04234	0	0.92347		0	
	P_{17}	w	0.34915	0.19680	−0.000355	0.00800	−1	−0.01524		1	
	P_5	2.44	1.18378	1.20777	0.007585	0.030895	0	−2.95198		0	
			−0.078113	−0.37185	−0.029872	−0.05421	0	1.11230		0	
			0.34915	0.19680	−0.000355	0.00800	0	−0.01524		0	
V.	P_1	2.40	0	0.39686	−0.02101	0.018987	−0.96473	2.02625			
	P_8	3.70	0	−0.41207	0.000450	−0.04507	0.34203	0.92868			
	P_9	2.60	1	0.56365	−0.001016	0.022913	−2.86412	−0.04365			
	P_5	2.44	0	0.54053	0.008788	0.003771	3.39050	−2.90031			
			0	−0.32782	−0.02995	−0.052423	−0.22372	1.10888			
			0	0	0	0	0	0			
VI.	P_3	2.18	0	0.42368	−0.02243	0.02027	−1.02992	2.16318			
	P_8	3.70	0	−0.44246	0.002058	−0.04653	0.41589	0.77356			
	P_9	2.60	1	0.57265	−0.001492	0.02334	−2.8860	0.00228			
	P_5	2.44	0	0.52729	0.009489	0.003138	3.4227	−2.96792			
			0	−0.47800	−0.02200	−0.05961	0.14135	0.34210			
VII.	P_3	2.18									
	P_{14}	0									
	P_9	2.60									
	P_5	2.44									
			0	−0.28233	−0.02241	−0.03903	−0.04257	0			

4. NETWORK FLOW PROBLEMS[1]

Let us first consider the mathematical formulation of the following problem. We are given a *transportation network* (pipeline system, railroad system, communication links) through which we wish to send a homogeneous commodity (oil, freight cars, message units) from a particular point of the network called the *source node* to a designated destination called the *sink node*. In addition to the source and sink nodes, the network consists of a set of *intermediate nodes* which are connected to each other or to the source and sink nodes by *arcs* or *links* of the network. These intermediate nodes can be interpreted as switching or transshipment points. We shall label the source node 0 and the sink node m and refer to the intermediate nodes by numbers or as node $i, j, k, \ldots$. We shall designate the arc connecting nodes i and j by the ordered pair (i,j) and assume that the flow of the commodity is directed from i to j; that is, an arc is a one-way street. If the flow can also go from j to i, the network would include both directed arcs (i,j) and (j,i). Each arc can accommodate a nonnegative flow, and we shall assume that each arc has a finite upper bound on its capacity which we shall designate f_{ij}. If we let x_{ij} be the unknown flow from node i to node j, then we have $0 \leq x_{ij} \leq f_{ij}$.

The flow of the commodity which originates at the source 0 is sent along the arcs to the intermediate nodes and then transshipped along additional arcs to other intermediate nodes or to the sink until all the commodity which began the trip at node 0 finally arrives at node m. That is, we impose upon the network the condition of *conservation of flow* at the intermediate nodes; what is shipped into a node is shipped out. The problem is to determine the maximum amount of flow f which can be sent from the source to the sink. We also, of course, wish to determine which arcs are used and to what extent of their capacity. A typical network is shown in Fig. 11.5. The number on each arc represents the capacity f_{ij} on arc (i,j).

We shall use the convention that if an arc (i,j) does not exist, we let $f_{ij} = 0$; that is, all arcs are possible, but flow is restricted to that set for which the corresponding $f_{ij} > 0$. Since we require the total flow into a

[1] This section is based on lectures by Prof. Robert Oliver, University of California, Berkeley, Calif. and Ford and Fulkerson [43*a*].

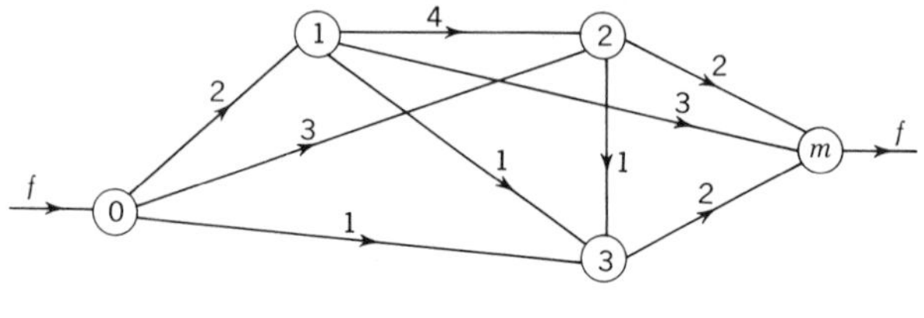

FIGURE 11.5

node to equal the total flow out of it, the mathematical model of the *maximal-flow network problem* is given by the following: Maximize

$$f$$

subject to

$$\sum_j (x_{0j} - x_{j0}) = f \tag{4.1}$$

$$\sum_j (x_{ij} - x_{ji}) = 0 \qquad i \neq 0,m \tag{4.2}$$

$$\sum_j (x_{mj} - x_{jm}) = -f \tag{4.3}$$

$$0 \leq x_{ij} \leq f_{ij} \tag{4.4}$$

Equation (4.1) represents the conservation of flow at the source; *i.e.*, the total amount shipped from node 0 minus the total amount shipped back to node 0 must equal the total flow in the network; Eqs. (4.2) represent the conservation at the intermediate nodes, while Eq. (4.3) represents the conservation equation at the sink. We note that the x_{ij} are non-negative variables with finite upper bounds and that $x_{ij} = 0, f = 0$ is a feasible solution. Hence, the maximal-flow problem is always feasible; i.e., the problem is a feasible linear-programming problem.

Let us write in explicit form the matrix of the network of Fig. 11.5. If we let the arcs correspond to the columns of the matrix and the nodes correspond to the rows, we obtain the following tableau (ignoring the capacity constraints):

nodes \ arcs	x_{01}	x_{02}	x_{03}	x_{12}	x_{13}	x_{1m}	x_{23}	x_{2m}	x_{3m}		f
0	1	1	1							=	1
1	−1			1	1	1				=	0
2		−1		−1			1	1		=	0
3			−1		−1		−1		1	=	0
m						−1		−1	−1	=	−1

(4.5)

The matrix of the equations is called the *node-arc incidence matrix*. For such matrices we note that one of the equations is redundant. We can, for example, drop the last equation which is given by the negative sum of the remaining equations. If we were to solve this problem by the simplex method we would couple m equations from (4.5) with the corresponding upper-bound constraints and apply the simplex algorithm for bounded variables. The objective function would be to maximize f. However, because of its simple structure, more efficient procedures, to be discussed below, are available. The reader will note that any problem which takes on the form of a node-arc incidence matrix can be interpreted as a network problem; i.e., each column has one $+1$ and no more than one -1. Before describing the procedure for solving the maximal-flow problem, we shall introduce some appropriate concepts from the theory of networks (Ford and Fulkerson [43*a*], Busacker and Saaty [8*b*], and Simonnard [90*b*]).

Formally, a *network* is a collection of elements i, j, k, . . . called nodes and a set of ordered pairs (i,j) called arcs. We shall discuss only *directed networks;* i.e., for an arc (i,j) the direction of flow is restricted to be from i to j. The network can also have the arc (j,i). A sequence of arcs (i,j), (j,k), . . . , (p,q) is called a *chain* which connects node i to node q. If $i = q$, then the chain is termed a *cycle* or *loop*. If we ignore the direction of an arc, then we call an *undirected chain* or *path* from i to q any sequence of arcs which form a connected set of arcs from i to q.

Let all the nodes of a given network be divided into two disjoint subsets S and $\bar{S}$; i.e., a node will be either in S or $\bar{S}$, but not in both. In addition, let $(S,\bar{S})$ denote the set of arcs that lead from those $i \in S$ to those $j \in \bar{S}$, where the symbol $\in$ means "contained in." For example, in Fig. 11.5, if $S = (0,1,2)$ and $\bar{S} = (3,m)$, then $(S,\bar{S}) = (0,3)$, $(1,3)$, $(2,3)$, $(1,m)$, $(2,m)$. A *cut* is the set of arcs $(S,\bar{S})$ with $0 \in S$ and $m \in \bar{S}$.† For Fig. 11.5, $S = (0,1,2)$, and $\bar{S} = (3,m)$, the cut is shown by dotted lines in Fig. 11.6. A network has a finite number of cuts. Another cut for the

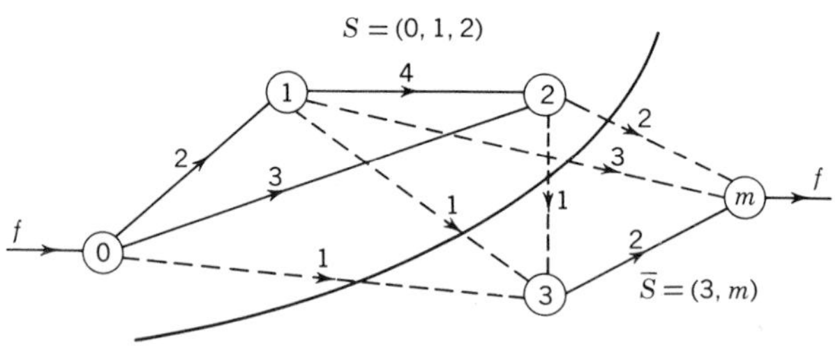

FIGURE 11.6

given network is $(0,1)$, $(0,2)$, $(0,3)$, as shown in Fig. 11.7; here $S = (0)$ and $\bar{S} = (1,2,3,m)$. We define the *capacity of a cut* $f(S,\bar{S})$ to be the sum of the individual capacities of all the arcs in the cut; that is, $f(S,\bar{S}) = \sum_{\substack{i \in S \\ j \in \bar{S}}} f_{ij}$.

The cut in Fig. 11.6 has a capacity of 8, while the cut in Fig. 11.7 has a

† An alternative definition of a cut is that it is a set of directed arcs such that every chain from the source node to the sink node contains at least one arc of the cut. Hence, if all the arcs of a cut were deleted from the network, there would be no chain from 0 to m.

capacity of 6. Since if we remove a cut from a network there would be no chains from source to sink, i.e., the maximal flow would be zero, the maximal flow cannot exceed the capacity of the minimal cut capacity. The cut of Fig. 11.6 limits the flow to no more than 8, while the cut of Fig. 11.7 limits the flow to no more than 6. The following theorem, the *max-flow min-cut theorem*, establishes the existence of the maximum flow whose capacity is exactly equal to the minimal cut capacity. For our example, this is a flow of 6.

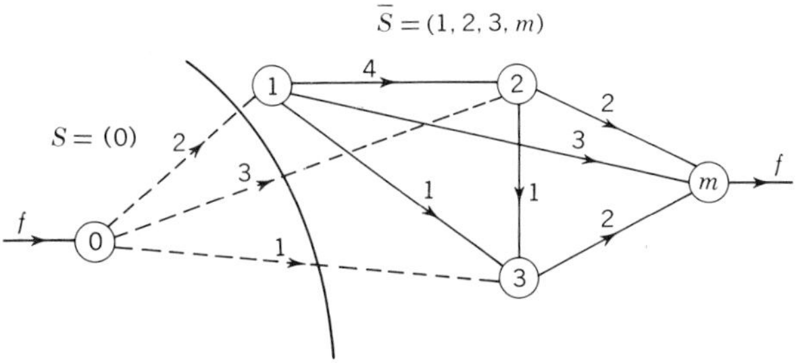

FIGURE 11.7

Max-flow min-cut theorem. *For any network the maximal-flow value from node 0 to node m is equal to the minimal cut capacity.*

Proof. The proof is given in two parts. We first show that the maximum flow is less than or equal to the minimum cut and then show that a flow exists which is equal to the minimal cut capacity.

For any cut $(S,\bar{S})$ we sum Eqs. (4.1) and (4.2) for all $i \in S$ to obtain

$$f = \sum_{i \in S} \left(\sum_{j} (x_{ij} - x_{ji}) \right)$$

This can be rewritten as

$$f = \sum_{\substack{i \in S \\ j \in S}} x_{ij} + \sum_{\substack{i \in S \\ j \in \bar{S}}} x_{ij} - \sum_{\substack{i \in S \\ j \in S}} x_{ji} - \sum_{\substack{i \in S \\ j \in \bar{S}}} x_{ji}$$

or

$$f = \sum_{\substack{i \in S \\ j \in \bar{S}}} x_{ij} - \sum_{\substack{i \in S \\ j \in \bar{S}}} x_{ji} + \left(\sum_{\substack{i \in S \\ j \in S}} x_{ij} - \sum_{\substack{i \in S \\ j \in S}} x_{ji} \right) \tag{4.6}$$

As both indices i and j of the term in parentheses range over the same set of values, this term will be equal to zero. We then have

$$f = \sum_{\substack{i \in S \\ j \in \bar{S}}} x_{ij} - \sum_{\substack{i \in S \\ j \in \bar{S}}} x_{ji} \tag{4.7}$$

Since each flow $x_{ji} \geq 0$ and each $x_{ij} \leq f_{ij}$, (4.7) becomes

$$f \leq \sum_{\substack{i \in S \\ j \in \bar{S}}} x_{ij} \leq \sum_{\substack{i \in S \\ j \in \bar{S}}} f_{ij} = f(S,\bar{S}) \tag{4.8}$$

and hence, the maximal flow is bounded above by the capacity of the arbitrary cut $(S,\bar{S})$, and thus f must be bounded above by the minimal cut capacity; that is, $f \leq \min f(S,\bar{S})$ for all possible cuts.

We next establish that for some flow and corresponding cut (to be defined), the equality of (4.8) holds. Since all arc capacities are finite, a maximum flow exists, although its value could be zero. Let the value of the maximal flow be f' and for this f' and corresponding arc flows x_{ij} we define a cut $(S',\bar{S}')$ as follows.

Let S' include the source node 0. For any $i \in S'$ if $x_{ij} < f_{ij}$, then $j \in S'$. In addition, if $i \in S'$ and $x_{ji} > 0$, then $j \in S'$. Let $\bar{S}'$ be all nodes not in S'. We first note that the set of nodes $\bar{S}'$ includes the sink node m. If $m \notin \bar{S}'$, that is, both nodes 0 and m are in S', then we can find a path from 0 to m such that for any arc in the path either $f_{ij} - x_{ij} > 0$ or $x_{ji} > 0$. We could then increase the flow from 0 to m by adding a small amount to those arcs for which $f_{ij} - x_{ij} > 0$ and subtracting the same amount from those arcs having $x_{ji} > 0$. Hence, we would have a flow from 0 to m which would be greater than f', contradicting the assumption that f' is the maximal flow. Therefore $m \in \bar{S}'$, and $(S',\bar{S}')$ is a cut. We next note that $x_{ij} = f_{ij}$ when $i \in S'$ and $j \in \bar{S}'$, and $x_{ji} = 0$ for $i \in S'$ and $j \in \bar{S}'$. This results from the definition of S' and $\bar{S}'$. Then for the cut $(S',\bar{S}')$, Eq. (4.7) becomes

$$f = \sum_{\substack{i \in S' \\ j \in \bar{S}'}} x_{ij} - \sum_{\substack{i \in S' \\ j \in \bar{S}'}} x_{ji} = \sum_{\substack{i \in S' \\ j \in \bar{S}'}} x_{ij} = \sum_{\substack{i \in S' \\ j \in \bar{S}'}} f_{ij} = f(S',\bar{S}')$$

But from (4.8) $f \leq f(S,\bar{S})$ for all cuts $(S,\bar{S})$ and since $f = f(S',\bar{S}')$, then $(S',\bar{S}')$ must yield the minimal cut capacity and $\max f = f' = \min f(S,\bar{S}) = f(S',\bar{S}')$.

At this point it is instructive to look at the dual problem of the maximal-flow problem. Recalling that f is a nonnegative variable and letting π_i be the dual variables associated with the ith node constraint given by (4.1), (4.2), and (4.3), and w_{ij} be the dual variables associated with the arc-capacity constraints (4.4), the dual problem as given by the node-arc incidence matrix is as follows: Minimize

$$\sum_i \sum_j f_{ij} w_{ij}$$

subject to

$$
\begin{aligned}
-\pi_0 + \pi_m \qquad\quad &\geq 1 \\
\pi_i - \pi_j + w_{ij} &\geq 0 \qquad \text{all } (i,j) \\
w_{ij} &\geq 0 \qquad \text{all } (i,j)
\end{aligned}
$$

The π_i are unrestricted as to sign as they correspond to primal equations. If we know a minimal cut $(S,\bar{S})$, we can construct the optimal solution to the dual problem as follows:

$$
\begin{aligned}
\pi_i &= 0 \qquad \text{for } i \in S \\
\pi_i &= 1 \qquad \text{for } i \in \bar{S} \\
w_{ij} &= 1 \qquad \text{for } (i,j) \in (S,\bar{S}) \\
w_{ij} &= 0 \qquad \text{for } (i,j) \notin (S,\bar{S})
\end{aligned}
\tag{4.9}
$$

This is a feasible solution, and the minimum value of the dual objective function is equal to the maximum value of the primal objective function. As pointed out in Ford and Fulkerson [43*a*], if we drop the redundant source equation, i.e., take $\pi_0 = 0$, it can be shown that all extreme-point solutions of the dual are of the form given by (4.9) for some S with $0 \in S$. Hence, the max-flow min-cut theorem would result from the duality theorem of linear programming.

We next describe an algorithm termed the *labeling method* for solving maximal-flow problems. The process converges in a finite number of iterations if the arc capacities f_{ij} are all rational numbers. Its validity is based on the max-flow min-cut theorem. The steps of the algorithm are given below and are designed to find a path over which a positive flow can be sent from source to sink. The steps are repeated until no such path can be found.

Maximal-flow algorithm

STEP 1. Find an initial feasible solution for the network. We may, for example, begin with all $x_{ij} = 0$.

STEP 2. Start with the source node 0 and give it a label $[-,\infty]$. The general label for any nodes i and j is indicated by $[i\pm,v_j]$, with v_j being a positive number representing a change in flow between i and j, $i+$ representing an increase of flow by the amount v_j from i to j, and $i-$ representing a decrease of flow by the amount v_j from j to i. The source label $[-,\infty]$ indicates that an unlimited amount of commodity is available at the source for shipment to the sink.

STEP 3. Select any unlabeled node i. Initially only node 0 is labeled.

a. For any unlabeled node j for which $x_{ij} < f_{ij}$, assign the label $[i+,v_j]$ to node j, where $v_j = \min(v_i, f_{ij} - x_{ij})$. This limits the amount sent from i to j either the amount already sent to i or the remaining capacity of the arc connecting i to j, whichever is smaller.

b. To all nodes j that are unlabeled and such that $x_{ji} > 0$, assign the label $[i-,v_j]$, where $v_j = \min\ (v_i,\ x_{ji})$. This enables us to reroute a flow going into i away from i. (For example, in Fig. 11.8 if we started with

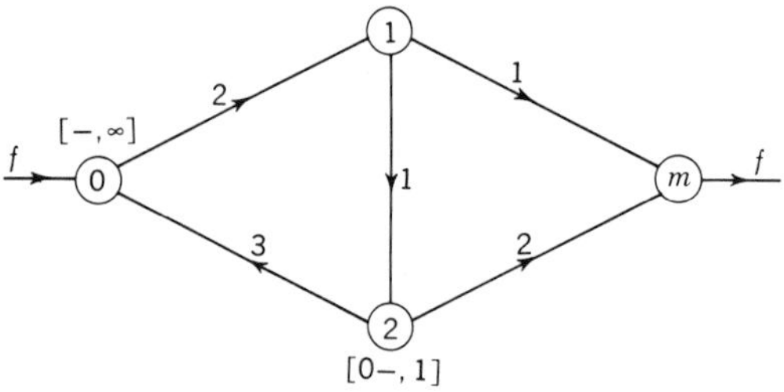

FIGURE 11.8

the initial solution $x_{01} = 2$, $x_{12} = 1$, $x_{20} = 1$, $x_{1m} = 1$, $x_{2m} = 0$, we have, starting with the origin label $[-,\infty]$, that the label for node 2 should be $[0-,1]$ since $v_2 = \min\ (\infty,\ x_{20} = 1) = 1$.)

As node j receives a label, it is processed through Step 3 until all labeled nodes have been looked at with respect to the unlabeled nodes connecting them. Either this process will bring us along a path to the sink node which then gets a label (Step 4), or we cannot find such a path which leads to the sink (Step 5).

STEP 4. The sink node has received a label $[i\ \pm,\ v_m]$ with $i \neq m$ and $v_m > 0$ by the selection rule for v_m. Hence we have found what is termed a flow-augmenting path and can add or subtract v_m to the arcs of the path leading from 0 to m. The new feasible solution is

$$x'_{ij} = x_{ij} + v_m \qquad \text{for } (i,j) \text{ in path and } j \text{ has label } [i+,v_j]$$
$$x'_{ji} = x_{ji} - v_m \qquad \text{for } (j,i) \text{ in path and } j \text{ has label } [i-,v_j]$$
$$x'_{ij} = x_{ij} \qquad \text{for } (i,j) \text{ not in path}$$

The labels are all erased, and we repeat the process starting with Step 2 and the new feasible flow x'_{ij}.

STEP 5. The process terminates in that we cannot find a path from source to sink. The maximal flow f is equal to the sum of all the v_m generated in the applications of Step 3, assuming initial solutions of all $x_{ij} = 0$. To show that this flow is maximal, we let the set of labeled nodes in the final iteration correspond to the set of nodes S' in the proof of the max-flow min-cut theorem.

We illustrate the algorithm by finding the value of the maximal flow for the network of Fig. 11.5. We let the initial feasible solution be $x_{ij} = 0$ for all i,j. We indicate the current feasible solution and the arc capacities by the ordered pair of numbers (x_{ij}, f_{ij}) attached to each arc of the network as shown in Fig. 11.9. The node labels are indicated by $[i\pm, v_j]$. Along the flow-augmenting path $(0,1,2,m)$ of Fig. 11.10 we can send an additional two units of flow ($v_m = 2$) which yields a new feasible solution as shown in Fig. 11.11. We again find an unsaturated path as shown by the labels

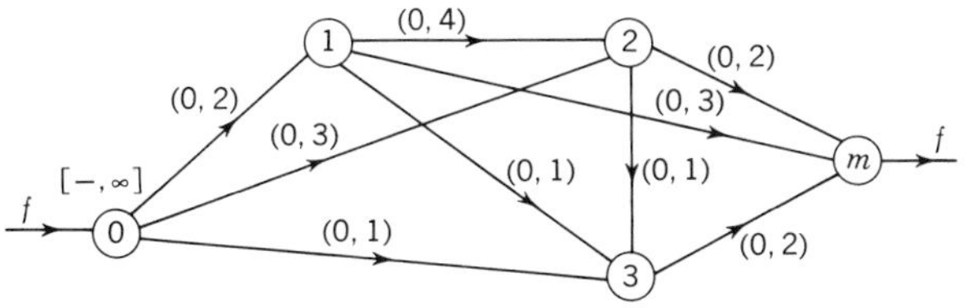

FIGURE 11.9

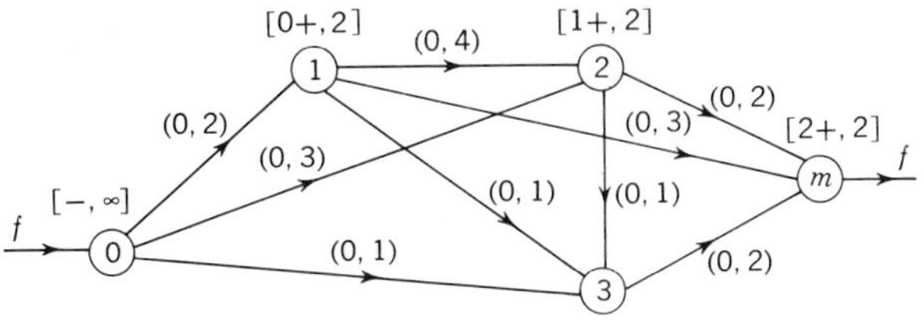

FIGURE 11.10

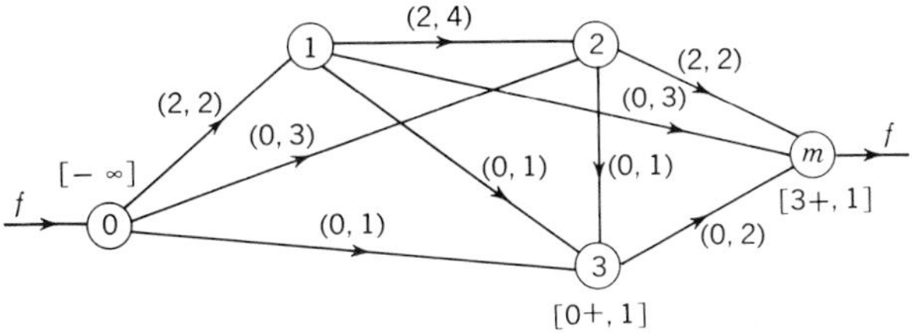

FIGURE 11.11

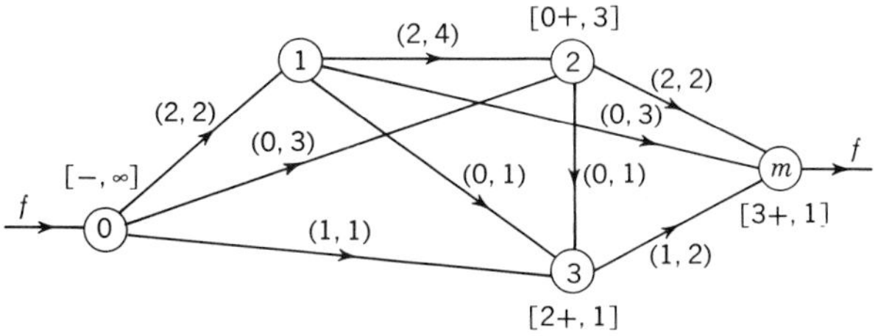

FIGURE 11.12

of Fig. 11.11. Here $v_m = 1$. The new feasible solution which sends a total of three units from source to sink is shown in Fig. 11.12, along with a new path over which an additional unit can be sent. The resulting feasible solution is shown in Fig. 11.13, along with a new unsaturated path for which $v_m = 2$. The reader will note that the finding of the labeled path of Fig. 11.13 required an application of the rerouting Step 3b. The corresponding feasible solution is shown in Fig. 11.14. This solution has a flow of six units and is optimum as we cannot label any node except the origin. The corresponding cut is given by arcs (0,1), (0,2), (0,3) and has a cut capacity of 6.

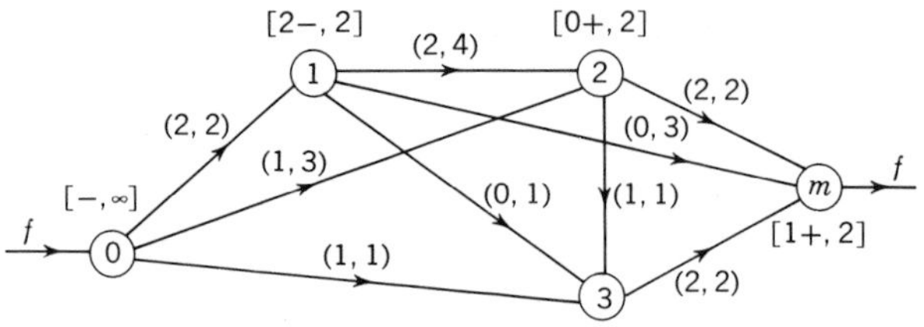

FIGURE 11.13

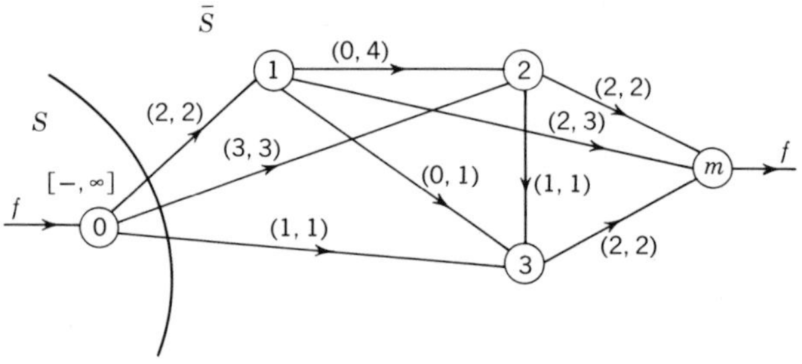

FIGURE 11.14

A number of modified maximal-flow network problems can be formulated in terms of the basic network model described above. For example, if a network has two (or more) sources, it can be recast into a single-source network by joining the given sources to a third source by artificial arcs which have infinite capacity, and similarly if we have multiple sinks. The added source or sink is identified by 0 or m, respectively. Another variation is found when in addition to the capacities on the arcs, we also have node capacities. These latter capacities could, for example, represent the maximum amount of the commodity that can be processed by the facilities at the node. We transform this problem into a maximal-flow problem by replacing node i with capacity k_i by two nodes i' and i'', where node i' is connected to the network by the same arcs going into i, nodes i' and i'' are connected by a directed arc from i' to i'' with capacity k_i (that is, $f_{i'i''} = k_i$), and node i'' is connected to the network by the arcs going out of i.

Another important class of network problems is the set of *minimal-cost network-flow problems*. As we shall see, this class of problems includes, among others, the transportation problem and the shortest-route problem.

We shall first develop the general formulation and then specialize it to particular problems.

For the minimal-cost flow problem we are given, as in the maximal-flow problem, a general network over which units of a homogeneous commodity are to be shipped from the source to the sink. Associated with each arc (i,j) is a cost c_{ij} of shipping one unit of the commodity from node i to node j.[1] We must ship a given quantity of F units from the source to the sink so as to minimize the total cost of shipping the F units to the sink. Assuming conservation of flow at the intermediate nodes and that the flow x_{ij} along an arc is bounded by both upper- and lower-capacity restrictions, $0 \leq l_{ij} \leq x_{ij} \leq f_{ij}$, the linear-programming model of the minimal-cost network problem is given by the following: Minimize

$$\sum_i \sum_j c_{ij} x_{ij}$$

subject to

$$\sum_j (x_{0j} - x_{j0}) = F \tag{4.10}$$

$$\sum_j (x_{ij} - x_{ji}) = 0 \qquad i \neq 0,m \tag{4.11}$$

$$\sum_j (x_{mj} - x_{jm}) = -F \tag{4.12}$$

$$0 \leq l_{ij} \leq x_{ij} \leq f_{ij} \tag{4.13}$$

A typical arc of a minimal-cost network is identified as shown in Fig. 11.15. We note that for a given F we have no assurance that the problem is

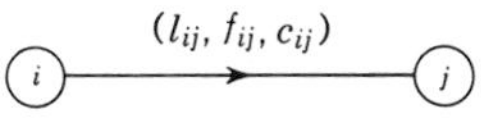

FIGURE 11.15

feasible. It is usually assumed that all coefficients are integers. We shall not describe any of the methods for solving this minimal-cost network problem as their development and explanation is rather involved. The most general algorithm, termed the *"out-of-kilter" algorithm,* is due to Fulkerson and is described in Ford and Fulkerson [43*a*] and Simonnard [90*b*].

If we let the network be a roadmap with the source being the origin city, the sink the destination city, and the c_{ij} being distances between cities (intermediate nodes), we can convert the above problem to that of finding the minimum distance between the origin and the destination by letting $F = 1$ and all $l_{ij} = 0$ and $f_{ij} = 1$. This problem can, of course, be

[1] We assume a linear cost relationship.

solved by the regular techniques of linear programming, but more efficient, specialized algorithms exist, for example, the simple combinatorial labeling scheme (Ford and Fulkerson [43*a*]; see also Dantzig [16*c*]). This procedure, described below, also yields the shortest routes from the source to all other nodes.

Shortest-route algorithm

STEP 1. Assign all nodes a label of the form $[-, \pi_i]$, where the first component indicates the preceding node in the shortest route and π_i indicates the shortest distance from node 0 to node i. Node 0 starts with a label of $\pi_0 = 0$, and its index is always $-$; all other nodes start with $\pi_i = \infty$.

STEP 2. For any arc (i,j) for which $\pi_i + c_{ij} < \pi_j$, change the label of node j to $[i, \pi_i + c_{ij}]$ and continue the process until no such arc can be found. In the latter situation the process is terminated, and the node labels indicate the shortest distance from node 0 to node j.

Ford and Fulkerson [43*a*] prove that the algorithm will terminate under the assumption that the sum of the costs around any directed cycle is nonnegative. The simple example of Fig. 11.16 illustrates the algorithm. The labels can be readily checked as being those which yield the minimum distances from the origin to the corresponding node. The minimum distance is 11.

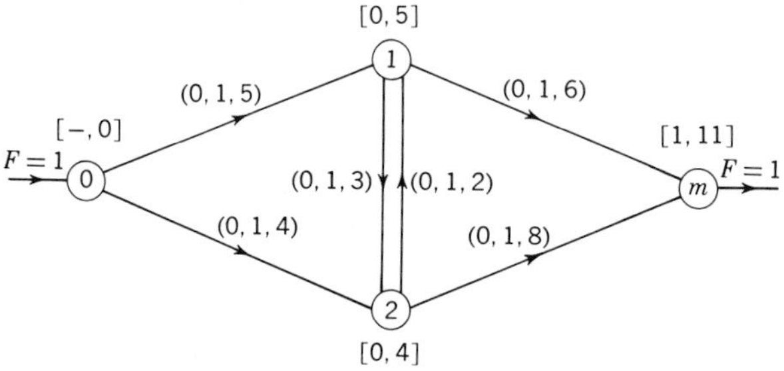

FIGURE 11.16

We next transform a transportation problem into a special network termed *bipartite network*.[1] For exposition purposes, let us consider a particular 2×3 transportation problem with costs c_{ij}, availabilities a_i, and requirements b_j, with $\Sigma a_i = \Sigma b_j$, as shown in Tableau 11.3 (see Chap. 10.) Here we use triangles and squares to distinguish between the original origins and the destinations. The minimal-flow network of Fig. 11.17

[1] A bipartite network is a network in which the nodes can be divided into two subsets such that the arcs of the network join the nodes of one subset to the other. In Fig. 11.17, we have a bipartite network if we omit nodes 0 and m.

represents an equivalent statement of the above transportation problem.

In addition to the method of Chap. 10, many special and computationally efficient algorithms for solving the transportation problem and the related assignment problem are available (Ford and Fulkerson [43*a*], Kuhn [67*a*], and Simonnard [90*b*]). Other special network problems

Tableau 11.3

	1	2	3	
1	c_{11}	c_{12}	c_{13}	a_1
2	c_{21}	c_{22}	c_{23}	a_2
	b_1	b_2	b_3	

include the following: (1) The *capacitated transportation problem*, which is a standard transportation problem with the added set of upper-bound constraints $0 \leq x_{ij} \leq f_{ij}$ for all shipments between origin i and destination j (Ford and Fulkerson [43*a*]). (2) The *multicommodity network-flow problem* in which more than one commodity can flow through the same net-

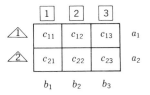

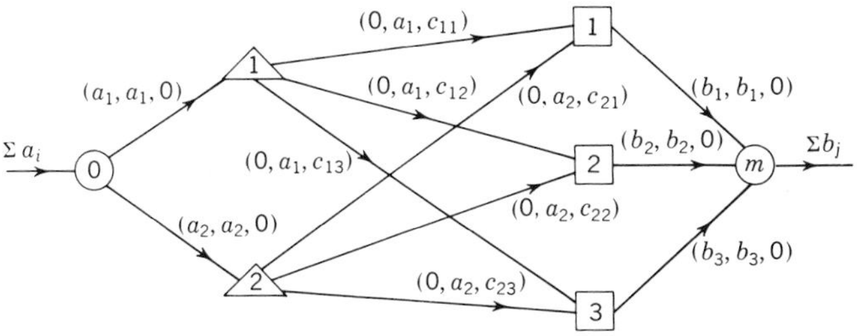

FIGURE 11.17

work. Each commodity has a separate origin and destination and the capacity of an arc cannot be exceeded by the sum of the individual flows of each commodity through that arc. In addition, each arc has specified capacity restrictions for each type of commodity. The object is to maximize the sum of the values of all the simultaneous flows (Ford and Fulkerson [43*a*] and Hu [61*b*]). (3) A *network with gains* is a single commodity network in which the flow x_{ij} sent from node i to node j along arc (i,j) is multiplied by a factor k_{ij} before it reaches node j; that is, $k_{ij}x_{ij}$ is the amount which arrives at node j. The multiplicative factor could represent the loss of a fluid due to evaporation or the gain of a product due to breeding (Jewell [62*b*]).

EXERCISES

1. Solve the following production-scheduling problem, using the reduced set of equations with the initial conditions $s_{-1} = 0$, $s_n = 0$, $r_0 = 0$. The objective function to be minimized is

$$16s_0 + \sum_{t=1}^{4} s_t + 15 \sum_{t=1}^{5} y_t$$

and requirements are $r_1 = 5$, $r_2 = 12$, $r_3 = 6$, $r_4 = 6$, $r_5 = 8$.

2. Formulate the following problem from Jacobs [62] as a linear-programming problem: A caterer knows that, in connection with the meals he has arranged to serve during the next n days, he will need $r_j \geq 0$ fresh napkins on the jth day, with $j = 1, 2, \ldots, n$. Laundering normally takes p days; i.e., a soiled napkin sent for laundering immediately after use on the jth day is returned in time to be used again on the $(j + p)$th day. However, the laundry also has a higher-cost service which returns the napkins in $q < p$ days (p and q integers). Having no usable napkins on hand or in the laundry, the caterer will meet his early needs by purchasing napkins at a cents each. Laundering expense is b and c cents per napkin for the normal and high-cost service, respectively. How does he arrange matters to meet his needs and minimize his outlays for the n days?

3. Given the following 3×3 input-output coefficient matrix:

$$\mathbf{A} = \begin{pmatrix} 0 & 0.4 & 0.3 \\ 0.2 & 0 & 0.5 \\ 0.1 & 0.2 & 0 \end{pmatrix}$$

Our three industries and their *unknown* production vectors $\mathbf{X}$, *unknown* bill-of-goods vectors $\mathbf{Y}$, *unknown* final-storage vectors $\mathbf{S}_1$, and known initial-storage vectors $\mathbf{S}_0$ are related by

$$(\mathbf{I} - \mathbf{A})\mathbf{X} = \mathbf{Y} + \mathbf{S}_1 - \mathbf{S}_0$$

The production vector is constrained by a *known* capacity vector $\mathbf{L}$, that is,

$$\mathbf{X} \leq \mathbf{L}$$

Here

$$\mathbf{S}_0 = \begin{pmatrix} 20 \\ 0 \\ 0 \end{pmatrix} \quad \text{and} \quad \mathbf{L} = \begin{pmatrix} 10 \\ 10 \\ 10 \end{pmatrix}$$

Write out the explicit equations of the system in terms of the three industries and determine values of $\mathbf{X}$, $\mathbf{Y}$, and $\mathbf{S}_1$ that maximize $\mathbf{cY}$, where $\mathbf{c} = (1,5,15)$. Here the problem is to determine not only a production vector but also a corresponding "optimal" bill of goods.

4. Formulate the following "trim problem" as a linear-programming problem and solve the exercise, both as a standard and an integer-programming problem. The description and numerical problem are from Eisemann [41*a*]. Also see Paull (Bibliography, Sec. 3*f*) and Gilmore and Gomory [49*a*].

> A problem of primary significance to a variety of industries is the suppression of trim losses in cutting rolls of paper, textiles, cellophane, metallic foil, or other material, for the execution of business orders. We define the measurement taken between the 2 circular ends of a roll as its "width," the measurement perpendicular

thereto when completely unwound as its "length." A number of cutting machines are at our disposal, the knives of which can be set for any combination of widths for which the combined total does not exceed the overall roll width. The knives then slice through the completely wound rolls much in the same way as a loaf of bread is sliced. Rolls are marketed in several standard widths. Orders specify desired widths and may prescribe either the number of rolls, or alternatively, the total length to be supplied for each width. In fitting the list of orders to the available rolls and machines, it is generally found that trimming losses due to odd end pieces are unavoidable; this wasted material represents a total loss, which may be somewhat alleviated by selling it as scrap. The problem then consists in fitting orders to rolls and machines in such a way as to subdue trimming losses to an absolute minimum.

As a numerical exercise, assume that two machines A and B are available; machine A is capable of cutting a standard roll of 100 in., and machine B is capable of cutting a standard roll of 80 in. The following orders of various widths must be cut: 45-in. roll, 862 rolls; 36-in. roll, 341 rolls; 31-in. roll, 87 rolls; and 14-in. roll, 216 rolls. We assume that as many standard rolls as necessary are available and that, excluding the unavoidable trim loss, only the widths (that is, 45, 36, 31, 14 in.) on order are cut. (HINT: Let the variables of the problem be equated to possible settings of the cutting blades; e.g., on machine A one setting of the blades would cut one 45-in. roll, one 31-in. roll, and one 14-in. roll, with a trim loss of 10 in. In order to ensure feasibility, the formulation must produce *at least* the number of each roll required.)

5. Formulate the following as a linear-programming problem (Magee [73]):

Consider a company that has one production line upon which it produces a single homogeneous commodity. We suppose that the commodity sells for a fixed unit price; that the costs for regular-time production, overtime production, and storage are known and vary between time periods; that the rate of production per unit time is known; and that an accurate sales forecast in the form of demand during each of a number of successive time periods is known. It is desired to formulate a production schedule that will meet the sales forecast and minimize the combined costs of production and storage.

t = number of time periods
r_i = number of units of finished product to be sold during ith time period
s_0 = initial inventory
m_i = maximum number of units that can be produced each time period on regular time
n_i = maximum number of units that can be produced each time period on overtime
a_i = cost of storage of one unit of product during time period i
c_i = cost of production of one unit on regular time during time period i
d_i = cost of production of one unit on overtime during time period i
x_i = regular-time production during ith time period
y_i = overtime production during ith time period

For concreteness, it will be assumed that each time period is 1 month in length and that the inventory for each month is taken on the last day of that month. This is equivalent to adding the month's production to inventory and withdrawing the month's sales from inventory at the end of each month.

6. *The multidimensional transportation problem.* Formulate the following example of a three-dimensional problem (Schell [90]). Consider a soap manufacturer who has

l factories in various parts of the country. Each of the l plants can manufacture m different types of soap. The soap is to be distributed from the factories to n different areas. Let

a_{ik} = required number of units to be shipped from factory i to area k
b_{jk} = required number of units of type j to be shipped to area k
d_{ij} = required number of units to be shipped from factory i of type j
x_{ijk} = amount of the jth type made in the ith plant shipped to the kth area
c_{ijk} = cost of shipping one unit of the jth type from the ith plant to the kth area
$x_{ijk} \geq 0$

Haley [52*e*] gives a computational procedure for solving this problem. There is no guarantee that a solution will be integral if the a_{ik}, b_{jk}, and d_{ij} are integral (Motzkin [76]).

7. Show how the caterer problem (Exercise 2) can be formulated and solved as a transportation problem (Garvin [45*a*] and Prager [85*b*]).

8. *The machine-loading problem.* Formulate the following problem. We are given m machines and n products and know the time a_{ij} required to process one unit of product j on machine i. Let x_{ij} be the number of units of product j produced on machine i, b_i the total time available on machine i, and d_j the number of units of product j which must be processed. We also know the cost (or time) c_{ij} of processing one unit of product j on machine i. (We assume each product needs to be processed on only one machine.) We wish to minimize the total cost. This problem can be looked at as a generalized transportation problem, and special algorithms exist for solving it (Eisemann and Lourie [41*b*] and Jewell [62*b*]).

9. Form the dual of the shortest-route problem of Fig. 11.16 and show that the final π_i of the Ford-Fulkerson algorithm represent an optimal solution to the dual.

10. Develop the equations for a two-commodity network with three nodes with a separate origin and destination for each commodity.

11. Construct an example to show that a multicommodity network with integer parameters can have a noninteger optimal solution.

12. Find the shortest-distance route from the origin to the destination in the following network:

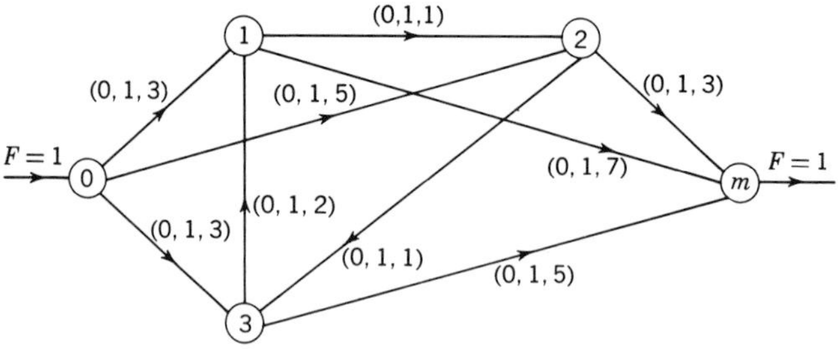

FIGURE E12

13. The transshipment problem is a generalized transportation problem in which each origin can ship to all other origins as well as to all destinations and a destination can in turn ship goods to all other destinations and all origins (see Exercise 8, Chap. 10). Show how to convert it to a minimum-cost network problem (Orden [85*a*] and Garvin [45*a*]).

14. Show how the caterer problem can be interpreted as a network problem (Ford and Fulkerson [43*a*]).

15. Write out the dual to the general minimum-cost network problem and develop the corresponding complementary slackness relationships.

16. *a.* Write out the explicit equations of the maximal-flow network shown below.
b. Solve the problem using the maximal-flow algorithm.

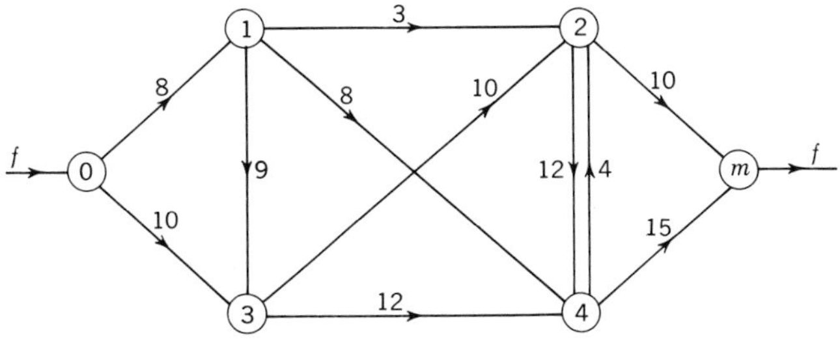

FIGURE E16

17. Develop the mathematical formulation of the network-with-gains problem (Jewell [62*b*]).

CHAPTER 12 / LINEAR PROGRAMMING AND THE THEORY OF GAMES

Like linear programming, the theory of games can be considered as a modern development in the field of mathematics. To the casual observer this would appear to be the only element which these areas have in common. For whereas, in the general linear-programming problem, we are concerned with the efficient use or allocation of limited resources to meet desired objectives, in the theory of games we are interested in developing a pattern or strategy of play for a given game which will enable us to win as much as possible. It is not until we develop the corresponding mathematical statement of a game-theory problem that the remarkable correspondence between these two problems is revealed. This relationship was first established by von Neumann and Dantzig. In this chapter we shall demonstrate the equivalence between the linear-programming model and the mathematical model of the important class of zero-sum two-person games.

1. INTRODUCTION TO THE THEORY OF GAMES

The main concern of the theory of games is the study of the following problem stated by von Neumann:[1] "If n players $P_1, P_2, \ldots, P_n$ play a given game Γ, how must the ith player, P_i, play to achieve the most favorable result?"

In interpreting this basic problem, we say that the term *game* refers to a set of rules and conventions for playing and a *play* refers to a particular possible realization of the rules, i.e., an individual contest. We shall assume that, at the end of each play of Γ, each of the players P_i will receive an amount of money v_i, called the *pay-off to player* P_i. We then assume that, in order to receive the most favorable result for himself, the only object of a player will be to maximize the amount of money he

[1] See Kuhn [67].

receives. In most parlor games, e.g., poker, the total amount of money lost by the losing players is equal to the total amount of money won by the winning players. Here we have

$$v_1 + v_2 + \cdots + v_n = 0$$

for all plays. Note that v_i can be positive, negative, or zero, where $v_i > 0$ represents a gain, $v_i < 0$ represents a loss, and $v_i = 0$ represents no pay-off. Games whose algebraic sum of the pay-offs is equal to zero are called *zero-sum games.* In zero-sum games, wealth is neither created nor destroyed by the players. An example of a non-zero-sum game would be poker in which a certain percentage of the pot is removed for the "house" before the final pay-off.

Games are also classified by the number of players and possible moves. Chess is a two-person game with a finite number of possible moves, and poker is a many-person game, also with a finite number of possible moves.[1] A duel in which the duelists may fire at any instant in a given time interval is a two-person game with an infinite number of possible moves. Games are further characterized by being cooperative or noncooperative. In the former the players have the ability to form coalitions and work as teams, while in the latter each player is concerned only with his own result. Two-person games are, of course, noncooperative. We shall concern ourselves only with finite, zero-sum, two-person games.

As an example of a finite, zero-sum, two-person game, let us consider the familiar game of matching pennies. To accomplish one move, which for this game is equivalent to a play, the first player P_1 selects either heads or tails, and the second player P_2, not knowing P_1's choice, also selects either heads or tails. After the choices have been made, P_2 pays P_1 1 unit if they match or -1 unit if they do not. The pay-off of -1 represents the giving of 1 unit by P_1 to P_2. The pay-off values of 1 and -1 are written in terms of a gain being a positive value to P_1 and a loss a negative value to P_1. When this is the case, we say that P_1 is the *maximizing player* and P_2 the *minimizing player*. We shall always consider P_1 as being the maximizing player and write the pay-off values accordingly.

The above statement of the problem can be summarized by the following diagram (Kuhn [67]):

		P_2 *Selections* Heads	P_2 *Selections* Tails
P_1 *Selections*	Heads	1	-1
	Tails	-1	1

[1] Here we include appropriate "stop rules" in the conventions of the games.

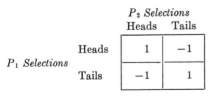

where at the end of a play P_2 pays P_1 the amount corresponding to the selections. We can further summarize the problem by means of the matrix

$$\begin{pmatrix} 1 & -1 \\ -1 & 1 \end{pmatrix}$$

Here the rows correspond to the possible choices for P_1 and the columns to the possible choices for P_2. As soon as P_1 selects a row and P_2 selects a column, the game has been played, and the pay-off value at the intersection of the selected row and column is paid to P_1.

After P_1 makes his selection, he must be careful to conceal his choice from P_2; otherwise P_2 will select the column that will yield a pay-off of -1. P_1 can hide his selection by using some chance device. For example, he could place two pieces of paper in a hat, one slip marked with an H and the other with a T. Before each play P_1 removes one of the marked slips from the hat, choosing heads if the slip has the H or tails if the slip has the T. The probability of P_1 choosing heads is $\frac{1}{2}$, and the probability of choosing tails is also $\frac{1}{2}$. Let us assume P_1 makes his choices in this manner. If P_2 selects heads, then P_1 has a mathematical expectation (i.e., the sum of the quantities obtained by multiplying the probability of choosing each row by the corresponding pay-off) of

$$\tfrac{1}{2}(1) + \tfrac{1}{2}(-1) = 0$$

and if P_2 selects tails,

$$\tfrac{1}{2}(-1) + \tfrac{1}{2}(1) = 0$$

Hence, P_1's expectation is zero. If fact, this is the only mode of play for P_1 that does not expose him to a negative expectation. To show this, suppose P_1 plays heads with probability x and tails with probability $1 - x$, where $0 \leq x \leq 1$. If P_2 plays heads, P_1's expectation is given by

$$E_H = x(1) + (1 - x)(-1) = 2x - 1$$

and if P_2 plays tails, P_1's expectation is

$$E_T = x(-1) + (1 - x)(1) = 1 - 2x$$

For $0 \leq x < \frac{1}{2}$, $E_H < 0$; and for $\frac{1}{2} < x \leq 1$, $E_T < 0$. Hence P_1 must select heads and tails with equal probabilities to obtain a zero expectation. The same situation is true for P_2.

Let us generalize the notions of the above discussion with the following definitions:

Definition 1. A *matrix game* Γ is determined by any $m \times n$ matrix

$$\mathbf{A} = \begin{pmatrix} a_{11} & a_{12} & \cdots & a_{1n} \\ a_{21} & a_{22} & \cdots & a_{2n} \\ \cdots & \cdots & \cdots & \cdots \\ a_{m1} & a_{m2} & \cdots & a_{mn} \end{pmatrix} = (a_{ij})$$

in which the elements a_{ij} are any real numbers. The matrix $\mathbf{A}$ is termed the pay-off matrix. The element a_{ij} represents the amount paid to P_1 by P_2 if P_1 selects the move associated with the ith row and P_2 selects the move associated with the jth column.

Definition 2. By a *mixed strategy for* P_1 we shall mean a row vector $\mathbf{X} = (x_1,x_2, \ldots ,x_m)$ of nonnegative numbers x_i such that

$$x_1 + x_2 + \cdots + x_m = 1$$

By a *mixed strategy for* P_2 we shall mean a column vector $\mathbf{Y} = (y_1,y_2, \ldots ,y_n)$ of nonnegative numbers y_j such that

$$y_1 + y_2 + \cdots + y_n = 1.$$

The elements x_i and y_j represent, respectively, the frequencies with which P_1 selects his ith move (row) and P_2 selects his jth move (column). For the matching-pennies game, a mixed strategy, i.e., the probabilities for selecting heads or tails, for P_1 could, for example, be any of the following: (1,0), (0,1), $(\frac{1}{4},\frac{3}{4})$, $(\frac{1}{2},\frac{1}{2})$. P_2 could employ similar mixed strategies.

Definition 3. For each $i = 1, 2, \ldots, m$, the mixed strategy which is 1 in the ith component and 0 elsewhere is called the *ith pure strategy for* P_1 and will be denoted by i.

Similarly, the *jth pure strategy for* P_2, denoted by j, is a mixed strategy for P_2 which is 1 in the jth conponent and 0 elsewhere.

Let us next consider the matrix game determined by the pay-off matrix

$$\mathbf{A} = \begin{pmatrix} a_{11} & a_{12} & \cdots & a_{1n} \\ a_{21} & a_{22} & \cdots & a_{2n} \\ \cdots & \cdots & \cdots & \cdots \\ a_{m1} & a_{m2} & \cdots & a_{mn} \end{pmatrix}$$

If P_1 chooses any pure strategy i, he is sure of being paid at least $\min_j a_{ij}$. For example, if the matrix game has the following 3×4 pay-off matrix:

$$\begin{pmatrix} 1 & 5 & 0 & 4 \\ 2 & 1 & 3 & 3 \\ 4 & 2 & -1 & 0 \end{pmatrix}$$

we have

$$\begin{aligned} &\min_j a_{1j} = a_{13} = 0 \qquad \min_j a_{2j} = a_{22} = 1 \\ &\min_j a_{3j} = a_{33} = -1 \end{aligned} \tag{1.1}$$

Since P_1 can select any i, he can, or course, select the i which *maximizes* his gain. That is, he can select the pure strategy that will make $\min_j a_{ij}$ as large as possible. By using this pure strategy, P_1 can win at least $\max_i \min_j a_{ij}$. For the 3×4 pay-off matrix we have from (1.1) that

$$\max_i \min_j a_{ij} = a_{22} = 1$$

If P_2 selects a j, the worst thing that can happen to him is to lose $\max_i a_{ij}$. P_2 can then select the pure strategy which *minimizes* his losses. By using this pure strategy, P_2 can keep P_1 from winning more than $\min_j \max_i a_{ij}$. For the 3×4 matrix we have

$$\begin{aligned} &\max_i a_{i1} = a_{31} = 4 \qquad \max_i a_{i2} = a_{12} = 5 \\ &\max_i a_{i3} = a_{23} = 3 \qquad \max_i a_{i4} = a_{14} = 4 \end{aligned}$$

and

$$\min_j \max_i a_{ij} = a_{23} = 3$$

If

$$\max_i \min_j a_{ij} = \min_j \max_i a_{ij} = v \tag{1.2}$$

then P_1 can be sure of winning v, and he can be prevented from winning more than v by P_2. Any matrix game for which (1.2) holds can best be played by P_1 and P_2 playing the corresponding pure strategies. Since any deviation from this strategy by P_1 lowers his opportunity for winning at least v and since any deviation from this strategy by P_2 raises P_1's opportunity for winning more than v, these pure strategies are called *optimal pure strategies.*

The following matrix game is one which has for its optimal pure strategies $\mathbf{X} = (1,0,0)$ and $\mathbf{Y} = (1,0,0)$:

$$\begin{pmatrix} 3 & 5 & 6 \\ 1 & 2 & 3 \\ 0 & 7 & 4 \end{pmatrix}$$

Here

$$\max_i \min_j a_{ij} = \min_j \max_i a_{ij} = a_{11} = 3$$

The reader will note that the element a_{11} is both the minimum element in its row and the maximum element in its column. Any such element is called a *saddle point.* If the element a_{kl} is a saddle point, then the pure strategies k and l for P_1 and P_2, respectively, are optimal pure strategies and $a_{kl} = v$.

Since not all matrix games can be optimally played by means of pure strategies (e.g., matching pennies), we need to consider the notion of optimal mixed strategies.

Definition 4. The *pay-off function* for P_1, i.e., the mathematical expectation of P_1, is defined to be

$$E(\mathbf{X},\mathbf{Y}) = \mathbf{XAY} = \sum_{i=1}^{m} \sum_{j=1}^{n} x_i a_{ij} y_j$$

where $\mathbf{X} = (x_1,x_2, \ldots ,x_m)$ and $\mathbf{Y} = (y_1,y_2, \ldots ,y_n)$ are any mixed strategies for P_1 and P_2, respectively.

To illustrate this definition, let the matrix game have the following pay-off matrix:

$$\begin{pmatrix} 0 & -1 & 1 \\ 1 & 0 & -1 \\ -1 & 1 & 0 \end{pmatrix}$$

If P_1 selects the mixed strategy $\mathbf{X} = (x_1,x_2,x_3)$ and P_2 the mixed strategy $\mathbf{Y} = (y_1,y_2,y_3)$, then $E(\mathbf{X},\mathbf{Y})$ is equal to the matrix product

$$(x_1,x_2,x_3) \begin{pmatrix} 0 & -1 & 1 \\ 1 & 0 & -1 \\ -1 & 1 & 0 \end{pmatrix} \begin{pmatrix} y_1 \\ y_2 \\ y_3 \end{pmatrix}$$

or

$$E(\mathbf{X},\mathbf{Y}) = (x_2 - x_3)y_1 + (-x_1 + x_3)y_2 + (x_1 - x_2)y_3$$

If $\mathbf{X} = (0.1,0.4,0.5)$ and $\mathbf{Y} = (0.3,0.3,0.4)$, then

$$E(\mathbf{X},\mathbf{Y}) = -0.03$$

Hence if P_1 and P_2 employ the above mixed strategies, P_1's expected pay-off is -0.03 of a unit.

Definition 5. A *solution* to a matrix game Γ is a pair of mixed strategies

$$\bar{\mathbf{X}} = (\bar{x}_1,\bar{x}_2, \ldots ,\bar{x}_m)$$
$$\bar{\mathbf{Y}} = (\bar{y}_1,\bar{y}_2, \ldots ,\bar{y}_n)$$

and a real number v such that

$$E(\bar{\mathbf{X}},j) \geq v \qquad \text{for the pure strategies } j = 1, 2, \ldots, n$$
$$E(i,\bar{\mathbf{Y}}) \leq v \qquad \text{for the pure strategies } i = 1, 2, \ldots, m$$

The $\bar{\mathbf{X}}$ and $\bar{\mathbf{Y}}$ are called *optimal strategies*, and the number v is called the *value* of the game.

From this definition we see that, if P_1 selects his moves as determined by the probabilities of his optimal strategy, then no matter what moves P_2 selects, P_1 can expect to win at least v units. A similar statement is true for P_2; i.e., he can expect to lose no more than v units. We should note that v can be positive, negative, or zero. For the game of matching pennies, whose pay-off matrix is

$$\begin{pmatrix} 1 & -1 \\ -1 & 1 \end{pmatrix}$$

the value of the game is 0 and the optimal strategies are $\bar{\mathbf{X}} = (\frac{1}{2},\frac{1}{2})$ and $\bar{\mathbf{Y}} = (\frac{1}{2},\frac{1}{2})$.

The following game from Kuhn [67], the skin game, is an example of a game which at first glance appears to be fair, i.e., has a value of $v = 0$. The two players are each provided with an ace of diamonds and an ace of clubs. P_1 is also given the two of diamonds and P_2 the two of clubs. In the first move, P_1 shows one of his cards, and P_2, ignorant of P_1's choice, shows one of his cards. P_1 wins if the suits match, and P_2 wins if they do not. The amount of the pay-off is the numerical value of the card shown by the winner. If the two deuces are shown, the pay-off is zero. The pay-off matrix is then

$$\begin{array}{c} \\ \blacklozenge \\ \clubsuit \\ 2\blacklozenge \end{array}\begin{array}{c} \begin{array}{ccc} \blacklozenge & \clubsuit & 2\clubsuit \end{array} \\ \begin{pmatrix} 1 & -1 & -2 \\ -1 & 1 & 1 \\ 2 & -1 & 0 \end{pmatrix} \end{array}$$

Since every element of the third row is greater than or equal to the corresponding element in the first row, there is no advantage to P_1 in playing his first strategy. We can then assume he will always play his third pure strategy in preference to his first, and hence $x_1 = 0$. We say that row 1 is *dominated* by row 3, and we can eliminate row 1 from our pay-off matrix. We now have

$$\begin{pmatrix} -1 & 1 & 1 \\ 2 & -1 & 0 \end{pmatrix}$$

P_2 can apply the same type of analysis to the reduced pay-off matrix, and since the elements of the second column are less than or equal to the corresponding elements of the third column, P_2 will not select his third pure strategy; that is, $y_3 = 0$. The final reduced pay-off matrix is

$$\begin{pmatrix} -1 & 1 \\ 2 & -1 \end{pmatrix}$$

In looking for a solution to the game defined by this reduced pay-off matrix, we should first determine whether there is a saddle point. Here there is none. If P_1 selects his second strategy with probability x and his third with probability $1 - x$ and if P_2 uses probability y for his first strategy and $1 - y$ for his second, then

$$E(\mathbf{X},\mathbf{Y}) = (-3x + 2)y + (2x - 1)(1 - y)$$

The reader can readily verify that an optimal solution to the original game is given by $\bar{\mathbf{X}} = (0,3/5,2/5)$, $\bar{\mathbf{Y}} = (2/5,3/5,0)$, and $E(\bar{\mathbf{X}},\bar{\mathbf{Y}}) = v = 1/5$. Hence the game is biased to P_1. The reader can also verify that, for any other possible mixed strategies, $\mathbf{X} = (0, x, 1 - x)$ and $\mathbf{Y} = (y, 1 - y, 0)$,

$$E(\mathbf{X},\bar{\mathbf{Y}}) \leq E(\bar{\mathbf{X}},\bar{\mathbf{Y}}) \leq E(\bar{\mathbf{X}},\mathbf{Y})$$

or

$$\max_{\mathbf{X}} \min_{\mathbf{Y}} E(\mathbf{X},\mathbf{Y}) = \min_{\mathbf{Y}} \max_{\mathbf{X}} E(\mathbf{X},\mathbf{Y}) = v\dagger$$

In other words, if P_1 plays his optimal mixed strategy, his value of the game will not be less than $E(\bar{\mathbf{X}},\bar{\mathbf{Y}})$, no matter what strategy $\mathbf{Y}$ is used by P_2. Similarly, if P_2 plays his optimal mixed strategy, he can expect to keep P_1 from winning more than $E(\bar{\mathbf{X}},\bar{\mathbf{Y}})$. If both play their optimal strategies, then the expected outcome can be predetermined. This mode of playing is a conservative one, in that it does not let the players take advantage of any deviations made from the optimal strategy by their opponents.

Definition 6. A symmetric game has a skew-symmetric matrix; that is, $a_{ij} = -a_{ji}$.

† See Kuhn [67] for derivations of these expressions. The first is called the saddle-point statement of the theory of games, while the second is the corresponding min-max statement by von Neumann and Morgenstern.

We next show that the value of a symmetric game is zero and that both players have identical optimal strategies. The pay-off function for P_1 is given by

$$E(\mathbf{X},\mathbf{Y}) = \mathbf{XAY} = \sum_{i=1}^{m} \sum_{j=1}^{n} x_i a_{ij} y_j$$

It is easy to verify, for a skew-symmetric $\mathbf{A}$ and for $\mathbf{X} = \mathbf{Y}$, that

$$E(\mathbf{X},\mathbf{Y}) = \mathbf{XAX} = 0$$

Hence, if both players use the same mixed strategy, then the expected value is zero. Let the respective optimum strategies be denoted by $\bar{\mathbf{X}}$ and $\bar{\mathbf{Y}}$. We have $\max_{\mathbf{X}} \min_{\mathbf{Y}} \mathbf{XAY} = \min_{\mathbf{Y}} \bar{\mathbf{X}}\mathbf{AY} = v$. If P_2 uses any mixed strategy, then $\bar{\mathbf{X}}\mathbf{AY} \geq v$. But we know, for $\mathbf{Y} = \bar{\mathbf{X}}$, that $\bar{\mathbf{X}}\mathbf{A}\bar{\mathbf{X}} = 0$. Hence, since P_2 can always make the value of the game no greater than zero by playing $\mathbf{Y} = \bar{\mathbf{X}}$, we must conclude that $v \leq 0$. Similarly, we have $\min_{\mathbf{Y}} \max_{\mathbf{X}} \mathbf{XAY} = \max_{\mathbf{X}} \mathbf{XA}\bar{\mathbf{Y}} = v$. If P_1 uses any mixed strategy, then $\mathbf{XA}\bar{\mathbf{Y}} \leq v$. For $\mathbf{X} = \bar{\mathbf{Y}}$ we have $\bar{\mathbf{Y}}\mathbf{A}\bar{\mathbf{Y}} = 0$, and hence P_1 can always make $v \geq 0$. Therefore, we must have $v = 0$, and both players have the same optimal strategies $\bar{\mathbf{X}} = \bar{\mathbf{Y}}$.

As an example of a symmetric game, let us develop the game matrix for the children's game of Stone, Paper, Scissors. In this zero-sum two-person game both players simultaneously call out stone, paper, or scissors. The combination of paper and stone is a win of one unit for the player calling paper (paper covers stone); stone and scissors is a win for stone (stone breaks scissors); and scissors and paper is a win for scissors (scissors cuts paper). A call of the same item represents no pay-off. The pay-off matrix in terms of P_1 as the maximizing player is

	Stone	Paper	Scissors
Stone	0	−1	1
Paper	1	0	−1
Scissors	−1	1	0

The optimal strategy for both players is $(\frac{1}{3},\frac{1}{3},\frac{1}{3})$.

In the next section we shall want to change the value v of a game by adding a fixed amount w to each element of the pay-off matrix (a_{ij}). In this connection we wish to show that, for the new pay-off matrix $(a_{ij} + w)$, the optimal strategies are the same as for the original game and the value of the new game is $v + w$.

By definition we have

$$E_1(\mathbf{X},\mathbf{Y}) = \sum_{i=1}^{m} \sum_{j=1}^{n} x_i a_{ij} y_j \tag{1.3}$$

for the original game and

$$E_2(\mathbf{X},\mathbf{Y}) = \sum_{i=1}^{m} \sum_{j=1}^{n} x_i(a_{ij} + w)y_j \tag{1.4}$$

for the modified game. Expanding (1.4), we have

$$E_2(\mathbf{X},\mathbf{Y}) = \sum_{i=1}^{m} \sum_{j=1}^{n} x_i a_{ij} y_j + w \sum_{i=1}^{m} \sum_{j=1}^{n} x_i y_j \tag{1.5}$$

Since $\Sigma x_i = \Sigma y_j = 1$, (1.5) becomes

$$E_2(\mathbf{X},\mathbf{Y}) = \sum_{i=1}^{m} \sum_{j=1}^{n} x_i a_{ij} y_j + w$$

From (1.3) we have

$$E_2(\mathbf{X},\mathbf{Y}) = E_1(\mathbf{X},\mathbf{Y}) + w$$

We see that the constant w does not play any role in the selection of an optimal strategy for the transformed game. In fact, the second game is the same as the original if, before the start of each play, P_2 gives P_1 w units. Hence the optimum strategies are the same, and

$$E_2(\bar{\mathbf{X}},\bar{\mathbf{Y}}) = E_1(\bar{\mathbf{X}},\bar{\mathbf{Y}}) + w = v + w$$

By selecting a suitable w we can make all elements of the pay-off matrix positive, thereby ensuring that the modified game has a positive value.

We next state, without proof, the following theorem:

The fundamental theorem of matrix games. *For every matrix game* $\max_{\mathbf{X}} \min_{\mathbf{Y}} E(\mathbf{X},\mathbf{Y})$ *and* $\min_{\mathbf{Y}} \max_{\mathbf{X}} E(\mathbf{X},\mathbf{Y})$ *exist and are equal. That is, every matrix game has a solution.*

The reader is referred to Kuhn [67], McKinsey [72], Luce and Raiffa [71*a*], von Neumann and Morgenstern [80], and Owen [85*aa*] for proofs of this theorem.

2. THE EQUIVALENCE OF THE MATRIX GAME AND THE PROBLEM OF LINEAR PROGRAMMING

We shall assume that we are given an arbitrary matrix game:

$$\mathbf{A} = \begin{pmatrix} a_{11} & a_{12} & \cdots & a_{1n} \\ a_{21} & a_{22} & \cdots & a_{2n} \\ \cdots & \cdots & \cdots & \cdots \\ a_{m1} & a_{m2} & \cdots & a_{mn} \end{pmatrix}$$

From Definitions 4 and 5, the problem for P_1 is to find a vector

$$\mathbf{X} = (x_1, x_2, \ldots, x_m)$$

and a number v such that

$$\begin{aligned}
a_{11}x_1 + a_{21}x_2 + \cdots + a_{m1}x_m &\geq v \\
a_{12}x_1 + a_{22}x_2 + \cdots + a_{m2}x_m &\geq v \\
\cdots\cdots\cdots\cdots\cdots\cdots\cdots & \\
a_{1n}x_1 + a_{2n}x_2 + \cdots + a_{mn}x_m &\geq v \\
x_1 + x_2 + \cdots + x_m &= 1 \\
x_1 &\geq 0 \\
x_2 &\geq 0 \\
\ddots \quad & \vdots \\
x_m &\geq 0
\end{aligned} \tag{2.1}$$

Similarly, for P_2 we have

$$\begin{aligned}
a_{11}y_1 + a_{12}y_2 + \cdots + a_{1n}y_n &\leq v \\
a_{21}y_1 + a_{22}y_2 + \cdots + a_{2n}y_n &\leq v \\
\cdots\cdots\cdots\cdots\cdots\cdots\cdots & \\
a_{m1}y_1 + a_{m2}y_2 + \cdots + a_{mn}y_n &\leq v \\
y_1 + y_2 + \cdots + y_n &= 1 \\
y_1 &\geq 0 \\
y_2 &\geq 0 \\
\ddots \quad & \vdots \\
y_n &\geq 0
\end{aligned} \tag{2.2}$$

Since every element of **A** can be made positive by the addition of a suitable constant to all the a_{ij}, we can assume that $v > 0$. Let us divide each of the relationships in (2.1) and (2.2) by v and let

$$x'_i = \frac{x_i}{v} \qquad \text{and} \qquad y'_i = \frac{y_j}{v}$$

Note that

$$\sum_i x'_i = \frac{1}{v} \sum_i x_i = \frac{1}{v}$$

and

$$\sum_j y'_j = \frac{1}{v} \sum_j y_j = \frac{1}{v}$$

Hence, by minimizing $\sum_i x'_i$, P_1 will maximize the value of the game, and by maximizing $\sum_j y'_j$, P_2 will minimize the value of the game. We can then restate (2.1) and (2.2) in terms of equivalent linear-programming problems and obtain the following symmetric dual problems:

The primal problem. Find a vector $\mathbf{X}' = (x'_1, x'_2, \ldots, x'_m)$ which minimizes

$$x'_1 + x'_2 + \cdots + x'_m$$

subject to

$$\begin{aligned} a_{11}x'_1 + a_{21}x'_2 + \cdots + a_{m1}x'_m &\geq 1 \\ a_{12}x'_1 + a_{22}x'_2 + \cdots + a_{m2}x'_m &\geq 1 \\ \cdots\cdots\cdots\cdots\cdots\cdots\cdots\cdots \\ a_{1n}x'_1 + a_{2n}x'_2 + \cdots + a_{mn}x'_m &\geq 1 \\ x'_i &\geq 0 \end{aligned}$$

The dual problem. Find a vector $\mathbf{Y}' = (y'_1, y'_2, \ldots, y'_n)$ which maximizes

$$y'_1 + y'_2 + \cdots + y'_n$$

subject to

$$\begin{aligned} a_{11}y'_1 + a_{12}y'_2 + \cdots + a_{1n}y'_n &\leq 1 \\ a_{21}y'_1 + a_{22}y'_2 + \cdots + a_{2n}y'_n &\leq 1 \\ \cdots\cdots\cdots\cdots\cdots\cdots\cdots\cdots \\ a_{m1}y'_1 + a_{m2}y'_2 + \cdots + a_{mn}y'_n &\leq 1 \\ y'_j &\geq 0 \end{aligned}$$

Since every game has a solution, optimum solutions to the above problems exist and

$$\min \sum_i x'_i = \max \sum_j y'_j = \frac{1}{v}\dagger$$

It should be noted that, if only the primal or only the dual is solved, the optimal strategy for the other problem is contained in the simplex tableau

† The set of x'_i and y'_j which satisfies the linear-programming problems must, of course, be converted to the optimal x_i and y_j that solve the game problems.

of the corresponding final solution. The optimal strategy corresponds to the $z_j - c_j$ elements for the slack vectors.

An alternate method for reducing the game problem to a linear-programming problem is the following: For P_1 the problem is again given by (2.1). Let us write the first n inequalities as equalities by subtracting a nonnegative variable from each to obtain

$$\begin{array}{l}
a_{11}x_1 + a_{21}x_2 + \cdots + a_{m1}x_m - x_{m+1} = v \\
a_{12}x_1 + a_{22}x_2 + \cdots + a_{m2}x_m \qquad - x_{m+2} = v \\
\cdots\cdots\cdots\cdots\cdots\cdots\cdots\cdots\cdots \\
a_{1n}x_1 + a_{2n}x_2 + \cdots + a_{mn}x_m \qquad - x_{m+n} = v \\
x_1 + x_2 + \cdots + x_m = 1 \\
x_i \geq 0
\end{array}$$

By subtracting the first equation from the succeeding $n - 1$ equations and using the first equation for the objective function, since it is equal to v, we then have the equivalent linear-programming problem of maximizing

$$a_{11}x_1 + a_{21}x_2 + \cdots + a_{m1}x_m - x_{m+1} = v$$

subject to

$$\begin{array}{l}
(a_{12} - a_{11})x_1 + (a_{22} - a_{21})x_2 + \cdots + (a_{m2} - a_{m1})x_m + x_{m+1} - x_{m+2} = 0 \\
\cdots\cdots\cdots\cdots\cdots\cdots\cdots\cdots\cdots \\
(a_{1n} - a_{11})x_1 + (a_{2n} - a_{21})x_2 + \cdots + (a_{mn} - a_{m1})x_m + x_{m+1} \qquad - x_{m+n} = 0 \\
x_1 + x_2 + \cdots + x_m = 1 \\
x_i \geq 0
\end{array}$$

Example. Here we are given a matrix game with the following payoff matrix:

$$\mathbf{A} = \begin{pmatrix} 3 & -2 & -4 \\ -1 & 4 & 2 \\ 2 & 2 & 6 \end{pmatrix}$$

The development of the corresponding linear-programming problem (using the above alternate method) is as follows: The initial system of constraints is

$$\begin{array}{l}
3x_1 - x_2 + 2x_3 \geq v \\
-2x_1 + 4x_2 + 2x_3 \geq v \\
-4x_1 + 2x_2 + 6x_3 \geq v \\
x_1 + x_2 + x_3 = 1 \\
x_1 \geq 0 \\
x_2 \geq 0 \\
x_3 \geq 0
\end{array}$$

Subtracting nonnegative slack variables, we have

$$
\begin{aligned}
3x_1 - x_2 + 2x_3 - x_4 \qquad\qquad &= v \\
-2x_1 + 4x_2 + 2x_3 \qquad - x_5 \qquad &= v \\
-4x_1 + 2x_2 + 6x_3 \qquad\qquad - x_6 &= v \\
x_1 + x_2 + x_3 \qquad\qquad\qquad &= 1 \\
x_i &\geq 0
\end{aligned}
$$

By subtracting the first equation from the second and third, we find that the corresponding linear-programming problem is to maximize

$$3x_1 - x_2 + 2x_3 - x_4$$

subject to

$$
\begin{aligned}
-5x_1 + 5x_2 \qquad + x_4 - x_5 \qquad &= 0 \\
-7x_1 + 3x_2 + 4x_3 + x_4 \qquad - x_6 &= 0 \\
x_1 + x_2 + x_3 \qquad\qquad\qquad &= 1 \\
x_i &\geq 0
\end{aligned}
$$

As the computational techniques for the direct solution of large matrix games are rather cumbersome, the efficient computational techniques of linear programming are usually employed to solve such games. There are, however, rapid iterative methods for solving games that yield good approximations to the optimal solution (Williams [101] and Luce and Raiffa [71*a*]).

The converse problem of expressing a given linear-programming problem as a matrix game can be readily accomplished by considering the problem and its corresponding dual. Let the original problem be given in terms of the inequalities

$$
\begin{aligned}
a_{11}x_1 + \cdots + a_{1n}x_n &\geq b_1 \\
\cdots\cdots\cdots\cdots\cdots & \qquad\qquad (2.3) \\
a_{m1}x_1 + \cdots + a_{mn}x_n &\geq b_m
\end{aligned}
$$

$$c_1x_1 + \cdots + c_nx_n \geq f \qquad (2.4)$$

where $x_j \geq 0$ and f is the minimum of the objective function (2.4). The associated dual is

$$
\begin{aligned}
a_{11}w_1 + \cdots + a_{m1}w_m &\leq c_1 \\
\cdots\cdots\cdots\cdots\cdots & \qquad\qquad (2.5) \\
a_{1n}w_1 + \cdots + a_{mn}w_m &\leq c_n
\end{aligned}
$$

$$b_1w_1 + \cdots + b_mw_m \leq g \qquad (2.6)$$

where $w_i \geq 0$ and g is the maximum of the objective function (2.6). From the duality theorem of Chap. 5 we have that, if a finite optimum solution to either problem exists, then a finite optimum solution for the other also exists and $f = g$. Let us assume that such solutions do exist for the problems (2.3), (2.4) and (2.5), (2.6). Let $(x_1, \ldots, x_n)$ and $(w_1, \ldots, w_m)$ be solutions to the former and latter problems, respectively. Multiplying the first inequality of (2.3) by w_1, the second by w_2, etc., and summing the m inequalities, we have

$$b_1w_1 + \cdots + b_mw_m \leq \left(\sum_{i=1}^{m} a_{i1}w_i\right) x_1 + \cdots + \left(\sum_{i=1}^{m} a_{in}w_i\right) x_n \tag{2.7}$$

From (2.5) we can replace the sums in (2.7) by their corresponding c_j to obtain the following inequality that is true for all solutions to the primal and dual:

$$b_1w_1 + \cdots + b_mw_m \leq c_1x_1 + \cdots + c_nx_n\dagger \tag{2.8}$$

This implies that $g \leq f$.

The reverse inequality of (2.8) is given by

$$b_1w_1 + \cdots + b_mw_m \geq c_1x_1 + \cdots + c_nx_n$$

or

$$b_1w_1 + \cdots + b_mw_m - (c_1x_1 + \cdots + c_nx_n) \geq 0 \tag{2.9}$$

This inequality restricts $g \geq f$. Hence, a simultaneous solution to the complete set of inequalities (2.3), (2.5), and (2.9) will yield optimum solutions for the primal and dual with, of course, $f = g$.

Let us multiply each inequality of the set (2.3), (2.5), and (2.9) by an unknown positive variable z to obtain

$$\begin{array}{rr}
a_{11}\bar{x}_1 + \cdots + a_{1n}\bar{x}_n - b_1z \geq 0 \\
\cdots\cdots\cdots\cdots\cdots\cdots \\
a_{m1}\bar{x}_1 + \cdots + a_{mn}\bar{x}_n - b_mz \geq 0 \\
-(a_{11}\bar{w}_1 + \cdots + a_{m1}\bar{w}_m) \qquad\qquad + c_1z \geq 0 \\
\cdots\cdots\cdots\cdots\cdots\cdots \\
-(a_{1n}\bar{w}_1 + \cdots + a_{mn}\bar{w}_m) \qquad\qquad + c_nz \geq 0 \\
(b_1\bar{w}_1 + \cdots + b_m\bar{w}_m) - (c_1\bar{x}_1 + \cdots + c_n\bar{x}_n) \qquad \geq 0
\end{array} \tag{2.10}$$

where $\bar{x}_j = zx_j$ and $\bar{w}_i = zw_i$. A solution to (2.10) with $z > 0$ will yield a solution to the original primal and dual problems. The above has the

† The inequality (2.8) is equivalent to formula (1.13) of Chap. 5.

following representation in terms of matrices:

$$(\bar{\mathbf{X}}\bar{\mathbf{W}}z)\begin{pmatrix} \mathbf{O} & \mathbf{A}' & -\mathbf{c}' \\ -\mathbf{A} & \mathbf{O} & \mathbf{b}' \\ \mathbf{c} & -\mathbf{b} & \mathbf{O} \end{pmatrix} \geq (\mathbf{O}) \tag{2.11}$$

Here $\mathbf{A}$ is an $m \times n$ matrix; $\mathbf{b}$, $\mathbf{c}$, $\bar{\mathbf{X}}$, and $\bar{\mathbf{W}}$ are row vectors; and the $\mathbf{O}$ matrices are of the right dimensions. We can consider the skew-symmetric matrix of (2.11) as defining a zero-sum two-person symmetric game and the row vector $(\bar{\mathbf{X}}\bar{\mathbf{W}}z)$ as the strategy vector for the maximizing player P_1. Since the value of a symmetric game is zero, we have from Definition 5 that a solution to inequalities (2.11) with

$$\Sigma\bar{x}_j + \Sigma\bar{w}_i + z = 1$$

is a solution to the corresponding game. By the fundamental theorem, a solution to this game always exists. However, the associated linear-programming problems may not have any solutions. This will be the case if the game solution yields an optimum strategy with $z = 0$. Hence a game solution with $z > 0$ will yield optimal solutions to the primal and dual linear-programming problems as defined by (2.3), (2.5), and (2.9), where $x_j = \bar{x}_j/z$ and $w_i = \bar{w}_i/z$ are the respective optimal solutions.

As an example of parametric programming techniques applied to the theory of games, let us next consider the matrix game with the following pay-off matrix:

$$\begin{pmatrix} \lambda & \lambda & \cdots & \lambda \\ a_{21} & a_{22} & \cdots & a_{2n} \\ \cdots & \cdots & \cdots & \cdots \\ a_{m1} & a_{m2} & \cdots & a_{mn} \end{pmatrix}$$

The parameter λ can assume any real value and represents the pay-off to player one (the maximizing player) when he employs his first strategy.

Player one wishes to determine a mixed strategy $\mathbf{X} = (x_1, x_2, \ldots, x_m)$ such that

$$\begin{aligned} \lambda x_1 + a_{21}x_2 + \cdots + a_{m1}x_m &\geq v \\ \lambda x_1 + a_{22}x_2 + \cdots + a_{m2}x_m &\geq v \\ \cdots\cdots\cdots\cdots\cdots\cdots \\ \lambda x_1 + a_{2n}x_2 + \cdots + a_{mn}x_m &\geq v \\ x_1 + \quad x_2 + \cdots + \quad x_m &= 1 \end{aligned}$$

where $x_i \geq 0$ and v, the value of the game, is to be a maximum.

Following the discussion above, the equivalent linear-programming problem is to maximize

$$\lambda x_1 + \qquad a_{21}x_2 + \cdots + \qquad a_{m1}x_m - x_{m+1} \qquad = v$$

subject to

$$
\begin{array}{lr}
(a_{22} - a_{21})x_2 + \cdots + (a_{m2} - a_{m1})x_m + x_{m+1} - x_{m+2} & = 0 \\
\cdots\cdots\cdots\cdots\cdots\cdots\cdots\cdots\cdots\cdots\cdots\cdots & \\
(a_{2n} - a_{21})x_2 + \cdots + (a_{mn} - a_{m1})x_m + x_{m+1} \qquad - x_{m+n} & = 0 \\
x_1 + \qquad x_2 + \cdots + \qquad x_m & = 1
\end{array}
$$

where $x_i \geq 0$ for $i = 1, 2, \ldots, m + n$.

This parametric linear-programming problem can be solved for all possible values of the parameter λ by the procedure described in Chap. 8.

Games which have more than one parametrized row, e.g.,

$$
\begin{pmatrix}
\lambda & \lambda & \cdots & \lambda \\
\mu & \mu & \cdots & \mu \\
a_{31} & a_{32} & \cdots & a_{3n} \\
\cdots & \cdots & \cdots & \cdots \\
a_{m1} & a_{m2} & \cdots & a_{mn}
\end{pmatrix}
$$

can also be transformed into equivalent linear-programming problems. However, efficient computational techniques for solving problems with more than one parameter are not available. For a discussion concerning the two-parameter objective function, see Gass and Saaty [47].

REMARKS

The introductory material on the theory of games is from Kuhn [67] and McKinsey [72]. The material in Sec. 2 is from Dantzig [19]. For additional reading the reader is referred to von Neumann and Morgenstern [80], Luce and Raiffa [71*a*], Owen [85*aa*], and Williams [101].

EXERCISES

1. Show that

$$\max_i \min_j a_{ij} \leq \min_j \max_i a_{ij}$$

in the arbitrary matrix

$$
\begin{pmatrix}
a_{11} & a_{12} & \cdots & a_{1n} \\
a_{21} & a_{22} & \cdots & a_{2n} \\
\cdots & \cdots & \cdots & \cdots \\
a_{m1} & a_{m2} & \cdots & a_{mn}
\end{pmatrix}
$$

2. Write the pay-off matrix and the equivalent linear-programming problem to the following game: Each player selects an integer from the set of integers (1,2,3). The player with the lower number wins 2 points unless his number is 1 lower, in which case he loses 4 points. When the numbers are equal, there is no pay-off.

3. Transform the following matrix games into their corresponding primal and dual linear-programming problems:

$$\begin{pmatrix} 2 & 1 & 0 & -2 \\ 1 & 0 & 3 & 2 \end{pmatrix}$$

$$\begin{pmatrix} 0 & -1 & 1 \\ 1 & 1 & -1 \\ 1 & -1 & 0 \end{pmatrix}$$

4. Write out the associated game matrix for the following linear-programming problems: Minimize

$$x_2 - x_3$$

subject to

$$\begin{aligned} x_1 + x_2 - x_3 &\geq 0 \\ -x_1 + x_2 + 3x_3 &\geq 1 \\ x_1 - 3x_2 + x_3 &\leq 1 \\ x_j &\geq 0 \end{aligned}$$

Maximize

$$w_1 + w_2 + w_3$$

subject to

$$\begin{aligned} w_1 + 4w_2 + 3w_3 &\leq 1 \\ 4w_1 - 4w_2 + 2w_3 &\leq 1 \\ 3w_1 - w_2 + 5w_3 &\leq 1 \\ w_i &\geq 0 \end{aligned}$$

5. Solve the following parametric game for all values of the parameter:

$$\mathbf{A} = \begin{pmatrix} \lambda & \lambda & \lambda \\ -1 & 4 & 0 \\ 3 & -1 & 2 \end{pmatrix}$$

6. By letting $v = v_1 - v_2$, that is, the difference of two nonnegative variables, in Eqs. (2.1), determine a third linear-programming formulation of the matrix game.

7. Solve the numerical example of Sec. 2.

8. Develop a graphical procedure for solving any $2 \times n$ matrix game. [Hint: Let v be one variable and x_1 the other variable of the two-dimensional graph, with $x_2 = 1 - x_1$, and use the formulation of Eqs. (2.1).]

9. Solve the skin game by the graphical method of Exercise 8.

PART FOUR / **NONLINEAR PROGRAMMING**

CHAPTER 13 / **NONLINEAR PROGRAMMING**

The linear-programming model, along with the implied concept of viewing complex systems in terms of interrelated activities,[1] has proved to be a major contribution to the field of scientific decision-making. Applications of linear programming exist in every category of industrial, business, and government endeavors. The versatility and adaptability of this deceptively simple mathematical model appears to have no bounds. However, early workers in the field recognized certain implied limitations of the linear model and began to develop theoretical, as well as computational, foundations for special and general nonlinear-programming problems.[2] One can, for example, trace such developments by studying the growth of programming applications within the petroleum industry (Charnes, Cooper, and Mellon [13], Manne [73*a*], Symonds [93*a*], Garvin et al [45*b*] and Catchpole [8*d*]). This industrial area yielded some of the first successful nonmilitary applications of linear programming. Extensions of these applications led to the need and development of many nonlinear-programming techniques. Nonlinearities in the programming model of an oil refinery arise in both the objective function and the system of constraints. As Catchpole [8*d*] and others point out, the cost of crude oil may increase with the quantity required, while the use of lead as an octane improver has a decreasing effect as more lead is used; i.e., the mathematical relations which describe these statements are not linear.

The consideration of these and other nonlinearities has advanced our ability to develop mathematical models of complex systems. We can now describe and deal with many generalizations of the linear model that

[1] See Dantzig [16*c*], Chap. 3, for a detailed discussion along these lines.

[2] The reader is referred to the work of, among others, Arrow, Hurwicz, and Uzawa [3], Barankin and Dorfman [3*d*], Beale [5*e*], Charnes and Lemke [15], Dantzig [16*c*], Dennis [32*f*], John [62*c*], Kuhn and Tucker [69], Markowitz [74*d*], Slater [90*d*], and Wolfe [105*a*].

take us into the field of nonlinear programming—or into the more inclusive field of *mathematical programming*.[1]

The purpose of this chapter is to present selected basic material necessary to the understanding of the foundations of nonlinear programming.[2] The material requires a broader mathematical background than does that covered in previous chapters. However, except for basic notions from the calculus, this chapter is essentially self-contained, assuming knowledge of the preceding discussions.

1. THE GENERAL PROBLEM OF MATHEMATICAL PROGRAMMING

As noted in Sec. 1 of Chap. 1, the mathematical model of the linear-programming problem can be described using relationships of the form

$$a_1x_1 + a_2x_2 + \cdots + a_jx_j + \cdots + a_nx_n = b$$

From a mathematical point of view this is a rather restrictive relationship in that it excludes a wide variety of possible interactions between the variables. For example, the output of an activity might be functionally related in a nonlinear fashion to the output of other activities or even to itself. Nonlinear expressions of the form

$$g(\mathbf{X}) = g(x_1,x_2, \ldots ,x_n) = a_1x_1^2 + a_2x_2x_5 + \cdots + a_nx_n^3 = b$$

and trigonometric, logarithmic, and other special functional forms can occur in the constraint set and objective function of practical, as well as theoretical, problems. Other nonlinearities include the restriction of a variable to integer values; for example, $x_1 = 0, 1, 2, 3, \ldots$, as discussed in Chap. 9.

For our purposes, we define the problem of mathematical programming to be:

Find a vector $\mathbf{X} = (x_1,x_2, \ldots ,x_n)$ which minimizes (maximizes) the objective function $f(\mathbf{X})$ subject to

$$g_i(\mathbf{X}) \leq 0 \qquad \text{for } i = 1,2, \ldots ,m \tag{1.1}$$

Here $f(\mathbf{X})$ and the $g_i(\mathbf{X})$ can be considered to be general functions of the variables $x_1, x_2, \ldots, x_n$. If a problem contains equality constraints or inequality constraints going in the opposite direction, these constraints can

[1] See the books of Abadie [1], Boot [6*c*], Saaty and Bram [88*b*], Hadley [52*d*], Kunzi and Krelle [69*a*], Vajda [97*a*], Zoutendijk [108], Fiacco and McCormick [42*c*], and the survey article by Dorn [40*a*].

[2] The reader interested in specific applications should consult the references given above and Bracken and McCormick [7*a*]. Although apparently not as plentiful and pervasive as linear programming, nonlinear applications include chemical equilibrium, least-square and Chebyshev approximations, stock portfolio selection, and many problems from mathematics and economics.

be expressed in the form (1.1), so there is no loss of generality in assuming the constraints to be of the above form.

For a linear-programming problem, problem (1.1) becomes: Minimize

$$f(\mathbf{X}) = c_1x_1 + \cdots + c_nx_n$$

subject to

$$\begin{aligned}
g_1(\mathbf{X}) &= a_{11}x_1 + \cdots + a_{1n}x_n - b_1 \leq 0 \\
&\cdots\cdots\cdots\cdots\cdots\cdots \\
g_m(\mathbf{X}) &= a_{m1}x_1 + \cdots + a_{mn}x_n - b_m \leq 0 \\
g_{m+1}(\mathbf{X}) &= x_1 \geq 0 \\
&\cdots\cdots\cdots\cdots\cdots\cdots \\
g_{m+n}(\mathbf{X}) &= x_n \geq 0
\end{aligned}$$

A simple example of a nonlinear problem involving logarithmic functions in the objective function (Dorn [40c]) is as follows: Minimize

$$f(x_1,x_2) = -\log x_1 - \log x_2$$

subject to

$$\begin{aligned}
x_1 + x_2 &\leq 2 \\
x_1 &\geq 0 \\
x_2 &\geq 0
\end{aligned}$$

The graph of the constraint set and the objective function, for various values of the objective function, is shown in Fig. 13.1. The solution space is the shaded triangle and the optimum occurs at the point (1,1),

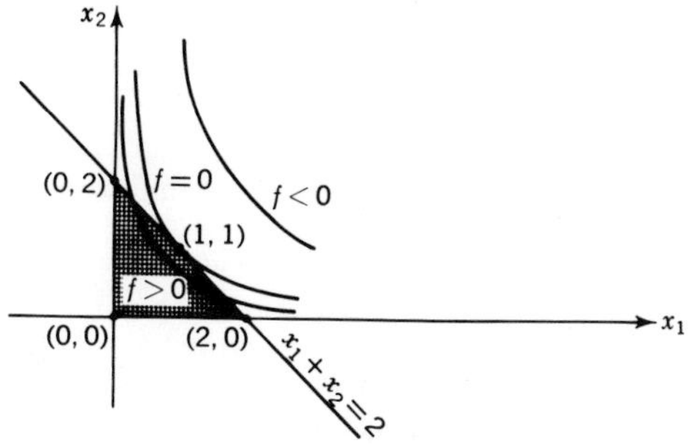

FIGURE 13.1

not at an extreme point of the solution space. When the constraint set can contain nonlinear functions, we are not necessarily limited to the nice mathematical situation found in linear programming, i.e., the search for an optimizing extreme point of a convex solution space. The mathematical-programming problem could have a nonconvex solution space, an interior optimum solution, and other characteristics which compound both theoretical and computational developments.

One of the main difficulties encountered in the realm of mathematical programming is the determination of a solution point which appears not only to optimize the objective function at that point, but to optimize the function over the complete range of the solution space. More precisely, we have:

Definition 1. A function $f(\mathbf{X})$ has a *global minimum* at a point $\mathbf{X}^0$ of a set of points $\mathbf{K}$ if and only if $f(\mathbf{X}^0) \leq f(\mathbf{X})$ for all $\mathbf{X}$ in $\mathbf{K}$.

Definition 2. A function $f(\mathbf{X})$ has a *local minimum* at a point $\mathbf{X}^0$ of a set of points $\mathbf{K}$ if and only if there exists a positive number ϵ such that $f(\mathbf{X}^0) \leq f(\mathbf{X})$ for all $\mathbf{X}$ in $\mathbf{K}$ at which $\|\mathbf{X}^0 - \mathbf{X}\| < \epsilon$.

In one dimension, it is easy to illustrate local and global minima for a function $f(\mathbf{X})$ as in Fig. 13.2. Here $\mathbf{K}$ is the set of all nonnegative x. The points $x = 0$ and $x = x_2$ are local minima, while $x = x_1$ is the global minimum. (A global minimum is also a local minimum.) The reader should note that global and local maxima can also be defined by obvious changes in Definitions 1 and 2.

Although nonlinear optimization problems call for the determination of a global minimum, computational procedures will, in general, lead to a solution which is only a local minimum. Moreover, it is usually impossible to determine if a local minimum is really a global minimum. Even if this could be done, computational procedures have no way of

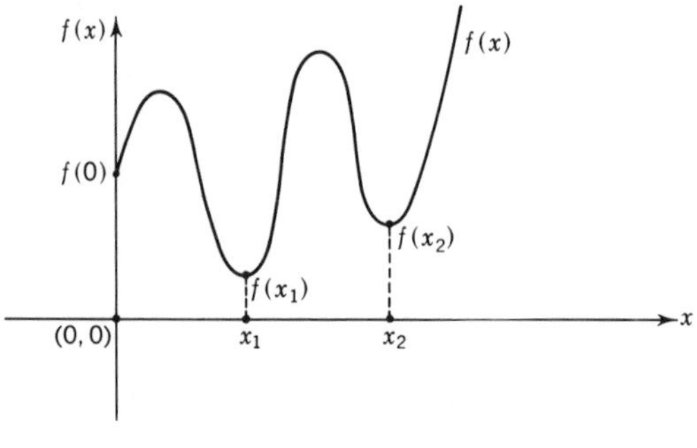

FIGURE 13.2

proceeding from a local minimum to a global minimum. This is not the case, of course, for the linear-programming problem. For that problem, we can show that the simplex algorithm arrives at a solution point which is not only a local minimum, but also a global minimum. This is a key characteristic of most computationally tractable nonlinear-programming problems. Thus, our discussions below will be directed toward problems for which we can demonstrate mathematically the coincidence of the local and global minima. To accomplish this end, we next describe and develop some necessary mathematical notions.

2. MATHEMATICAL BACKGROUND

a. Quadratic forms. In some mathematical-programming problems, the objective function is restricted to a specialized expression, a positive-definite quadratic form.

Definition 3. A *quadratic form* $f(\mathbf{X})$ in n variables is an expression of second degree which can be written as $f(\mathbf{X}) = \sum_{i=1}^{n} \sum_{j=1}^{n} c_{ij}x_i x_j$, or

$$f(\mathbf{X}) = c_{11}x_1x_1 + c_{12}x_1x_2 + \cdots + c_{1n}x_1x_n + \cdots + c_{n1}x_nx_1 + \cdots + c_{nn}x_nx_n$$

For $n = 2$ we have

$$f(x_1,x_2) = c_{11}x_1^2 + c_{12}x_1x_2 + c_{21}x_1x_2 + c_{22}x_2^2$$

If $\mathbf{C} = (c_{ij})$, then the quadratic form is expressed as $f(\mathbf{X}) = \mathbf{X'CX}$, where $\mathbf{X} = (x_1, \ldots ,x_n)$ is an $n \times 1$ column vector and $\mathbf{C}$ is an $n \times n$ matrix. For $n = 2$,

$$\begin{aligned} f(x) &= \begin{pmatrix} x_1 \\ x_2 \end{pmatrix}' \mathbf{C} \begin{pmatrix} x_1 \\ x_2 \end{pmatrix} = (x_1x_2) \begin{pmatrix} c_{11} & c_{12} \\ c_{21} & c_{22} \end{pmatrix} \begin{pmatrix} x_1 \\ x_2 \end{pmatrix} \\ &= (c_{11}x_1 + c_{21}x_2 \quad c_{12}x_1 + c_{22}x_2) \begin{pmatrix} x_1 \\ x_2 \end{pmatrix} \\ &= c_{11}x_1^2 + c_{12}x_1x_2 + c_{21}x_2x_1 + c_{22}x_2^2 \\ &= c_{11}x_1^2 + (c_{12} + c_{21})x_1x_2 + c_{22}x_2^2 \end{aligned}$$

Noting that for any $\mathbf{X}$ and $\mathbf{C}$ the product $\mathbf{X'CX}$ is a scalar and thus $\mathbf{X'CX} = (\mathbf{X'CX})' = (\mathbf{CX})'(\mathbf{X}')' = \mathbf{X'C'X}$, we can write

$$\mathbf{X'CX} = \tfrac{1}{2}[\mathbf{X'CX} + \mathbf{X'C'X}] = \frac{\mathbf{X'(C + C')X}}{2}$$

The matrix $\frac{1}{2}(\mathbf{C} + \mathbf{C}')$ is a symmetric matrix with the general element $\frac{1}{2}(c_{ij} + c_{ji})$. In this form, the coefficient of $x_i x_j$ is $\frac{1}{2}(c_{ij} + c_{ji})$ and of $x_j x_i$ is $\frac{1}{2}(c_{ji} + c_{ij})$. However, for the total product term of x_i and x_j, the coefficient is $(c_{ij} + c_{ji})$, as it is in the original form $\mathbf{X}'\mathbf{C}\mathbf{X}$. For any $\mathbf{C}$ we can redefine the coefficients of $x_i x_j$ to be $\frac{1}{2}(c_{ij} + c_{ji})$ and of $x_j x_i$ to also be $\frac{1}{2}(c_{ij} + c_{ji})$ so that the matrix representation of the quadratic form is in terms of a symmetric matrix. In what follows, we shall always assume that the matrix $\mathbf{C}$ is symmetric.

Definition 4. A quadratic form is called *positive definite* if $\mathbf{X}'\mathbf{C}\mathbf{X} > 0$ for $\mathbf{X} \neq 0$. For example $f(\mathbf{X}) = x_1^2 + x_2^2$ is positive definite.

Definition 5. A quadratic form is called *positive semidefinite* if $\mathbf{X}'\mathbf{C}\mathbf{X} \geq 0$ for all $\mathbf{X}$ and there exists at least one vector $\mathbf{X} \neq 0$ such that $\mathbf{X}'\mathbf{C}\mathbf{X} = 0$. The quadratic form

$$f(\mathbf{X}) = x_1^2 - x_1 x_2 - x_2 x_1 + x_2^2 = x_1^2 - 2x_1 x_2 + x_2^2 = (x_1 - x_2)^2 \geq 0$$

for all x_1 and x_2 and equals zero for $x_1 = x_2 = 1$, and thus $f(\mathbf{X})$ is positive semidefinite.

Negative definite and *negative semidefinite* forms can also be defined by appropriate reversal of the inequality signs in the above definitions.

For quadratic forms of small dimension we can test for positive definiteness by the following theorem (Hadley [52*b*]):

Theorem 1. *The quadratic form* $f(\mathbf{X}) = \mathbf{X}'\mathbf{C}\mathbf{X}$ *is positive definite if and only if the principal minors, i.e., the minors obtained by taking successively larger minors along the principal diagonal (see below), are all strictly positive.* Thus for $n = 4$ we must have

$$c_{11} > 0 \qquad \begin{vmatrix} c_{11} & c_{12} \\ c_{21} & c_{22} \end{vmatrix} > 0 \qquad \begin{vmatrix} c_{11} & c_{12} & c_{13} \\ c_{21} & c_{22} & c_{23} \\ c_{31} & c_{32} & c_{33} \end{vmatrix} > 0 \qquad |\mathbf{C}| > 0$$

b. Convex functions. The main mathematical concept required in our development of nonlinear programming is that of a convex function. Given that the constraints $g_i(\mathbf{X})$ and the objective function $f(\mathbf{X})$ are convex functions we can, as shall be shown below, prove that a local minimum is also a global minimum.

Definition 6. A function $f(\mathbf{X}) = f(x_1,x_2, \ldots ,x_n)$ defined over a set of points which lie in a convex set $\mathbf{K}$ in $\mathbf{E}_n$ is called a *convex function* if for any two points $\mathbf{X}_1$ and $\mathbf{X}_2$ in $\mathbf{K}$ and any $0 \leq \lambda \leq 1$ we have

$$f[\lambda\mathbf{X}_1 + (1 - \lambda)\mathbf{X}_2] \leq \lambda f(\mathbf{X}_1) + (1 - \lambda)f(\mathbf{X}_2) \tag{2.1}$$

If strict inequality holds in (2.1), $f(\mathbf{X})$ is said to be *strictly convex.* Similarly, a function $f(\mathbf{X})$ is a *concave function* if (2.1) holds with the inequality reversed, that is, $-f(\mathbf{X})$ is convex.

A convex function $f(\mathbf{X})$ in one variable is pictured in Fig. 13.3. The reader will note that the expression $\lambda f(x_1) + (1 - \lambda) f(x_2)$ is the ordinate corresponding to an abscissa of $\lambda x_1 + (1 - \lambda) x_2$ for the line segment joining the points $[x_1,f(x_1)]$ and $[x_2,f(x_2)]$. Thus, a function is convex if every point on each such line segment overestimates the corresponding point on the function; i.e., the line segment lies above the function. A linear function is both convex and concave; the sum of convex functions is also a convex function; i.e., if $f_i(\mathbf{X})$, $i = 1, \ldots, k$ are convex functions, then $f(\mathbf{X}) = \sum_{i=1}^{k} f_i(\mathbf{X})$ is convex.

Theorem 2. *The positive semidefinite quadratic form* $f(\mathbf{X}) = \mathbf{X'CX}$ *is a convex function for all* $\mathbf{X}$ *in* $\mathbf{E}_n$.

Proof. We wish to show for $0 \leq \lambda \leq 1$ and all $\mathbf{X}_1$, $\mathbf{X}_2$ in $\mathbf{E}_n$ that

$$f[\lambda\mathbf{X}_1 + (1 - \lambda)\mathbf{X}_2] \leq \lambda f(\mathbf{X}_1) + (1 - \lambda)f(\mathbf{X}_2)$$

or

$$f[\lambda\mathbf{X}_1 + (1 - \lambda)\mathbf{X}_2] - \lambda f(\mathbf{X}_1) - (1 - \lambda)f(\mathbf{X}_2) \leq 0 \tag{2.2}$$

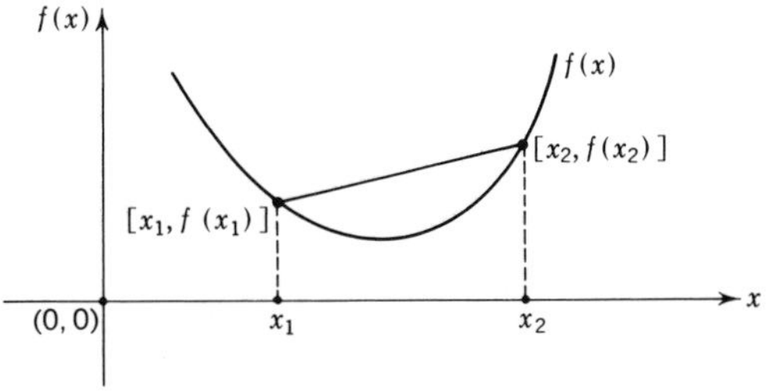

FIGURE 13.3

Noting that $\mathbf{X}_2'\mathbf{C}\mathbf{X}_1 = \mathbf{X}_1'\mathbf{C}\mathbf{X}_2$, the left-hand side of (2.2) can be rewritten as

$$\begin{aligned}
&[\lambda\mathbf{X}_1 + (1-\lambda)\mathbf{X}_2]'\mathbf{C}[\lambda\mathbf{X}_1 + (1-\lambda)\mathbf{X}_2] \\
&\qquad\qquad - \lambda\mathbf{X}_1'\mathbf{C}\mathbf{X}_1 - (1-\lambda)\mathbf{X}_2'\mathbf{C}\mathbf{X}_2 \\
&= \lambda^2\mathbf{X}_1'\mathbf{C}\mathbf{X}_1 + (1-\lambda)^2\mathbf{X}_2'\mathbf{C}\mathbf{X}_2 + 2\lambda(1-\lambda)\mathbf{X}_1'\mathbf{C}\mathbf{X}_2 \\
&\qquad\qquad - \lambda\mathbf{X}_1'\mathbf{C}\mathbf{X}_1 - (1-\lambda)\mathbf{X}_2'\mathbf{C}\mathbf{X}_2 \\
&= (\lambda^2-\lambda)\mathbf{X}_1'\mathbf{C}\mathbf{X}_1 + (1-\lambda)[(1-\lambda)\mathbf{X}_2'\mathbf{C}\mathbf{X}_2 - \mathbf{X}_2'\mathbf{C}\mathbf{X}_2] \\
&\qquad\qquad + 2\lambda(1-\lambda)\mathbf{X}_1'\mathbf{C}\mathbf{X}_2 \\
&= \lambda(\lambda-1)\mathbf{X}_1'\mathbf{C}\mathbf{X}_1 + \lambda(\lambda-1)\mathbf{X}_2'\mathbf{C}\mathbf{X}_2 - 2\lambda(\lambda-1)\mathbf{X}_1'\mathbf{C}\mathbf{X}_2 \\
&= \lambda(\lambda-1)[\mathbf{X}_1'\mathbf{C}\mathbf{X}_1 + \mathbf{X}_2'\mathbf{C}\mathbf{X}_2 - 2\mathbf{X}_1'\mathbf{C}\mathbf{X}_2] \\
&= \lambda(\lambda-1)[(\mathbf{X}_1-\mathbf{X}_2)'\mathbf{C}(\mathbf{X}_1-\mathbf{X}_2)] \qquad (2.3)
\end{aligned}$$

Since the form $\mathbf{X}'\mathbf{C}\mathbf{X} \geq 0$ for any $\mathbf{X}$, for (2.3) we have $[(\mathbf{X}_1-\mathbf{X}_2)'\mathbf{C}(\mathbf{X}_1-\mathbf{X}_2)] \geq 0$. Since $\lambda(\lambda-1) < 0$ for $0 < \lambda < 1$ and $\lambda(\lambda-1) = 0$ for $\lambda = 0$ or 1, then (2.3) will always be less than or equal to zero for all vectors $\mathbf{X}_1$ and $\mathbf{X}_2$, which was to be shown. Thus $f(\mathbf{X}) = \mathbf{X}'\mathbf{C}\mathbf{X}$ is convex for all $\mathbf{X}$ in $\mathbf{E}_n$.

Theorem 3. *If $f(\mathbf{X})$ is convex on a convex set $\mathbf{K}$, then $f(\mathbf{X})$ has at most one local minimum. If there is such a minimum, it is a global minimum and is attained on a convex set.*

Proof. Suppose there is a local minimum at a point $\mathbf{X}^0$. For *any* $\mathbf{X} = \bar{\mathbf{X}}$ in $\mathbf{K}$, by definition of a local minimum and convexity of $f(\mathbf{X})$, we have

$$f(\mathbf{X}^0) \leq f[(1-\lambda)\mathbf{X}^0 + \lambda\bar{\mathbf{X}}] \leq (1-\lambda)f(\mathbf{X}^0) + \lambda f(\bar{\mathbf{X}}) \qquad (2.4)$$

for λ a sufficiently small positive number. From the extremes of (2.4) we have

$$\begin{aligned}
f(\mathbf{X}^0) &\leq (1-\lambda)f(\mathbf{X}^0) + \lambda f(\bar{\mathbf{X}}) \\
\lambda f(\mathbf{X}^0) &\leq \lambda f(\bar{\mathbf{X}})
\end{aligned}$$

and since $\lambda > 0$, then $f(\mathbf{X}^0) \leq f(\bar{\mathbf{X}})$, which implies $f(\mathbf{X}^0)$ is a global minimum by definition, as $\bar{\mathbf{X}}$ is any point in $\mathbf{K}$.

If $\mathbf{X}^0$ and $\mathbf{X}^1$ are two points at which $f(\mathbf{X})$ attains its minimum value z_0, then for $0 \leq \lambda \leq 1$

$$z_0 \leq f[(1-\lambda)\mathbf{X}^0 + \lambda\mathbf{X}^1] \leq (1-\lambda)f(\mathbf{X}^0) + \lambda f(\mathbf{X}^1) = z_0$$

Hence, $f(\mathbf{X})$ also attains its minimum at $\mathbf{X} = (1-\lambda)\mathbf{X}^0 + \lambda\mathbf{X}^1$ and thus the set of solutions is convex.

Theorem 4. *Let* $\mathbf{K}$ *be the set of points of* $\mathbf{E}_n$ *which satisfy the constraints*

$$g_i(\mathbf{X}) \leq 0 \qquad i = 1,2, \ldots, m$$
$$\mathbf{X} \geq \mathbf{O}$$

If the $g_i(\mathbf{X})$ *are convex functions, then* $\mathbf{K}$ *is a convex set. (We assume* $\mathbf{K}$ *is not empty.)*

Proof. Let $\mathbf{X}_1$ and $\mathbf{X}_2$ be any two points in $\mathbf{K}$ and define

$$\bar{\mathbf{X}} = \lambda \mathbf{X}_1 + (1 - \lambda)\mathbf{X}_2$$

for $0 \leq \lambda \leq 1$. We need to show that $\bar{\mathbf{X}}$ satisfies the constraints.

$\bar{\mathbf{X}} \geq \mathbf{O}$ as it is the nonnegative sum of nonnegative quantities. For any i, we have

$$g_i(\bar{\mathbf{X}}) = g_i[\lambda \mathbf{X}_1 + (1 - \lambda)\mathbf{X}_2] \leq \lambda g_i(\mathbf{X}_1) + (1 - \lambda) g_i(\mathbf{X}_2)$$

But $g_i(\mathbf{X}_1) \leq 0$ and $g_i(\mathbf{X}_2) \leq 0$ and for $0 \leq \lambda \leq 1$ we must have

$$g_i(\bar{\mathbf{X}}) \leq \lambda g_i(\mathbf{X}_1) + (1 - \lambda) g_i(\mathbf{X}_2) \leq 0$$

N.B. The preceding two theorems state that for the general programming problem of minimizing $f(\mathbf{X})$ subject to $g_i(\mathbf{X}) \leq 0 (i = 1, \ldots, m)$ and $\mathbf{X} \geq \mathbf{O}$, if $f(\mathbf{X})$ and all $g_i(\mathbf{X})$ are convex functions, then a local minimum of $f(\mathbf{X})$ subject to the constraint set is also a global minimum.

c. The gradient vector and saddle point. The following concepts will enable us to develop necessary and sufficient conditions for a point $\mathbf{X}^0$ to be a solution of certain restricted mathematical-programming problems.

Definition 7. If a function $f(\mathbf{X})$ and all its first derivatives are continuous over some subset of $\mathbf{E}_n$, then for each point $\mathbf{X}^0$ in this subset we define the n-component column vector $\nabla f(\mathbf{X}^0)$, termed the gradient vector of $f(\mathbf{X})$ at $\mathbf{X}^0$, to be

$$\nabla f(\mathbf{X}^0) = \begin{pmatrix} \dfrac{\partial f(\mathbf{X}^0)}{\partial x_1} \\ \cdot \\ \cdot \\ \cdot \\ \dfrac{\partial f(\mathbf{X}^0)}{\partial x_n} \end{pmatrix}$$

$\nabla f(\mathbf{X}^0)$ is a vector perpendicular to the contour of $f(\mathbf{X})$ which passes through $\mathbf{X}^0$. Its direction is the direction of maximum increase of $f(\mathbf{X})$, and its length is the magnitude of that maximum rate of increase. This direction is often referred to as the direction of steepest ascent.

From the calculus we note that a necessary condition for $\mathbf{X}^0$ to be a minimum for the general differentiable function $f(x_1, \ldots, x_n) = f(\mathbf{X})$ is that

$$\left[\frac{\partial f(\mathbf{X})}{\partial x_j}\right]_{\mathbf{X}^0} = 0$$

for $\mathbf{X}^0$ an interior point of the region of definition (Buck [8*a*] and Widder [100*c*]). To find the minimum of $f(\mathbf{X})$ with $a_j \leq x_j \leq b_j$, we must investigate the boundary points as well as the interior points. For example, if $f(\mathbf{X})$ is a convex function of the single variable x, we could have the three cases illustrated in Fig. 13.4*a* to *c*.

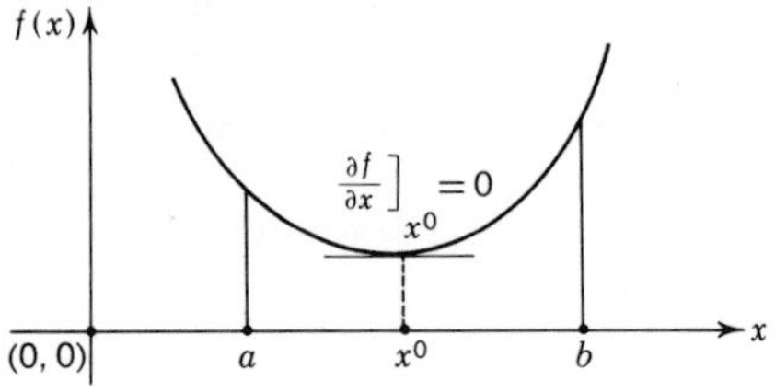

FIGURE 13.4a

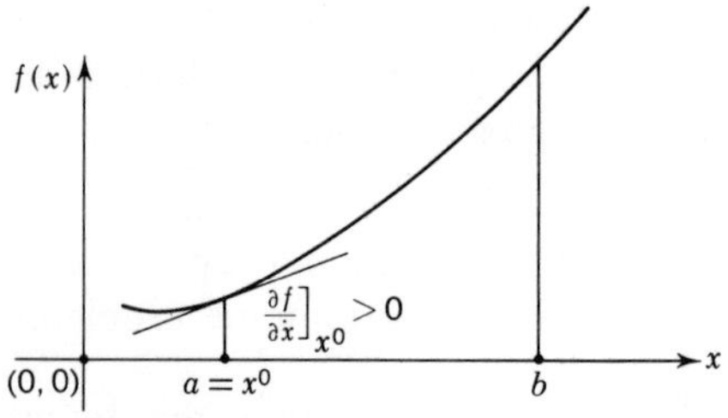

FIGURE 13.4b

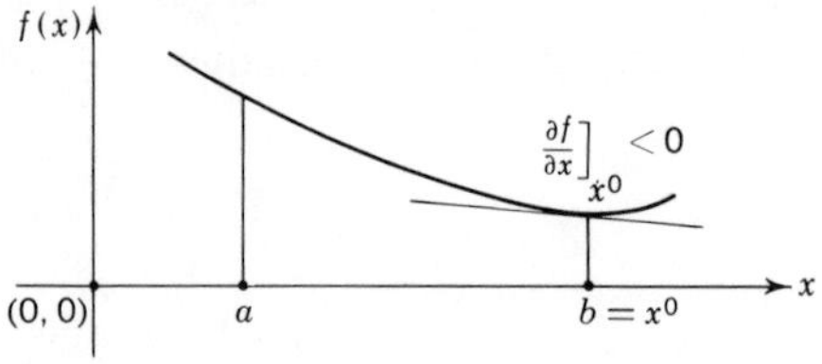

FIGURE 13.4c

The next two theorems relate the notions of convexity, the gradient and conditions for a point $\mathbf{X}^0$ to minimize a convex function over a convex set.

Theorem 5. *If $f(\mathbf{X})$ is defined on an open convex set $\mathbf{K}$ and $f(\mathbf{X})$ is differentiable in $\mathbf{X}$, then $f(\mathbf{X})$ is convex if and only if*

$$f(\mathbf{X}_1) - f(\mathbf{X}_2) \geq (\mathbf{X}_1 - \mathbf{X}_2)'\nabla f(\mathbf{X}_2)$$

for all $\mathbf{X}_1$ and $\mathbf{X}_2$ in $\mathbf{K}$.

Proof.

(*a*) We first prove if $f(\mathbf{X}_1) - f(\mathbf{X}_2) \geq (\mathbf{X}_1 - \mathbf{X}_2)'\nabla f(\mathbf{X}_2)$, then $f(\mathbf{X})$ is convex (sufficiency).

For $\mathbf{X}_1$ and $\mathbf{X}_2$ in $\mathbf{K}$ and $0 \leq \lambda \leq 1$, let $\mathbf{X}_3 = \lambda\mathbf{X}_1 + (1 - \lambda)\mathbf{X}_2$. We note that $\mathbf{X}_3$ is also in $\mathbf{K}$. For $\mathbf{X}_1$ and $\mathbf{X}_3$ we have

$$f(\mathbf{X}_1) - f(\mathbf{X}_3) \geq (\mathbf{X}_1 - \mathbf{X}_3)'\nabla f(\mathbf{X}_3) \tag{2.5}$$

and for $\mathbf{X}_2$ and $\mathbf{X}_3$ we have

$$f(\mathbf{X}_2) - f(\mathbf{X}_3) \geq (\mathbf{X}_2 - \mathbf{X}_3)'\nabla f(\mathbf{X}_3) \tag{2.6}$$

Multiplying (2.5) by λ and (2.6) by $(1 - \lambda)$ and adding, we obtain

$$\begin{aligned}\lambda f(\mathbf{X}_1) - \lambda f(\mathbf{X}_3) + (1 - \lambda)f(\mathbf{X}_2) - (1 - \lambda)f(\mathbf{X}_3)& \\ \geq \lambda(\mathbf{X}_1 - \mathbf{X}_3)'\nabla f(\mathbf{X}_3) + (1 - \lambda)(\mathbf{X}_2 - \mathbf{X}_3)'\nabla f(\mathbf{X}_3)&\end{aligned}$$

or

$$\begin{aligned}\lambda f(\mathbf{X}_1) + (1 - \lambda)f(\mathbf{X}_2) \geq f(\mathbf{X}_3) + [\lambda\mathbf{X}_1' + (1 - \lambda)\mathbf{X}_2']\nabla f(\mathbf{X}_3)& \\ - \mathbf{X}_3'\nabla f(\mathbf{X}_3)&\end{aligned} \tag{2.7}$$

Applying the definition of $\mathbf{X}_3$, (2.7) becomes

$$\lambda f(\mathbf{X}_1) + (1 - \lambda)f(\mathbf{X}_2) \geq f[\lambda\mathbf{X}_1 + (1 - \lambda)\mathbf{X}_2]$$

which implies that $f(\mathbf{X})$ is convex.

(*b*) We next prove that if $f(\mathbf{X})$ is convex we must also have that $f(\mathbf{X}_1) - f(\mathbf{X}_2) \geq (\mathbf{X}_1 - \mathbf{X}_2)'\nabla f(\mathbf{X}_2)$ (necessity).

For $0 < \lambda \leq 1$ we have for any $\mathbf{X}_1$ and $\mathbf{X}_2$ in $\mathbf{K}$ that

$$\lambda f(\mathbf{X}_1) + (1 - \lambda)f(\mathbf{X}_2) \geq f[\lambda\mathbf{X}_1 + (1 - \lambda)\mathbf{X}_2]$$

Regrouping terms,

$$\lambda f(\mathbf{X}_1) - \lambda f(\mathbf{X}_2) \geq f[\lambda\mathbf{X}_1 + (1 - \lambda)\mathbf{X}_2] - f(\mathbf{X}_2)$$

or

$$f(\mathbf{X}_1) - f(\mathbf{X}_2) \geq \frac{f[\mathbf{X}_2 + \lambda(\mathbf{X}_1 - \mathbf{X}_2)] - f(\mathbf{X}_2)}{\lambda}$$

Taking the limit of the right-hand side as λ approaches zero, we have the result

$$f(\mathbf{X}_1) - f(\mathbf{X}_2) \geq (\mathbf{X}_1 - \mathbf{X}_2)'\nabla f(\mathbf{X}_2)$$

Theorem 6. *Let $f(\mathbf{X})$ be a convex continuously differentiable function defined on a convex set $\mathbf{K}$. Let $\mathbf{X}^0$ be in $\mathbf{K}$. Then $f(\mathbf{X}^0) \leq f(\mathbf{X})$ for all $\mathbf{X}$ in $\mathbf{K}$ (that is, $\mathbf{X}^0$ minimizes $f(\mathbf{X})$ for all $\mathbf{X}$ in $\mathbf{K}$) if and only if $(\mathbf{X} - \mathbf{X}^0)'\nabla f(\mathbf{X}^0) \geq 0$ for all $\mathbf{X}$ in $\mathbf{K}$.*

Proof.

(*a*) (Necessity.) For $f(\mathbf{X}^0) \leq f(\mathbf{X})$, the minimum point $\mathbf{X}^0$ is either an interior point of $\mathbf{K}$ or a boundary point of $\mathbf{K}$. If $\mathbf{X}^0$ is interior, by the calculus we must have $\nabla f(\mathbf{X}^0) = 0$ and, of course, $(\mathbf{X} - \mathbf{X}^0)'\nabla f(\mathbf{X}^0) = 0$. For $\mathbf{X}^0$ *any* minimum point, we have for $f(\mathbf{X})$ convex and $\mathbf{K}$ a convex set (see Theorem 3)

$$f(\mathbf{X}^0) \leq f[\lambda\mathbf{X} + (1 - \lambda)\mathbf{X}^0]$$

for all $\mathbf{X}$ in $\mathbf{K}$ and $0 \leq \lambda \leq 1$. For $\lambda > 0$, then

$$\frac{f[\mathbf{X}^0 + \lambda(\mathbf{X} - \mathbf{X}^0)] - f(\mathbf{X}^0)}{\lambda} \geq 0$$

Taking the limit as λ approaches zero, we have

$$(\mathbf{X} - \mathbf{X}^0)'\nabla f(\mathbf{X}^0) \geq 0$$

(*b*) (Sufficiency.) Given that $(\mathbf{X} - \mathbf{X}^0)'\nabla f(\mathbf{X}^0) \geq 0$ for all $\mathbf{X}$ in $\mathbf{K}$, by convexity of $f(\mathbf{X})$ we have from Theorem 5 that

$$f(\mathbf{X}) - f(\mathbf{X}^0) \geq (\mathbf{X} - \mathbf{X}^0)'\nabla f(\mathbf{X}^0) \geq 0$$

Hence $f(\mathbf{X}) \geq f(\mathbf{X}^0)$ for all $\mathbf{X}$ in $\mathbf{K}$ and $\mathbf{X}^0$ minimizes $f(\mathbf{X})$ over $\mathbf{K}$.

From a geometric point of view, the above theorem states that $\mathbf{X}^0$ is a minimum point if the angle between the vectors $\nabla f(\mathbf{X}^0)$ and $(\mathbf{X} - \mathbf{X}^0)'$ for all $\mathbf{X}$ in $\mathbf{K}$ is less than or equal to 90°. See Fig. 13.5.

For a point $\mathbf{X}_1$ not the minimum and if for some $\mathbf{X}_2$ we have $f(\mathbf{X}_1) \geq f(\mathbf{X}_2)$, then from Theorem 5

$$0 \geq f(\mathbf{X}_2) - f(\mathbf{X}_1) \geq (\mathbf{X}_2 - \mathbf{X}_1)'\nabla f(\mathbf{X}_1)$$

i.e., the angle between the gradient at $\mathbf{X}_1$ and $(\mathbf{X}_2 - \mathbf{X}_1)'$ is no less than 90°.

In the following section, we shall need the concept of a saddle point of a function. The reader will recognize the extension of the saddle point of a matrix developed for the theory of games in Chap. 12.

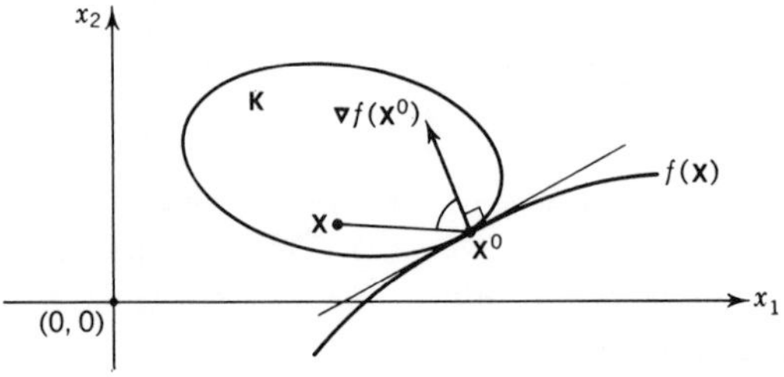

FIGURE 13.5

Definition 8. The point $(\mathbf{X}^0,\pi^0)$ is said to be a *saddle point* of a function $F(\mathbf{X},\pi)$ if there exists an $\epsilon > 0$ such that for all $\mathbf{X}$ and π in an ϵ neighborhood of $\mathbf{X}^0$ and π^0, that is, with $\|\mathbf{X} - \mathbf{X}^0\| < \epsilon$ and $\|\pi - \pi^0\| < \epsilon$, we have

$$F(\mathbf{X}^0,\pi) \leq F(\mathbf{X}^0,\pi^0) \leq F(\mathbf{X},\pi^0) \tag{2.8}$$

If (2.8) holds for all $\mathbf{X}$ and π then $(\mathbf{X}^0,\pi^0)$ is a global saddle point. π^0 is an m-component vector.

The last theorem of this section is stated without proof (Valentine [97*c*]).

Theorem 7. *Theorem of the separating hyperplane. If two convex sets* $\mathbf{K}_1$ *and* $\mathbf{K}_2$ *have, at most, common boundary points, there exists a hyperplane* $\mathbf{aX} = b$, $\mathbf{a} \neq \mathbf{O}$ *which separates* $\mathbf{K}_1$ *and* $\mathbf{K}_2$ *in the following sense:*

$$\mathbf{aX}_1 \leq b \qquad \text{for all } \mathbf{X}_1 \text{ in } \mathbf{K}_1 \text{ and}$$
$$\mathbf{aX}_2 \geq b \qquad \text{for all } \mathbf{X}_2 \text{ in } \mathbf{K}_2$$

Here $\mathbf{a}$ *is an n-component row vector and b is a scalar.*

3. THE CONVEX-PROGRAMMING PROBLEM[1]

We can now specialize our discussion to the important class of nonlinear problems termed *convex programs* and develop necessary and sufficient conditions for solution to such problems, the *Kuhn-Tucker conditions.* In this section we consider

The convex-programming problem. Minimize

$$f(\mathbf{X})$$

subject to

$$g_i(\mathbf{X}) \leq 0 \qquad i = 1, 2, \ldots, m \tag{3.1}$$
$$\mathbf{X} \geq \mathbf{O}$$

where $f(\mathbf{X})$ and all the $g_i(\mathbf{X})$ are convex, continuously differentiable functions.

For these functions, define the *Lagrangian function*

$$F(\mathbf{X},\pi) = f(\mathbf{X}) + \sum_{i=1}^{m} \pi_i g_i(\mathbf{X})$$

[1] This section follows the development of Kunzi and Krelle [69*a*]. See also Karlin [63*a*] and Vajda [97*a*].

where the m variables π_i are called *Lagrange multipliers.* We can now describe a new problem whose solution has a direct relationship to the convex-programming problem.

The saddle-point problem. Find nonnegative vectors $\mathbf{X}^0$, $\boldsymbol{\pi}^0$ for the Lagrangian function $F(\mathbf{X},\boldsymbol{\pi})$ such that

$$F(\mathbf{X}^0,\boldsymbol{\pi}) \leq F(\mathbf{X}^0,\boldsymbol{\pi}^0) \leq F(\mathbf{X},\boldsymbol{\pi}^0) \tag{3.2}$$

for all $\mathbf{X} \geq \mathbf{O}$ and $\boldsymbol{\pi} \geq \mathbf{O}$. The point $(\mathbf{X}^0,\boldsymbol{\pi}^0)$ is termed a nonnegative saddle point. We can interpret (3.2) to read

$$F(\mathbf{X}^0,\boldsymbol{\pi}^0) = \max_{\boldsymbol{\pi}\geq\mathbf{O}} [\min_{\mathbf{X}\geq\mathbf{O}} F(\mathbf{X},\boldsymbol{\pi})] = \min_{\mathbf{X}\geq\mathbf{O}} [\max_{\boldsymbol{\pi}\geq\mathbf{O}} F(\mathbf{X},\boldsymbol{\pi})]\dagger$$

† This relationship can be developed as follows: From (3.2) we have

$$F(\mathbf{X}^0,\boldsymbol{\pi}) \leq F(\mathbf{X}^0,\boldsymbol{\pi}^0) \tag{3.2a}$$

$$F(\mathbf{X}^0,\boldsymbol{\pi}^0) \leq F(\mathbf{X},\boldsymbol{\pi}^0) \tag{3.2b}$$

From (3.2*a*) we have

$$\max_{\boldsymbol{\pi}\geq\mathbf{O}} F(\mathbf{X}^0,\boldsymbol{\pi}) \leq F(\mathbf{X}^0,\boldsymbol{\pi}^0)$$

and from (3.2*b*)

$$F(\mathbf{X}^0,\boldsymbol{\pi}^0) \leq \min_{\mathbf{X}\geq\mathbf{O}} F(\mathbf{X},\boldsymbol{\pi}^0)$$

so that

$$\max_{\boldsymbol{\pi}\geq\mathbf{O}} F(\mathbf{X}^0,\boldsymbol{\pi}) \leq F(\mathbf{X}^0,\boldsymbol{\pi}^0) \leq \min_{\mathbf{X}\geq\mathbf{O}} F(\mathbf{X},\boldsymbol{\pi}^0) \tag{3.2c}$$

Since

$$\min_{\mathbf{X}\geq\mathbf{O}} \max_{\boldsymbol{\pi}\geq\mathbf{O}} F(\mathbf{X},\boldsymbol{\pi}) \leq \max_{\boldsymbol{\pi}\geq\mathbf{O}} F(\mathbf{X}^0,\boldsymbol{\pi})$$

and

$$\min_{\mathbf{X}\geq\mathbf{O}} F(\mathbf{X},\boldsymbol{\pi}^0) \leq \max_{\boldsymbol{\pi}\geq\mathbf{O}} \min_{\mathbf{X}\geq\mathbf{O}} F(\mathbf{X},\boldsymbol{\pi})$$

we conclude from (3.2*c*) that

$$\min_{\mathbf{X}\geq\mathbf{O}} \max_{\boldsymbol{\pi}\geq\mathbf{O}} F(\mathbf{X},\boldsymbol{\pi}) \leq F(\mathbf{X}^0,\boldsymbol{\pi}^0) \leq \max_{\boldsymbol{\pi}\geq\mathbf{O}} \min_{\mathbf{X}\geq\mathbf{O}} F(\mathbf{X},\boldsymbol{\pi}) \tag{3.2d}$$

But for $F(\mathbf{X},\boldsymbol{\pi})$ we have for any $\mathbf{X}$ and $\boldsymbol{\pi}$

$$\min_{\mathbf{X}\geq\mathbf{O}} F(\mathbf{X},\boldsymbol{\pi}) \leq F(\mathbf{X},\boldsymbol{\pi})$$

and

$$F(\mathbf{X},\boldsymbol{\pi}) \leq \max_{\boldsymbol{\pi}\geq\mathbf{O}} F(\mathbf{X},\boldsymbol{\pi})$$

and hence

$$\min_{\mathbf{X}\geq\mathbf{O}} F(\mathbf{X},\boldsymbol{\pi}) \leq \max_{\boldsymbol{\pi}\geq\mathbf{O}} F(\mathbf{X},\boldsymbol{\pi}) \tag{3.2e}$$

The point $(\mathbf{X}^0,\boldsymbol{\pi}^0)$ of the saddle-point problem and the optimal solution to the convex-programming problem are related by

Theorem 8. *The Kuhn-Tucker theorem.* $\mathbf{X}^0$ *is a solution to the convex-programming problem* (3.1) *if and only if a vector* $\boldsymbol{\pi}^0$ *exists such that* $(\mathbf{X}^0,\boldsymbol{\pi}^0)$ *is a solution to the saddle-point problem.*

The proof of this theorem enables us to evolve the necessary and sufficient conditions which a vector $\mathbf{X}^0$ must satisfy in order to be a solution to (3.1). The knowledge of such conditions leads to the development of appropriate algorithms and justifies their validity.

Proof. We first show that the saddle-point conditions are sufficient to ensure that $\mathbf{X}^0$ is a solution to the convex-programming problem. For the saddle point $(\mathbf{X}^0,\boldsymbol{\pi}^0)$ and $F(\mathbf{X},\boldsymbol{\pi}) = f(\mathbf{X}) + \sum_{i=1}^{m} \pi_i g_i(\mathbf{X})$ we have from (3.2)

$$f(\mathbf{X}^0) + \sum_{i=1}^{m} \pi_i g_i(\mathbf{X}^0) \leq f(\mathbf{X}^0) + \sum_{i=1}^{m} \pi_i^0 g_i(\mathbf{X}^0) \leq f(\mathbf{X}) + \sum_{i=1}^{m} \pi_i^0 g_i(\mathbf{X}) \tag{3.3}$$

for all $\mathbf{X} \geq \mathbf{O}$ and $\boldsymbol{\pi} \geq \mathbf{O}$. For the left-hand inequality of (3.3) to hold for all $\boldsymbol{\pi} \geq \mathbf{O}$, we must have both $g_i(\mathbf{X}^0) \leq 0$ for all i and $\sum_{i=1}^{m} \pi_i^0 g_i(\mathbf{X}^0) = 0$. We show this as follows. The left-hand inequality is

$$f(\mathbf{X}^0) + \sum_{i=1}^{m} \pi_i g_i(\mathbf{X}^0) \leq f(\mathbf{X}^0) + \sum_{i=1}^{m} \pi_i^0 g_i(\mathbf{X}^0)$$

or

$$\sum_{i=1}^{m} \pi_i g_i(\mathbf{X}^0) \leq \sum_{i=1}^{m} \pi_i^0 g_i(\mathbf{X}^0) \tag{3.4}$$

If some $g_i(\mathbf{X}^0) > 0$, then a corresponding $\pi_i > 0$ can be chosen large enough that the inequality (3.4) does not hold. Therefore $g_i(\mathbf{X}^0) \leq 0$ for all i. Since $\mathbf{X}^0 \geq \mathbf{O}$ we have shown that $\mathbf{X}^0$ satisfies the constraints

---

Since the left-hand side of (3.2*e*) is independent of $\mathbf{X}$, we have

$$\min_{\mathbf{X}\geq\mathbf{O}} F(\mathbf{X},\boldsymbol{\pi}) \leq \min_{\mathbf{X}\geq\mathbf{O}} \max_{\boldsymbol{\pi}\geq\mathbf{O}} F(\mathbf{X},\boldsymbol{\pi}) \tag{3.2f}$$

Also, since the right-hand side (3.2*f*) is independent of $\boldsymbol{\pi}$, we have

$$\max_{\boldsymbol{\pi}\geq\mathbf{O}} \min_{\mathbf{X}\geq\mathbf{O}} F(\mathbf{X},\boldsymbol{\pi}) \leq \min_{\mathbf{X}\geq\mathbf{O}} \max_{\boldsymbol{\pi}\geq\mathbf{O}} F(\mathbf{X},\boldsymbol{\pi}) \tag{3.2g}$$

Relationships (3.2*d*) and (3.2*g*) yield the desired result.

(3.1) of the convex-programming problem. Also as (3.4) must hold for $\boldsymbol{\pi} = \mathbf{O}$, we have $0 \leq \sum_{i=1}^{m} \pi_i^0 g_i(\mathbf{X}^0)$. But as $\boldsymbol{\pi}^0 \geq \mathbf{O}$, all $g_i(\mathbf{X}^0) \leq 0$; then $\sum_{i=1}^{m} \pi_i^0 g_i(\mathbf{X}^0) \leq 0$, which implies $\sum_{i=1}^{m} \pi_i^0 g_i(\mathbf{X}^0) = 0$. Thus either $\pi_i^0 = 0$, $g_i(\mathbf{X}^0) = 0$, or both equal zero.

From the right-hand inequality of (3.3) we now can write

$$f(\mathbf{X}^0) \leq f(\mathbf{X}) + \Sigma\pi_i^0 g_i(\mathbf{X}) \qquad \text{for all } \mathbf{X} \geq \mathbf{O}$$

Since $\boldsymbol{\pi}^0 \geq \mathbf{O}$, $f(\mathbf{X}^0) \leq f(\mathbf{X})$ for all $\mathbf{X} \geq \mathbf{O}$ for which $g_i(\mathbf{X}) \leq \mathbf{O}$. Hence, as $g_i(\mathbf{X}^0) \leq 0$, $\mathbf{X}^0$ is a solution to the convex-programming problem.

To prove necessity, i.e., that a solution to the convex-programming problem (3.1) solves the saddle-point problem, is a more complex task. Here we have to introduce an apparently mild assumption concerning the convex region of feasible solutions to (3.1), that is, that the interior of this convex feasible region is not empty. This regularity or constraint qualification, which is due to Slater [90*d*], takes the place of the assumption dealing with the differentiability of the functions required by Kuhn and Tucker in their original proof [69]. The Kuhn-Tucker qualification rules out certain peculiarities on the boundary of the feasible-solution space. For our purposes, then, we require the following assumption: For the constraints (3.1) there exists an $\mathbf{X} \geq \mathbf{O}$ such that all $g_i(\mathbf{X}) < 0$.[1]

To continue with the proof, let $\mathbf{X}^0$ be a solution to (3.1). We shall show that a $\boldsymbol{\pi}^0 \geq \mathbf{O}$ exists such that $(\mathbf{X}^0,\boldsymbol{\pi}^0)$ satisfies the saddle-point problem. For this purpose we construct in $\mathbf{E}_{m+1}$ two point sets $\mathbf{K}_1$ and $\mathbf{K}_2$ with vectors $\mathbf{Y} = (y_0, y_1, \ldots, y_m)$. $\mathbf{K}_1$ is the set of all points $\mathbf{Y}$ for which there exists at least one $\mathbf{X} \geq \mathbf{O}$ with $f(\mathbf{X}) \leq y_0$ and $g_i(\mathbf{X}) \leq y_i$ for $i = 1, \ldots, m$; that is,

$$\mathbf{K}_1 = \{\mathbf{Y} \mid f(\mathbf{X}) \leq y_0;\ g_i(\mathbf{X}) \leq y_i,\ i = 1, \ldots, m, \text{ for at least one } \mathbf{X} \geq \mathbf{O}\}$$

$\mathbf{K}_2$ is defined in terms of $\mathbf{X}^0$ to be

$$\mathbf{K}_2 = \{\mathbf{Y} \mid f(\mathbf{X}^0) > y_0;\ 0 > y_i,\ i = 1, \ldots, m\}$$

Since $f(\mathbf{X})$ and all $g_i(\mathbf{X})$ are convex, $\mathbf{K}_1$ is a convex set. Also, $\mathbf{K}_2$ is convex as it represents an open subset of $\mathbf{E}_{m+1}$ which is bounded by planes parallel to the coordinate axes of $\mathbf{E}_{m+1}$. It consists of the

[1] This assumption apparently rules out linear-equality restrictions such as $\mathbf{AX} - \mathbf{b} = \mathbf{O}$. However, it can be shown that the assumption is not required for those $g_i(\mathbf{X})$ which are linear. The Kuhn-Tucker theorem is true without restriction if all constraints are linear (Karlin [63*a*] and Arrow, Hurwicz, and Uzawa [3]).

interior of an orthant with vertex at $[f(\mathbf{X}^0),\mathbf{O}]$. As $\mathbf{X}^0$ minimizes $f(\mathbf{X})$ for all $\mathbf{X} \geq \mathbf{O}$, $\mathbf{K}_1$ and $\mathbf{K}_2$ have no points in common; i.e., the intersection of $\mathbf{K}_1$ and $\mathbf{K}_2$ is empty.

From the above and the theorem of the existence of a hyperplane which separates two nonintersecting convex sets, there exists a vector $\mathbf{a} \neq \mathbf{O}$, a constant b, such that the hyperplane $\mathbf{aY} = b$ separates $\mathbf{K}_1$ from $\mathbf{K}_2$; that is,

$$\mathbf{aY}_1 \geq \mathbf{aY}_2 \tag{3.5}$$

for all $\mathbf{Y}_1$ in $\mathbf{K}_1$ and $\mathbf{Y}_2$ in $\mathbf{K}_2$. Since the components of $\mathbf{Y}_2$ are allowed to be arbitrarily small, we can conclude that $\mathbf{a} \geq \mathbf{O}$. For if some component $a_i < 0$, then as the corresponding component of $\mathbf{Y}_2$ can be made as small as desired, the separating hyperplane inequality (3.5) would not be satisfied.

If we choose $\mathbf{Y}_1 = [f(\mathbf{X}),g_1(\mathbf{X}), \ldots ,g_m(\mathbf{X})]$ and $\mathbf{Y}_2 = [f(\mathbf{X}^0), 0, \ldots , 0]$, since (3.5) is satisfied for the boundary points of $\mathbf{K}_2$, we have

$$a_0f(\mathbf{X}) + a_1g_1(\mathbf{X}) + \cdots + a_mg_m(\mathbf{X}) \geq a_0f(\mathbf{X}^0) \tag{3.6}$$

for all $\mathbf{X} \geq \mathbf{O}$. If $a_0 = 0$, then from (3.6) for all $\mathbf{X} \geq \mathbf{O}$,

$$a_1g_1(\mathbf{X}) + \cdots + a_mg_m(\mathbf{X}) \geq 0 \tag{3.7}$$

and all $a_i \geq 0$ and some $a_i \neq 0$. But by the regularity assumption there exists some $\mathbf{X} \geq \mathbf{O}$, say $\bar{\mathbf{X}}$, for which all $g_i(\bar{\mathbf{X}}) < 0$. For $\bar{\mathbf{X}}$, (3.7) would be strictly less than zero. Thus, we can conclude that $a_0 > 0$. Dividing (3.6) by a_0 we obtain

$$f(\mathbf{X}) + \pi_1^0g_1(\mathbf{X}) + \cdots + \pi_m^0g_m(\mathbf{X}) \geq f(\mathbf{X}^0) \tag{3.8}$$

with $\pi_i^0 = a_i/a_0 \geq 0$. This can be rewritten as

$$F(\mathbf{X},\boldsymbol{\pi}^0) \geq f(\mathbf{X}^0) \tag{3.9}$$

for all $\mathbf{X} \geq \mathbf{O}$. Letting $\mathbf{X} = \mathbf{X}^0$ in (3.8) we have

$$F(\mathbf{X}^0,\boldsymbol{\pi}^0) \geq f(\mathbf{X}^0)$$

or

$$F(\mathbf{X}^0,\boldsymbol{\pi}^0) = f(\mathbf{X}^0) + \pi_1^0g_1(\mathbf{X}^0) + \cdots + \pi_m^0g_m(\mathbf{X}^0) \geq f(\mathbf{X}^0)$$

or

$$\pi_1^0g_1(\mathbf{X}^0) + \cdots + \pi_m^0g_m(\mathbf{X}^0) \geq 0$$

But as all $g_i(\mathbf{X}^0) \leq 0$ and all $\pi_i^0 \geq 0$ we must have

$$\pi_1^0g_1(\mathbf{X}^0) + \cdots + \pi_m^0g_m(\mathbf{X}^0) = 0$$

and thus

$$F(\mathbf{X}^0,\pi^0) = f(\mathbf{X}^0) \tag{3.10}$$

Also, for all $\boldsymbol{\pi} \geq \mathbf{O}$, as $g_i(\mathbf{X}^0) \leq 0$ for all i, it follows that

$$f(\mathbf{X}^0) \geq f(\mathbf{X}^0) + \sum_{i=1}^{m} \pi_i g_i(\mathbf{X}^0) = F(\mathbf{X}^0,\boldsymbol{\pi}) \tag{3.11}$$

Combining (3.9) to (3.11) we obtain

$$F(\mathbf{X}^0,\boldsymbol{\pi}) \leq f(\mathbf{X}^0) = F(\mathbf{X}^0,\boldsymbol{\pi}^0) = f(\mathbf{X}^0) \leq F(\mathbf{X},\boldsymbol{\pi}^0)$$

or

$$F(\mathbf{X}^0,\boldsymbol{\pi}) \leq F(\mathbf{X}^0,\boldsymbol{\pi}^0) \leq F(\mathbf{X},\boldsymbol{\pi}^0)$$

This completes the proof. For purposes of reference we summarize the above in the following:

Theorem 9. *The equivalence of the saddle-point and convex-programming problems. Let $f(\mathbf{X})$ and $g_i(\mathbf{X})$ be convex functions defined on the convex set $\mathbf{X} \geq \mathbf{O}$ of $\mathbf{E}_n$, with the property that there exists some $\bar{\mathbf{X}} \geq \mathbf{O}$ for which $g_i(\bar{\mathbf{X}}) < 0$ for all $i = 1, \ldots, m$. If $\mathbf{X}^0$ is a point at which $f(\mathbf{X})$ achieves its minimum subject to $g_i(\mathbf{X}) \leq 0$, $\mathbf{X} \geq \mathbf{O}$, then there exists some $\boldsymbol{\pi}^0 \geq \mathbf{O}$ such that for $F(\mathbf{X},\boldsymbol{\pi}) = f(\mathbf{X}) + \sum_{i=1}^{m} \pi_i g_i(\mathbf{X})$, we have*

$$F(\mathbf{X}^0,\boldsymbol{\pi}) \leq F(\mathbf{X}^0,\boldsymbol{\pi}^0) \leq F(\mathbf{X},\boldsymbol{\pi}^0) \tag{3.12}$$

for all $\mathbf{X} \geq \mathbf{O}$, $\boldsymbol{\pi} \geq \mathbf{O}$, and $\Sigma \pi_i^0 g_i(\mathbf{X}^0) = 0$. Conversely, if $(\mathbf{X}^0,\boldsymbol{\pi}^0)$ satisfies (3.12), then $\mathbf{X}^0$ minimizes $f(\mathbf{X})$ subject to all $\mathbf{X}$ satisfying $\mathbf{X} \geq \mathbf{O}$ and $g_i(\mathbf{X}) \leq 0$ for all i.

If $f(\mathbf{X})$ and all the $g_i(\mathbf{X})$ are continuously differentiable as well as convex functions, then the saddle-point inequalities are equivalent to certain conditions which are necessary and sufficient to provide a solution to the convex-programming problem. These are given by the following theorem:

Theorem 10. *For $\mathbf{X} \geq \mathbf{O}$ and $f(\mathbf{X})$ and all $g_i(\mathbf{X})$ continuously differentiable convex functions, (3.13) and (3.14) are necessary and sufficient conditions for a point $(\mathbf{X}^0,\boldsymbol{\pi}^0)$ to satisfy the constraints (3.2) of the saddle-point problem.*

The Kuhn-Tucker conditions:

$$\frac{\partial F(\mathbf{X}^0,\boldsymbol{\pi}^0)}{\partial x_j} = \frac{\partial f(\mathbf{X}^0)}{\partial x_j} + \frac{\sum_{i=1}^{m} \pi_i^0 \, \partial g_i(\mathbf{X}^0)}{\partial x_j} \geq 0 \tag{3.13a}$$

$$(\mathbf{X}^0)' \left[\frac{\partial F(\mathbf{X}^0,\boldsymbol{\pi}^0)}{\partial \mathbf{X}}\right] = \sum_{j=1}^{n} x_j^0 \left\{\frac{\partial f(\mathbf{X}^0)}{\partial x_j} + \frac{\sum_{i=1}^{m} \pi_i^0 \, \partial g_i(\mathbf{X}^0)}{\partial x_j}\right\} = 0 \tag{3.13b}$$

$$\mathbf{X}^0 = \{x_j^0\} \geq \mathbf{O} \tag{3.13c}$$

and

$$\frac{\partial F(\mathbf{X}^0,\boldsymbol{\pi}^0)}{\partial \pi_i} = g_i(\mathbf{X}^0) \leq 0 \qquad (3.14a)$$

$$\boldsymbol{\pi}^0 \left[\frac{\partial F(\mathbf{X}^0,\boldsymbol{\pi}^0)}{\partial \boldsymbol{\pi}}\right] = \sum_{i=1}^{m} \pi_i^0 g_i(\mathbf{X}^0) = 0 \qquad (3.14b)$$

$$\boldsymbol{\pi}^0 = \{\pi_i^0\} \geq \mathbf{O} \qquad (3.14c)$$

From the nonnegativity implications of (3.13*a* and *c*) we have that the individual terms of (3.13*b*)

$$x_j^0 \left\{\frac{\partial f(\mathbf{X}^0)}{\partial x_j} + \frac{\sum_{i=1}^{m} \pi_i^0 \, \partial g_i(\mathbf{X}^0)}{\partial x_j}\right\} = 0$$

Similarly, from (3.14*a* and *c*) we have that the individual terms of (3.14*b*) $\pi_i^0 g_i(\mathbf{X}^0) = 0$.

Proof. For necessity, i.e., assuming $(\mathbf{X}^0,\boldsymbol{\pi}^0)$ is a saddle-point solution, (3.13) and (3.14) assert that $F(\mathbf{X},\boldsymbol{\pi}^0)$ has a local minimum at $F(\mathbf{X}^0,\boldsymbol{\pi}^0)$ and $F(\mathbf{X}^0,\boldsymbol{\pi})$ has a local maximum at $F(\mathbf{X}^0,\boldsymbol{\pi}^0)$, respectively. This must be the case for $F(\mathbf{X}^0,\boldsymbol{\pi}) \leq F(\mathbf{X}^0,\boldsymbol{\pi}^0) \leq F(\mathbf{X},\boldsymbol{\pi}^0)$. For first assume (3.13*a* and *b*) do not hold with $\mathbf{X}^0 \geq \mathbf{O}$. We fix $\boldsymbol{\pi} = \boldsymbol{\pi}^0$ and note that $F(\mathbf{X},\boldsymbol{\pi}^0)$ is a convex function of $\mathbf{X}$. If some $x_j^0 > 0$, that is, we have an interior point, and $\partial F(\mathbf{X}^0,\boldsymbol{\pi}^0)/\partial x_j \neq 0$, then we could decrease $F(\mathbf{X}^0,\boldsymbol{\pi}^0)$ by taking a different value of $\mathbf{X}$, say $\bar{\mathbf{X}}$, such that $F(\bar{\mathbf{X}},\boldsymbol{\pi}^0) < F(\mathbf{X}^0,\boldsymbol{\pi}^0)$. See Fig. 13.6. If $x_j^0 = 0$, that is, a boundary point, and $\partial F(\mathbf{X}^0,\boldsymbol{\pi}^0)/\partial x_j < 0$, then again we could find $\bar{\mathbf{X}} > 0$ such that $F(\bar{\mathbf{X}},\boldsymbol{\pi}^0) < F(\mathbf{X}^0,\boldsymbol{\pi}^0)$. See Fig. 13.7. These situations cannot arise as $F(\mathbf{X},\boldsymbol{\pi})$ is minimized by $\mathbf{X}^0$ for all $\boldsymbol{\pi}$. Thus (3.13*a* and *b*) must hold, and, of course, $\mathbf{X}^0 \geq \mathbf{O}$, which is (3.13*c*).[1]

[1] From Theorem 6 it is possible to prove that $\partial f(\mathbf{X}^0,\boldsymbol{\pi}^0)/\partial x_j = 0$ when $x_j^0 > 0$ and $\partial F(\mathbf{X}^0,\boldsymbol{\pi}^0)/\partial x_j \geq 0$ when $x_j^0 = 0$.

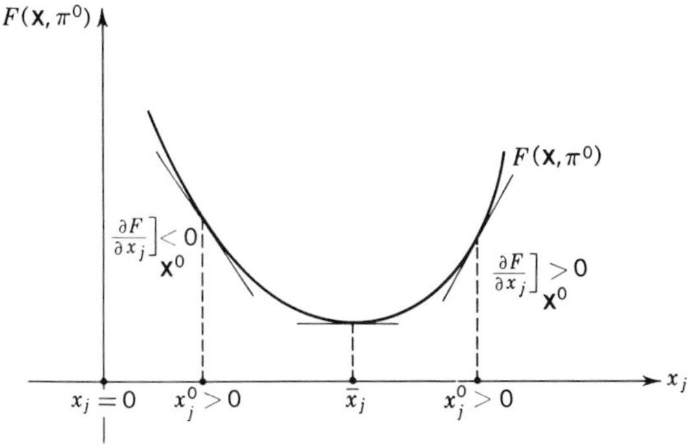

FIGURE 13.6

Similarly, for $\mathbf{X} = \mathbf{X}^0$, we note that $F(\mathbf{X}^0,\boldsymbol{\pi})$ is a linear function of the π_i. From (3.4) we have for all $\boldsymbol{\pi} \geq \mathbf{O}$ and $(\mathbf{X}^0,\boldsymbol{\pi}^0)$,

$$\sum_{i=1}^{m} \pi_i g_i(\mathbf{X}^0) \leq \sum_{i=1}^{m} \pi_i^0 g_i(\mathbf{X}^0) \tag{3.15}$$

Since $\partial F(\mathbf{X}^0,\boldsymbol{\pi}^0)/\partial\pi_i = g_i(\mathbf{X}^0)$, if $\partial F(\mathbf{X}^0,\boldsymbol{\pi}^0)/\partial\pi_i > 0$, then $g_i(\mathbf{X}^0) > 0$. From (3.15), if a $g_i(\mathbf{X}^0) > 0$, we could select π_i very large such

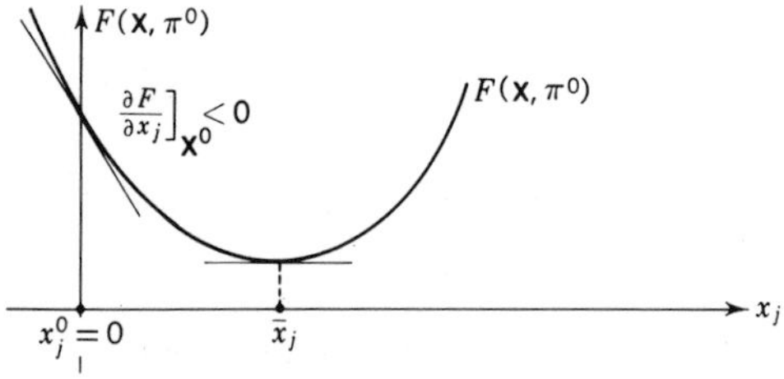

FIGURE 13.7

that this condition could not hold. Therefore, we must have all $\partial F(\mathbf{X}^0,\boldsymbol{\pi}^0)/\partial\pi_i = g_i(\mathbf{X}^0) \leq 0$. We have already shown that $\sum_{i=1}^{m} \pi_i^0 g_i(\mathbf{X}^0) = 0$ (Theorem 8) and, of course, $\boldsymbol{\pi}^0 \geq \mathbf{O}$. Thus, (3.14) holds.

For the sufficiency, i.e., given that the Kuhn-Tucker conditions hold, we have from Theorem 5 and the fact that $F(\mathbf{X},\boldsymbol{\pi}^0)$ is convex in $\mathbf{X}$,

$$F(\mathbf{X},\boldsymbol{\pi}^0) \geq F(\mathbf{X}^0,\boldsymbol{\pi}^0) + (\mathbf{X} - \mathbf{X}^0)' \left[\frac{\partial F(\mathbf{X}^0,\boldsymbol{\pi}^0)}{\partial \mathbf{X}}\right]$$

or

$$F(\mathbf{X},\boldsymbol{\pi}^0) \geq F(\mathbf{X}^0,\boldsymbol{\pi}^0) + \mathbf{X}' \left[\frac{\partial F(\mathbf{X}^0,\boldsymbol{\pi}^0)}{\partial \mathbf{X}}\right] - (\mathbf{X}^0)' \left[\frac{\partial F(\mathbf{X}^0,\boldsymbol{\pi}^0)}{\partial \mathbf{X}}\right] \tag{3.16}$$

From (3.13*b*) the last term of (3.16) is zero. Also, from (3.13*a*) and $\mathbf{X} \geq \mathbf{O}$ the middle term on the right-hand side of (3.16) is nonnegative. We thus have $F(\mathbf{X},\boldsymbol{\pi}^0) \geq F(\mathbf{X}^0,\boldsymbol{\pi}^0)$. Since $F(\mathbf{X}^0,\boldsymbol{\pi})$ is also convex, i.e., linear, in $\boldsymbol{\pi}$ we have

$$F(\mathbf{X}^0,\boldsymbol{\pi}) = F(\mathbf{X}^0,\boldsymbol{\pi}^0) + (\boldsymbol{\pi} - \boldsymbol{\pi}^0) \left[\frac{\partial F(\mathbf{X}^0,\boldsymbol{\pi}^0)}{\partial \boldsymbol{\pi}}\right]$$

or

$$F(\mathbf{X}^0,\boldsymbol{\pi}) = F(\mathbf{X}^0,\boldsymbol{\pi}^0) + \boldsymbol{\pi}\left[\frac{\partial F(\mathbf{X}^0,\boldsymbol{\pi}^0)}{\partial \boldsymbol{\pi}}\right] - \boldsymbol{\pi}^0\left[\frac{\partial F(\mathbf{X}^0,\boldsymbol{\pi}^0)}{\partial \boldsymbol{\pi}}\right] \quad (3.17)$$

From (3.14a) and (3.14b) we have for $\boldsymbol{\pi} \geq \mathbf{O}$ that the last term of (3.17) is zero and the middle term is nonpositive and thus $F(\mathbf{X}^0,\boldsymbol{\pi}) \leq F(\mathbf{X}^0,\boldsymbol{\pi}^0)$. Together the results yield that $F(\mathbf{X}^0,\boldsymbol{\pi}) \leq F(\mathbf{X}^0,\boldsymbol{\pi}^0) \leq F(\mathbf{X},\boldsymbol{\pi}^0)$, that is, that $\mathbf{X}^0 \geq \mathbf{O}$ and $\boldsymbol{\pi}^0 \geq \mathbf{O}$ satisfy the saddle-point constraints. This completes the proof.

To recapitulate, we have shown the equivalence of the optimum solution $\mathbf{X}^0$ to the convex-programming problem and the $\mathbf{X}^0$ of the point $(\mathbf{X}^0,\boldsymbol{\pi}^0)$ which solves the saddle-point problem under the assumption of a nonempty interior of the solution space to the convex-programming problem. Then, assuming differentiability of the convex functions, necessary and sufficient conditions for a point $(\mathbf{X}^0,\boldsymbol{\pi}^0)$ to solve the saddle-point problem were developed using arguments from the calculus and the properties of convex functions. A number of computational procedures exist based on the above developments which enable us to solve convex-programming problems. We shall discuss below only one such procedure for solving *quadratic-programming problems*, i.e., problems in which the objective function is a quadratic form and the constraints are linear. The reader is referred to Kunzi and Krelle [69a] and Hadley [52d] for discussions concerning more general situations and algorithms, e.g., gradient methods and approximation methods. The reader should recognize that the field of nonlinear programming is indeed more difficult in an algorithmic sense than linear programming. New approaches are constantly being reported in the literature, but much work still needs to be done in this area to make the finding of a solution to a nonlinear-programming problem as efficient and as sure as finding one to a linear problem.

A number of special convex-programming problems are of interest in terms of their associated Kuhn-Tucker conditions. The complete developments are left as exercises for the reader, but we shall next discuss a few points. Suppose our problem is the primal linear program: Minimize $\mathbf{cX}$, subject to $\mathbf{AX} \geq \mathbf{b}$, $\mathbf{X} \geq \mathbf{O}$. We let $f(\mathbf{X}) = \mathbf{cX}$, $[g_i(\mathbf{X})] = \mathbf{b} - \mathbf{AX} \leq \mathbf{O}$, with each $g_i(\mathbf{X})$ a linear function. Then $F(\mathbf{X},\boldsymbol{\pi}) = \mathbf{cX} + \boldsymbol{\pi}[\mathbf{b} - \mathbf{AX}]$. The Kuhn-Tucker conditions for this problem lead to the complementary slackness conditions for the optimal solutions to the primal and dual problems, and the equality of the extreme values of the primal and dual objective function. Here the π_i are the dual variables.

For the linear program: minimize $\mathbf{cX}$, $\mathbf{AX} = \mathbf{b}$, $\mathbf{X} \geq \mathbf{O}$, we transform the equalities to $[\mathbf{AX} \leq \mathbf{b},\ \mathbf{AX} \geq \mathbf{b}]$ and let $[g(\mathbf{X})] = \begin{bmatrix}\mathbf{AX} - \mathbf{b} \\ \mathbf{b} - \mathbf{AX}\end{bmatrix} \leq \mathbf{O}$. We now have $2m$ inequalities. For this problem the $\boldsymbol{\pi}$ are unrestricted.

This will be true if the more general convex problem is stated in terms of $g_i(\mathbf{X}) = 0$ and we convert the equality to two inequalities. Thus, if the ith constraint is an equality, the corresponding π_i will be unrestricted as to sign; i.e., it can be represented by the difference of two nonnegative variables. If the $\mathbf{X}$ are unrestricted in the convex-programming problem, then the Kuhn-Tucker conditions (3.13) reduce to

$$\frac{\partial f(\mathbf{X}^0)}{\partial x_j} + \frac{\sum_{i=1}^{m} \pi_i^0 \, \partial g_i(\mathbf{X}^0)}{\partial x_j} = 0$$

This can be shown by writing $\mathbf{X}$ as the difference of two nonnegative variables and direct application of (3.13).

As an example of the above, consider the following problem (Dorn [40*c*]): Minimize $f(\mathbf{X}) = -\log x_1 - \log x_2$ subject to $x_1 + x_2 - 2 \leq 0$ and $x_j \geq 0$. Here $F(\mathbf{X},\boldsymbol{\pi}) = -\log x_1 - \log x_2 + \pi_1(x_1 + x_2 - 2)$ and the Kuhn-Tucker conditions for optimal $\mathbf{X}^0$ and $\boldsymbol{\pi}^0$ are

$$-\frac{1}{x_1^0} + \pi_1^0 \geq 0 \qquad -\frac{1}{x_2^0} + \pi_1^0 \geq 0$$

$$x_1^0\left[-\frac{1}{x_1^0} + \pi_1^0\right] + x_2^0\left[-\frac{1}{x_2^0} + \pi_1^0\right] = 0$$

$$x_j^0 \geq 0$$

$$x_1^0 + x_2^0 - 2 \leq 0$$

$$\pi_1^0(x_1^0 + x_2^0 - 2) = 0$$

$$\pi_1^0 \geq 0$$

An optimal solution is $\mathbf{X}^0 = (1,1)$ and $\pi_1^0 = 1$.

As in the case for linear programming, primal-dual problem definitions exist for nonlinear programming along with theorems relating the solutions of the two problems. A duality theorem for convex programs with linear constraints is given by Dorn [40*c*]. More general theorems exist, and the reader is referred to the papers by Dantzig and Cottle, Kuhn, and Whinston in Abadie [1] and Moeske [74*g*].

4. QUADRATIC PROGRAMMING

An important class of convex-programming problems arises when the usual linear-constraint set of $\mathbf{AX} = \mathbf{b}$, $\mathbf{X} \geq \mathbf{O}$ is coupled with a convex objective function. This class includes the separable convex-programming problem and the quadratic-programming problem. The solution of both these problems can be accomplished by adaptations of the simplex method. We shall discuss only the quadratic problem and leave the separable case for the reader as an exercise (see Exercise 8).

For our purposes we shall state the quadratic-programming problem as follows:[1] Minimize

$$f(\mathbf{X}) = \mathbf{pX} + \mathbf{X'CX} = \sum_{j=1}^{n} p_j x_j + \sum_{i=1}^{n}\sum_{j=1}^{n} c_{ij}x_i x_j$$

subject to

$$\begin{aligned} \mathbf{AX} &= \mathbf{b} \\ \mathbf{X} &\geq \mathbf{O} \end{aligned} \tag{4.1}$$

where $\mathbf{p}$ is an n-component row vector of given constants. We assume the matrix $\mathbf{C}$ is symmetric and the quadratic form $\mathbf{X'CX}$ is positive semidefinite. From Theorem 2 we note that $\mathbf{X'CX}$ is a convex function for all $\mathbf{X}$ in $\mathbf{E}_n$ and, since $\mathbf{pX}$ is linear, $f(\mathbf{X})$ is a convex function. We next develop the Kuhn-Tucker conditions for problem (4.1).

We rewrite the equations of (4.1) as inequalities in order to formulate the problem in the convex-programming form of (3.1); that is, we have the following: Minimize

$$f(\mathbf{X}) = \mathbf{pX} + \mathbf{X'CX}$$

subject to

$$\begin{aligned} \mathbf{AX} - \mathbf{b} &\leq \mathbf{O} \\ -\mathbf{AX} + \mathbf{b} &\leq \mathbf{O} \\ \mathbf{X} &\geq \mathbf{O} \end{aligned} \tag{4.2}$$

If we let $\boldsymbol{\pi}_1$ and $\boldsymbol{\pi}_2$ be m-component row vectors representing the set of Lagrange multipliers π_{1i} and π_{2i} for the first and second set of inequalities, respectively, we can write the Lagrangian function for (4.2) as

$$F(\mathbf{X},\boldsymbol{\pi}_1,\boldsymbol{\pi}_2) = \mathbf{pX} + \mathbf{X'CX} + \boldsymbol{\pi}_1(\mathbf{AX} - \mathbf{b}) + \boldsymbol{\pi}_2(-\mathbf{AX} + \mathbf{b}) \tag{4.3}$$

Noting that $\partial\mathbf{X'CX}/\partial\mathbf{X} = 2\mathbf{CX}$, we have for (4.3) the following derivatives:

$$\frac{\partial F(\mathbf{X},\boldsymbol{\pi}_1,\boldsymbol{\pi}_2)}{\partial x_j} = p_j + 2\sum_{i=1}^{m} c_{ij}x_i + \sum_{i=1}^{m} \pi_{1i}a_{ij} - \sum_{i=1}^{m} \pi_{2i}a_{ij}$$

$$\frac{\partial F(\mathbf{X},\boldsymbol{\pi}_1,\boldsymbol{\pi}_2)}{\partial \pi_{1i}} = \sum_{j=1}^{n} a_{ij}x_j - b_i$$

$$\frac{\partial F(\mathbf{X},\boldsymbol{\pi}_1,\boldsymbol{\pi}_2)}{\partial \pi_{2i}} = -\sum_{j=1}^{n} a_{ij}x_j + b_i$$

[1] Applications which lead to quadratic-programming formulations include stock portfolio selection, quadratic diet problem, chemical equilibrium, allocation of resources, electrical networks, structural mechanics, and regression analysis (Boot [6c], Dorn [40a], and Saaty and Bram [88b]).

or by grouping terms and writing the above as column vectors

$$\frac{\partial F(\mathbf{X},\boldsymbol{\pi}_1,\boldsymbol{\pi}_2)}{\partial \mathbf{X}} = \mathbf{p}' + 2\mathbf{CX} + \mathbf{A}'\boldsymbol{\pi}_1' - \mathbf{A}'\boldsymbol{\pi}_2'$$

$$\frac{\partial F(\mathbf{X},\boldsymbol{\pi}_1,\boldsymbol{\pi}_2)}{\partial \boldsymbol{\pi}_1} = \mathbf{AX} - \mathbf{b}$$

$$\frac{\partial F(\mathbf{X},\boldsymbol{\pi}_1,\boldsymbol{\pi}_2)}{\partial \boldsymbol{\pi}_2} = -\mathbf{AX} + \mathbf{b}$$

From (3.13) and (3.14) for $\mathbf{X}^0$, $\boldsymbol{\pi}_1^0$, $\boldsymbol{\pi}_2^0$ to solve the saddle-point problem we have the following Kuhn-Tucker conditions for the solution to the quadratic-programming problem:

$$\mathbf{p}' + 2\mathbf{CX}^0 + \mathbf{A}'\boldsymbol{\pi}_1^{0\prime} - \mathbf{A}'\boldsymbol{\pi}_2^{0\prime} \geq \mathbf{O} \tag{4.4a}$$

$$(\mathbf{X}^0)'[\mathbf{p}' + 2\mathbf{CX}^0 + \mathbf{A}'\boldsymbol{\pi}_1^{0\prime} - \mathbf{A}'\boldsymbol{\pi}_2^{0\prime}] = 0 \tag{4.4b}$$

$$\mathbf{X}^0 \geq \mathbf{O} \tag{4.4c}$$

$$\left.\begin{aligned} \mathbf{AX}^0 - \mathbf{b} &\leq \mathbf{O} \\ -\mathbf{AX}^0 + \mathbf{b} &\leq \mathbf{O} \end{aligned}\right. \tag{4.5a}$$

$$[\boldsymbol{\pi}_1^0,\boldsymbol{\pi}_2^0][\mathbf{AX}^0 - \mathbf{b},\ -\mathbf{AX}^0 + \mathbf{b}] = 0 \tag{4.5b}$$

$$\left.\begin{aligned} \boldsymbol{\pi}_1^0 &\geq \mathbf{O} \\ \boldsymbol{\pi}_2^0 &\geq \mathbf{O} \end{aligned}\right. \tag{4.5c}$$

Inequalities (4.5a) are just $\mathbf{AX}^0 = \mathbf{b}$; Eq. (4.5$b$) is always true since it can be rewritten as $(\boldsymbol{\pi}_1^0 - \boldsymbol{\pi}_2^0)(\mathbf{AX}^0 - \mathbf{b}) = \mathbf{O}$; and from inequalities (4.5c) we can let $\boldsymbol{\pi}^0 = (\boldsymbol{\pi}_1^0 - \boldsymbol{\pi}_2^0)'$, where the $m \times 1$ column vector $\boldsymbol{\pi}^0$ is a vector of multipliers π_i^0 which are unrestricted as to sign. The conditions (4.4a to c) and (4.5a to c) can now be rewritten as

$$\begin{aligned} \mathbf{p}' + 2\mathbf{CX}^0 + \mathbf{A}'\boldsymbol{\pi}^0 &\geq \mathbf{O} \\ (\mathbf{X}^0)'[\mathbf{p}' + 2\mathbf{CX}^0 + \mathbf{A}'\boldsymbol{\pi}^0] &= 0 \\ \mathbf{X}^0 &\geq \mathbf{O} \\ \mathbf{AX}^0 &= \mathbf{b} \\ (\mathbf{AX}^0 - \mathbf{b})'\boldsymbol{\pi}^0 &= 0 \\ \boldsymbol{\pi}^0 \text{ unrestricted} & \end{aligned}$$

Let $\mathbf{V}^0$ be an $n \times 1$ nonnegative column vector of slack variables v_j^0, that is, $\mathbf{p}' + 2\mathbf{CX}^0 + \mathbf{A}'\boldsymbol{\pi}^0 - \mathbf{IV}^0 = \mathbf{O}$ or $\mathbf{V}^0 = \mathbf{p}' + 2\mathbf{CX}^0 + \mathbf{A}'\boldsymbol{\pi}^0$. The Kuhn-

Tucker conditions are now

$$\mathbf{p}' + 2\mathbf{C}\mathbf{X}^0 + \mathbf{A}'\boldsymbol{\pi}^0 - \mathbf{I}\mathbf{V}^0 = \mathbf{O} \qquad (4.6a)$$

$$(\mathbf{X}^0)'\mathbf{V}^0 = 0 \qquad (4.6b)$$

$$\left.\begin{array}{r}\mathbf{X}^0 \geq \mathbf{O} \\ \mathbf{V}^0 \geq \mathbf{O}\end{array}\right. \qquad (4.6c)$$

$$\mathbf{A}\mathbf{X}^0 = \mathbf{b} \qquad (4.7a)$$

$$(\mathbf{A}\mathbf{X}^0 - \mathbf{b})'\boldsymbol{\pi}^0 = 0 \qquad (4.7b)$$

$$\boldsymbol{\pi}^0 \text{ unrestricted} \qquad (4.7c)$$

Based on Theorem 8, the Kuhn-Tucker theorem on the equivalence of the solutions to the saddle-point problem and the convex-programming problem, we can restate the above in the following theorem:

Theorem 11. *The vector* $\mathbf{X}^0 \geq \mathbf{O}$ *solves the quadratic-programming problem if and only if* $\mathbf{A}\mathbf{X}^0 = \mathbf{b}$ *and vectors* $\mathbf{V}^0 \geq \mathbf{O}$ *and* $\boldsymbol{\pi}^0$ *unrestricted exist such that*

$$\mathbf{p}' + 2\mathbf{C}\mathbf{X}^0 + \mathbf{A}'\boldsymbol{\pi}^0 - \mathbf{I}\mathbf{V}^0 = \mathbf{O} \qquad (4.8)$$

$$(\mathbf{X}^0)'\mathbf{V}^0 = 0 \qquad (4.9)$$

The reader will note that the conditions of the theorem are now linear except for (4.9), which states that the vector sum $\sum_{j=1}^{n} x_j^0 v_j^0 = 0$; that is, since $x_j^0 \geq 0$, $v_j^0 \geq 0$, then each term $x_j^0 v_j^0 = 0$. If $x_j^0 > 0$, then $v_j^0 = 0$, and conversely. The total number of linear constraints is $m + n$, m from $\mathbf{A}\mathbf{X}^0 = \mathbf{b}$ and n from (4.8). The total number of variables to be considered is $2n$ nonnegative variables, $\mathbf{X}^0$ and $\mathbf{V}^0$, and m unrestricted variables $\boldsymbol{\pi}^0$. If we look at the equations which must be satisfied by the optimum $\mathbf{X}^0$, $\mathbf{V}^0$, $\boldsymbol{\pi}^0$, we have (dropping the superscript notation)

$$\mathbf{A}\mathbf{X} = \mathbf{b} \qquad (4.10)$$

$$2\mathbf{C}\mathbf{X} - \mathbf{I}\mathbf{V} + \mathbf{A}'\boldsymbol{\pi} = -\mathbf{p}' \qquad (4.11)$$

with $\mathbf{X} \geq \mathbf{O}$ and $\mathbf{V} \geq \mathbf{O}$. The side condition $\sum_{j=1}^{n} x_j v_j$ states that no more than n of the $2n$ variables x_j and v_j can be positive. Thus the possible solutions to the equations are restricted to those with no more than $n + m$ of the variables $\mathbf{X}$, $\mathbf{V}$, $\boldsymbol{\pi}$ nonzero; i.e., the solution set is restricted to the set of basic solutions to (4.10) and (4.11), with the implied nonnegativity of $\mathbf{X}$ and $\mathbf{V}$. Using these concepts, Wolfe [105a] developed a finite computational procedure—based on the simplex method—to determine the optimum basic solution to these equations. Wolfe's scheme is given in two forms, the short form for $\mathbf{p} = \mathbf{O}$ or $\mathbf{X'CX}$ positive definite and the long form for $\mathbf{p} \neq \mathbf{O}$ or $\mathbf{X'CX}$ positive semidefinite. We shall next describe these procedures.[1]

Wolfe's quadratic-programming problem algorithm. SHORT FORM. Here we assume $\mathbf{p} = \mathbf{O}$ or $\mathbf{X'CX}$ is positive definite. We wish to find an $\mathbf{X} \geq \mathbf{O}$, $\mathbf{V} \geq \mathbf{O}$, and $\boldsymbol{\pi}$ unrestricted which satisfies the constraints

$$\begin{aligned} \mathbf{AX} \qquad\qquad\quad &= \mathbf{b} \\ 2\mathbf{CX} - \mathbf{IV} + \mathbf{A}'\boldsymbol{\pi} &= -\mathbf{p}' \end{aligned}$$

and such that $\mathbf{X'V} = 0$. We next find a starting basic solution to these equations by a variation of the artificial-basis procedure of the simplex method. Assuming $\mathbf{b} \geq \mathbf{O}$, let $\mathbf{W} = (w_1, w_2, \ldots, w_m) \geq \mathbf{O}$, $\mathbf{Z}_1 = (z_{11}, z_{12}, \ldots, z_{1n}) \geq \mathbf{O}$, $\mathbf{Z}_2 = (z_{21}, z_{22}, \ldots, z_{2n}) \geq \mathbf{O}$ be artificial nonnegative variables and rewrite the equations of the system in the form

$$\mathbf{AX} \qquad\qquad + \mathbf{IW} \qquad\qquad = \mathbf{b} \tag{4.12}$$

$$2\mathbf{CX} - \mathbf{IV} + \mathbf{A}'\boldsymbol{\pi} \qquad + \mathbf{IZ}_1 - \mathbf{IZ}_2 = -\mathbf{p}' \tag{4.13}$$

Here, we take $w_i = b_i$ and either $z_{j1} = -p_j$ or $z_{j2} = p_j$ depending on whether $-p_j \geq 0$ or $-p_j < 0$, respectively. This yields a set of nonnegative values for the components of $\mathbf{W}$, $\mathbf{Z}_1$, and $\mathbf{Z}_2$ which correspond to an artificial basic feasible solution to (4.12) and (4.13). Keeping $\mathbf{V} = \mathbf{O}$ and $\boldsymbol{\pi} = \mathbf{O}$, we next solve (4.12) and (4.13) as a linear-programming problem with an objective function of minimizing $\sum_{i=1}^{m} w_i$. This process will yield a feasible solution to $\mathbf{AX} = \mathbf{b}$, if one exists, and vectors $\mathbf{Z}_1$ and $\mathbf{Z}_2$ satisfying (4.13). (Of course, if a basic feasible solution to (4.12) is known with $\mathbf{W} = \mathbf{O}$, then it can be utilized immediately by introducing it into (4.12) and (4.13), with appropriate changes made in the right-hand side of (4.13).)

At this point we define a new artificial nonnegative vector $\mathbf{Z} = (z_1, z_2, \ldots, z_n) \geq \mathbf{O}$ such that the components of $\mathbf{Z}$ are equal to the n nonnegative components of $\mathbf{Z}_1$ and $\mathbf{Z}_2$ which form the basic feasible solution to (4.13). That is, we now have a basic feasible solution $(\mathbf{X}, \mathbf{Z})$ to the equations

$$\mathbf{AX} \qquad\qquad\qquad = \mathbf{b} \tag{4.14}$$

$$2\mathbf{CX} - \mathbf{IV} + \mathbf{A}'\boldsymbol{\pi} + \bar{\mathbf{I}}\mathbf{Z} = -\mathbf{p}' \tag{4.15}$$

[1] There are a number of other procedures for solving the quadratic-programming problem, most of which are described in Kunzi and Krelle [69*a*].

where the diagonal elements of $\bar{\mathbf{I}}$ are positive or negative ones, according to the sign of the corresponding $-p_j$ and all other elements of $\bar{\mathbf{I}}$ are zero. Next, we need to find a solution to (4.14) and (4.15) such that the artificial variables $\mathbf{Z}$ are eliminated with $\mathbf{X} \geq \mathbf{O}$, $\mathbf{V} \geq \mathbf{O}$, $\boldsymbol{\pi}$ unrestricted and $\mathbf{X}'\mathbf{V} = 0$. We note that our present artificial solution ($\mathbf{X},\mathbf{Z},\mathbf{V} = \mathbf{O}$, $\boldsymbol{\pi} = \mathbf{O}$) satisfies $\mathbf{X}'\mathbf{V} = 0$.

To accomplish the above we apply the simplex method to (4.14) and (4.15) with the objective function of minimizing $\sum_{j=1}^{n} z_j$ and modify the method to account for the nonlinear constraint of $\mathbf{X}'\mathbf{V} = 0$. This modification calls for the following rule: If for any j, x_j is a basic variable, then the corresponding v_j is not allowed to become basic; similarly if v_j is basic, x_j is not allowed to become basic. Under this modification, Wolfe has shown that for $\mathbf{X}'\mathbf{C}\mathbf{X}$ positive definite or $\mathbf{p} = \mathbf{O}$, the simplex algorithm will find an optimum solution in a finite number of iterations with $\sum_{j=1}^{n} z_j = 0$, and hence, each $z_j = 0$. The corresponding $\mathbf{X}$ is an optimum solution to the quadratic-programming problem, and the final $\mathbf{X}$, $\mathbf{V}$ and $\boldsymbol{\pi}$ satisfy the Kuhn-Tucker conditions as given in (4.6a to c) and (4.7a to c).

In order to apply the simplex method to the constraints (4.14) and (4.15), we must convert the unrestricted variables $\boldsymbol{\pi}$ to nonnegative variables. Thus we let $\boldsymbol{\pi} = \boldsymbol{\pi}_1 - \boldsymbol{\pi}_2$, with $\boldsymbol{\pi}_1 \geq \mathbf{O}$, $\boldsymbol{\pi}_2 \geq \mathbf{O}$ and rewrite the starting problem as follows: Minimize

$$\sum_{j=1}^{n} z_j$$

subject to

$$\mathbf{AX} = \mathbf{b} \tag{4.16}$$

$$2\mathbf{CX} - \mathbf{IV} + \mathbf{A}'\boldsymbol{\pi}_1 - \mathbf{A}'\boldsymbol{\pi}_2 + \bar{\mathbf{I}}\mathbf{Z} = -\mathbf{p}' \tag{4.17}$$

with $\mathbf{X}'\mathbf{V} = 0$ and $\mathbf{X} \geq \mathbf{O}$, $\mathbf{V} \geq \mathbf{O}$, $\boldsymbol{\pi}_1 \geq \mathbf{O}$, $\boldsymbol{\pi}_2 \geq \mathbf{O}$, $\mathbf{Z} \geq \mathbf{O}$. This problem is solved by direct application of the simplex method as modified above.

LONG FORM. Here we consider the general quadratic-programming problem with $\mathbf{p} \neq \mathbf{O}$ and $\mathbf{X}'\mathbf{C}\mathbf{X}$ positive semidefinite. To accomplish the optimization we consider the parametric objective function $f(\mathbf{X}) = \lambda\mathbf{pX} + \mathbf{X}'\mathbf{C}\mathbf{X}$ subject to $\mathbf{AX} = \mathbf{b}$, $\mathbf{X} \geq \mathbf{O}$, $\lambda \geq 0$. We desire, of course, the optimum solution for $\lambda = 1$. The problem equivalent to (4.16) and (4.17) is as follows: Minimize

$$\sum_{j=1}^{n} z_j$$

subject to

$$\mathbf{AX} = \mathbf{b} \tag{4.18}$$
$$2\mathbf{CX} - \mathbf{IV} + \mathbf{A}'\boldsymbol{\pi}_1 - \mathbf{A}'\boldsymbol{\pi}_2 + \bar{\mathbf{I}}\mathbf{Z} = -\lambda\mathbf{p}' \tag{4.19}$$

with $\mathbf{X}'\mathbf{V} = 0$, $(\mathbf{X}, \mathbf{V}, \boldsymbol{\pi}_1, \boldsymbol{\pi}_2, \mathbf{Z}, \lambda) \geq \mathbf{O}$. This problem is first solved for $\lambda = 0$ by applying the short-form algorithm. Next, restricting all $z_j = 0$, we apply the parametric simplex algorithm (see Chap. 8) with the modification rule of the short form which restricts $\mathbf{X}'\mathbf{V} = 0$. The parametrization, to find solutions for $\lambda > 0$, is accomplished by using the new objective function minimize $-\lambda$, that is, maximize λ. Wolfe [105*a*] proves the following:

1. If no change of basis can be made starting with the short-form solution and the objective function minimize $-\lambda$, then the original quadratic-programming objective function $f(\mathbf{X})$ is unbounded for $\lambda > 0$.
2. Otherwise a sequence of solutions $\mathbf{X}_0, \mathbf{X}_1, \ldots, \mathbf{X}_t$ corresponding to parameter values $0 = \lambda_0, \lambda_1, \ldots, \lambda_t$ can be found such that for $\lambda_k \leq \lambda \leq \lambda_{k+1}$, the quadratic-programming problem for this range of λ is solved by

$$\mathbf{X} = \frac{\lambda_{k+1} - \lambda}{\lambda_{k+1} - \lambda_k}\mathbf{X}_k + \frac{\lambda - \lambda_k}{\lambda_{k+1} - \lambda_k}\mathbf{X}_{k+1}$$

and for $\lambda \geq \lambda_t$

$$\mathbf{X} = \mathbf{X}_t + (\lambda - \lambda_t)\mathbf{X}_\infty$$

where $\mathbf{X}_\infty$ is the final solution of the iterative process.

The reader is referred to Dantzig [16*c*], Beale [5*e*], and Van de Panne and Whinston [97*e*] for other variations of the simplex method for handling both the positive definite and positive semidefinite cases.

To establish a primal-dual relationship for quadratic programs, we define the following problems:

Primal: Minimize

$$f(\mathbf{X}) = \mathbf{pX} + \mathbf{X}'\mathbf{CX}$$

subject to

$$\mathbf{AX} = \mathbf{b}$$
$$\mathbf{X} \geq \mathbf{O}$$

Dual: Maximize

$$g(\mathbf{X},\boldsymbol{\pi}) = \mathbf{b}'\boldsymbol{\pi} - \mathbf{X}'\mathbf{CX}$$

subject to

$$-2\mathbf{CX} + \mathbf{A}'\boldsymbol{\pi} \leq \mathbf{p}'$$

We then have the following duality theorem of quadratic programming, Dorn [40*b*].

Theorem 12. *If there exists a vector* $\mathbf{X}^0$ *which minimizes* $f(\mathbf{X})$ *in the primal, then there exists a* $\boldsymbol{\pi}^0$ *for which* $(\mathbf{X}^0,\boldsymbol{\pi}^0)$ *is an optimal solution to the dual. Conversely, if* $(\mathbf{X}^0,\boldsymbol{\pi}^0)$ *maximizes* $g(\mathbf{X},\boldsymbol{\pi})$ *of the dual, then* $\mathbf{X}^0$ *is an optimal solution to the primal. In either case,* $f(\mathbf{X}^0) = g(\mathbf{X}^0,\boldsymbol{\pi}^0)$*; that is, the minimum value for the primal problem is equal to the maximum value for the dual problem.*

We shall not prove this theorem, but the reader should be able to verify its validity by the derivation and discussion of the Kuhn-Tucker conditions. If $\mathbf{C} = \mathbf{O}$, the above quadratic dual statements reduce to the linear-programming unsymmetric dual problems. For the symmetric case, that is, $\mathbf{AX} \geq \mathbf{b}$, $\mathbf{X} \geq \mathbf{O}$, the quadratic dual constraints would be $-2\mathbf{CX} + \mathbf{A}'\boldsymbol{\pi} \leq \mathbf{p}'$, $\boldsymbol{\pi} \geq \mathbf{O}$, with the same objective functions.

5. SPECIAL NONLINEAR PROBLEMS

Computational algorithms exist for solving many other variations of the convex-programming problem: *gradient methods* for linear and nonlinear constraints with convex objective function, Rosen [87*b*], Zoutendijk [108], Hadley [52*d*], and Kunzi and Krelle [69*a*]; *cutting-plane methods,* Kelley [64*b*]; *sequential unconstrained minimization,* Fiacco and McCormick [42*c*]; *separable convex programming,* Charnes and Lemke [15], Dantzig [16*c*], and Hadley [52*d*]; and *generalized programming,* Dantzig [16*c*]. See Saaty and Bram [88*b*] and Wilde and Beightler [100*cc*] for general discussions on nonlinear mathematics and optimization.

In addition, a number of other nonlinear-programming problems have been formulated and solved. Chief among these is the class of integer-programming problems and its variations (see Chap. 9, Sec. 3). Like integer and quadratic programs, other nonlinear or related problems can be solved by variations of the simplex method. A discussion of some of these problems follows.

A *linear fractional-programming problem* has the form: Maximize

$$f(\mathbf{X}) = \frac{\alpha + \mathbf{cX}}{\beta + \mathbf{dX}}$$

subject to

$$\mathbf{AX} \leq \mathbf{b}$$
$$\mathbf{X} \geq \mathbf{O}$$

where α and β are scalars, $\mathbf{c}$ and $\mathbf{d}$ are n-component row vectors of constants. Charnes and Cooper [10*a*] show how to transform this problem into equivalent linear programs, the solution of which solves the fractional program. Other approaches also exist (Dorn [40*d*] and Isbell and Marlow [61*c*]). For this problem it can be shown that $f(\mathbf{X})$ is neither convex nor concave, that a local maximum is a global maximum, and if the maximum occurs at a finite point, it occurs also at a basic (primal) feasible solution of the constraints.

Many problems dealing with statistical and other approximation procedures can be converted to linear programs (Charnes, Cooper, and Ferguson [11*a*], Kelley [64*a*], Stiefel [91*a*], Ward [99*f*], Wagner [99*c*], and Zukhovitskiy and Avdeyeva [109]). In addition Dantzig [27*h*], Rosen [87*d*], Torng [94*a*], Whalen [100*b*], and Zadeh [107] have related linear-programming procedures to the area of optimal-control theory. We shall next discuss two basic formulations which have proved of value in these fields.

The problem of minimizing the sum of absolute deviations can be handled as follows (Wagner [99*c*]). Let a_{ij}, $i = 1, 2, \ldots, k$ and $j = 1, 2, \ldots, p$ denote a set of k observational measurements on p independent variables and b_i, $i = 1, 2, \ldots, k$ denote the associated measurements on the dependent variable. The problem is to find the regression coefficients x_j such that we minimize $\sum_i \left| \sum_j a_{ij}x_j - b_i \right|$; that is, we wish to find values of the regression coefficients such that the sum of the absolute differences $\left| \sum_j a_{ij}x_j - b_i \right|$ is a minimum. Let $z_i' - z_i'' = b_i - \sum_j a_{ij}x_j$, $z_i' \geq 0$, $z_i'' \geq 0$. Since for any set of x_j the expression $b_i - \sum_j a_{ij}x_j$ can be positive or negative, we represent this difference as the difference of two nonnegative numbers. We can now rewrite the problem as follows: Minimize

$$\sum_i (z_i' + z_i'')$$

subject to

$$\sum_j a_{ij}x_j + z_i' - z_i'' = b_i$$

$$z_i' \geq 0$$

$$z_i'' \geq 0$$

with the variables x_j unrestricted as to sign. The x_j must, of course, be written as the difference of two nonnegative variables to put the problem into the standard linear-programming format. Since in a basic feasible solution z_i' and z_i'' cannot both be positive, the optimum basic solution will select a set of x_j which minimizes the sum of the absolute differences.

In a similar vein, the Chebyshev problem for a set of observations is to find a set of coefficients x_j such that we minimize $\{\underset{x_j}{\text{minimize}}\ \underset{i}{\text{maximum}}|\Sigma a_{ij}x_j - b_i|\}$; that is, we wish to find a set of x_j such that the maximum deviation between the corresponding observations is a minimum. Let $z \geq \left|\sum_j a_{ij}x_j - b_i\right|$ for all i. The variable z is nonnegative and we wish to have z a minimum. This inequality in absolute terms can be rewritten for each i as two inequalities; i.e., we have that the value of $\sum_j a_{ij}x_j - b_i$ can lie between z and $-z$, or $-z \leq \sum_j a_{ij}x_j - b_i \leq z$. The Chebyshev problem is now as follows: Minimize

$$z$$

subject to

$$\sum_j a_{ij}x_j - b_i - z \leq 0$$

$$\sum_j a_{ij}x_j - b_i + z \geq 0$$

$$z \geq 0$$

and x_j unrestricted as to sign.

EXERCISES

1. Construct and graph two-dimensional programming problems with the following characteristics:

a. an interior point which is optimal

b. a nonconvex solution space

2. Prove that a linear function is both convex and concave.

3. Prove that the set of points $(\mathbf{X},\mathbf{Y})$ for which $\mathbf{Y} \geq f(\mathbf{X})$, $f(\mathbf{X})$ a convex function, is a convex set, i.e., the region above and bounded by a convex function is convex.

4. Prove that any nonnegative combination of convex functions is convex.

5. Determine if the following functions are convex:

(*a*) $|x|$ for all x

(*b*) e^x for all x

(*c*) $(x - 1)^3$ for $1 \leq x < \infty$

(*d*) $(x - 1)^3$ for $0 \leq x < \infty$

(*e*) $\log x$ for $0 < x < \infty$

(*f*) e^{-x^2} for $0 \leq x < \infty$

6. Develop the Kuhn-Tucker conditions for the following problems:

a. Minimize $\mathbf{cX}$, subject to $\mathbf{AX} \geq \mathbf{b}$, $\mathbf{X} \geq \mathbf{O}$.

b. Minimize $\mathbf{cX}$, subject to $\mathbf{AX} = \mathbf{b}$, $\mathbf{X} \geq \mathbf{O}$.

c. Minimize $f(\mathbf{X})$, subject to $g_i(\mathbf{X}) \leq 0$, $i = 1, \ldots, m$ and $f(\mathbf{X})$ and all $g_i(\mathbf{X})$ are convex.

d. Minimize $f(\mathbf{X})$ subject to $\mathbf{AX} = \mathbf{b}$, $\mathbf{X} \geq \mathbf{O}$, $f(\mathbf{X})$ a convex function.

7. Develop the Kuhn-Tucker conditions for the following convex-programming problems:

a. Minimize

$$a_1e^{-b_1x_1} + a_2e^{-b_2x_2}$$

subject to

$$\begin{aligned} x_1 + x_2 - 1 &\leq 0 \\ x_j &\geq 0 \end{aligned}$$

and $(a_1, a_2, b_1, b_2) > \mathbf{O}$ (Carr and Howe [8*c*]).

b. Minimize

$$x_1^2 + 2x_2^2 + e^{x_3}$$

subject to

$$\begin{aligned} x_1x_2 + x_2 \qquad\quad - 1 &\leq 0 \\ x_2 + x_3^2 - 2 &\leq 0 \end{aligned}$$

and all $x_j \geq 0$.

c. Minimize

$$-x_1 - x_2 - x_3 + \tfrac{1}{2}(x_1^2 + x_2^2 + x_3^2)$$

subject to

$$\begin{aligned} x_1 + x_2 + x_3 &\leq 1 \\ 4x_1 + 2x_2 \qquad &\leq \tfrac{7}{3} \end{aligned}$$

all $x_j \geq 0$ (Moeske [74*g*]). (Note that the objective function is the sum of a linear function and a convex function.)

8. Separable convex programming (Dantzig [16*c*], Hadley [52*d*]): Consider the problem of minimizing $f(\mathbf{X}) = \sum_{j=1}^{n} f_j(x_j)$ subject to $\mathbf{AX} = \mathbf{b}$, $\mathbf{X} \geq \mathbf{O}$, where the objective function is the sum of separable convex functions $f_j(x_j)$. (A function is *separable* in its n variables if it can be written as a sum of n functions $f_j(x_j)$, each of which is a function of only a single variable; for example, $f(\mathbf{X}) = x_1^2 + x_2^2 + x_3$.) Develop a linear-programming procedure which utilizes linear approximations for each $f_j(x_j)$ by taking advantage of the convexity of the $f_j(x_j)$. Apply the approximation procedure to the objective function of Exercise 7*b*.

9. Show that for $\mathbf{X'CX}$ positive definite the quadratic-programming problem cannot have an unbounded solution (Hadley [52*d*]).

10. Develop the necessary changes in Wolfe's algorithm for $\mathbf{AX} \leq \mathbf{b}$, $\mathbf{X} \geq \mathbf{O}$.

11. Solve the following problems:

a. Wolfe [105*a*]. Minimize

$$\lambda(x_1 - 2x_3) + \tfrac{1}{2}(x_1^2 + x_2^2 + x_3^2)$$

subject to

$$\begin{aligned} x_1 - x_2 + x_3 &= 1 \\ x_j &\geq 0 \end{aligned}$$

Solution $\quad \lambda = 1,\ x_1 = 0,\ x_2 = \tfrac{1}{2},\ x_3 = \tfrac{3}{2},\ f(\mathbf{X}) = -\tfrac{7}{4}$

b. Hadley [52*d*]. Minimize

$$-2x_1 - x_2 + x_1^2$$

subject to

$$2x_1 + 3x_2 + x_3 \qquad = 6$$

$$2x_1 + \ x_2 \qquad + x_4 = 4$$

$$x_j \geq 0$$

Solution $\quad x_1 = 2/3,\ x_2 = 14/9,\ x_3 = 0$

$$x_4 = 10/9,\ f(\mathbf{X}) = -22/9$$

c. Carr and Howe [8*c*]. Minimize

$$-10x_1 - 20x_2 - x_1x_2 - 2x_1^2 - 2x_2^2$$

subject to

$$x_2 + x_3 \qquad = 8$$

$$x_1 + x_2 \qquad + x_4 = 10$$

$$x_j \geq 0$$

Solution $\quad x_1 = 4,\ x_2 = 6,\ x_3 = 0,\ x_4 = 0,\ f(\mathbf{X}) = 288$

d. Beale [5*e*]. Minimize

$$-6x_1 + 2x_1^2 - 2x_1x_2 + 2x_2^2$$

subject to

$$x_1 + x_2 \leq 2$$

$$x_j \geq 0$$

Solution $\quad x_1 = 3/2,\ x_2 = 1/2,\ f(\mathbf{X}) = -11/2$

12. Solve *b*, *c*, and *d* of Exercise 11 graphically.

13. Transform the following problem into a linear-constraint problem:

$$\mathbf{AX} = \mathbf{b}$$

$$|x_j| \leq 1,\ x_j \text{ unrestricted}$$

14. Transform the following problem into a linear-programming problem: Minimize $z_i = \sum_j |x_j|$ subject to $\mathbf{AX} = \mathbf{b}$, $\mathbf{X}$ unrestricted.

15. Prove the sufficiency of Theorem 11; that is, if $\mathbf{X}^0$, $\mathbf{V}^0$, π^0 satisfy $\mathbf{AX}^0 = \mathbf{b}$, (4.8), and (4.9), then $f(\mathbf{X}^0)$ is the minimum solution. HINT: Need to show $f(\mathbf{X}^0) \leq f(\mathbf{X})$ for all other $\mathbf{X}$. Use the fact that $(\mathbf{X} - \mathbf{X}^0)'\mathbf{C}(\mathbf{X} - \mathbf{X}^0) \geq 0$ of Theorem 2.

16. Show for $\mathbf{X}'\mathbf{C}\mathbf{X}$ positive semidefinite that $\mathbf{X}'\mathbf{C}\mathbf{X} = 0$ implies that $\mathbf{C}\mathbf{X} = \mathbf{O}$.

BIBLIOGRAPHY OF LINEAR-PROGRAMMING APPLICATIONS

The material that formed the survey of linear-programming applications of Sec. 3 in Chap. 1 was taken from the "Bibliography on Linear Programming and Related Techniques" by Vera Riley and Saul I. Gass, Johns Hopkins Press, Baltimore, 1958. This comprehensive bibliography includes over 1,000 items with abstracts arranged by topic, e.g., introductory survey, computational techniques, game theory, convex sets and linear inequalities, applications, etc. The list of applications cited below is only a small part of the total references on linear-programming applications.[1]

1. Agricultural Applications

Boles, James N.: Linear Programming and Farm Management Analysis, *Journal of Farm Economics*, vol. 37, no. 1, pp. 1–24, February, 1955.

Candler, Wilfred: A Modified Simplex Solution for Linear Programming with Variable Capital Restriction, *Journal of Farm Economics*, vol. 38, no. 4, pp. 940–955, November, 1956.

Fisher, Walter D., and Leonard W. Shruben: Linear Programming Applied to Feed-mixing under Different Price Conditions, *Journal of Farm Economics*, vol. 35, no. 4, pp. 471–483, November, 1953.

Fox, Karl A., and Richard C. Taeuber: Spatial Equilibrium Models of the Livestock-feed Economy, *The American Economic Review*, vol. 45, no. 4, pp. 584–608, September, 1955.

———: "A Spatial Equilibrium Model of the Livestock-feed Economy in the United States." Paper presented before the meeting of the Econometric Society, Dec. 27, 1952, Chicago, Ill.; published in *Econometrica*, vol. 21, no. 4, pp. 547–566 (including references), October, 1953.

*Heady, E. O., and A. C. Egbert: Regional Programming of Efficient Agricultural Production Patterns, *Econometrica*, vol. 32, no. 3, July, 1964.

Hildreth, Clifford G.: "Economic Implications of Some Cotton Fertilizer Experiments." Paper presented at a joint meeting of the Econometric Society

[1] References with an asterisk (*) are not included in the 1958 bibliography as they were published after that date.

and the American Farm Economic Association, December, 1953, and at a Cowles Commission staff meeting January, 1954; published in *Econometrica*, vol. 23, no. 1, pp. 88–98, January, 1955.

——— and Stanley Reiter: On the Choice of a Crop Rotation Plan, chap. 11 (pp. 177–188) in Tjalling C. Koopmans (ed.), "Activity Analysis of Production and Allocation," *Cowles Commission Monograph* 13 (proceedings of a Conference on Linear Programming held in Chicago, Ill., by the Cowles Commission for Research in Economics, June 20–24, 1949), John Wiley & Sons, Inc., New York, 1951.

King, Richard A.: Use of Economic Models: Some Applications of Activity Analysis in Agricultural Economics, North Carolina Agricultural Experiment Station Journal Paper 508, *Journal of Farm Economics*, vol. 35, no. 5, pp. 823–833, December, 1953.

*Muto, Kazuo: Application of Linear Programming to Planning in Agriculture, *OR, JUSE* (Japan), vol. 6, no. 4, December, 1961.

*Pierson, D. R.: Farm Profits Up by 40 Percent, *Automatic Data Processing*, vols. 4, 5, May, 1962.

Swanson, Earl R.: Integrating Crop and Livestock Activities in Farm Management Activity Analysis, *Journal of Farm Economics*, vol. 37, no. 5, pp. 1249–1258, December, 1955.

*Swanson, L. W., and J. G. Woodruff: A Sequential Approach to the Feed-mix Problem, *Operations Research*, vol. 12, no. 1, January–February, 1964.

2. Contract Awards

Percus, Jerome, and Leon Quinto: The Application of Linear Programming to Competitive Bond Bidding, *Econometrica*, vol. 24, no. 4, pp. 314–428 (including references), October, 1956.

*Waggener, H. A., and G. Suzuki: Bid Evaluation for Procurement of Aviation Fuel at DFSC: A Case History, *Naval Research Logistics Quarterly*, vol. 14, no. 1, March, 1967.

3. Industrial Applications

a. Chemical Industry

Arnoff, E. Leonard: "The Application of Linear Programming." Paper presented at the Case Institute of Technology, Cleveland, Ohio, Jan. 20–22, 1954; published in *Proceedings of Conference on Operations Research in Production and Inventory Control*, pp. 47–52, Case Institute of Technology, Cleveland, Ohio, 1954.

Dannerstedt, Gunnar, and Hrand Saxenian: Machine Scheduling to Meet Seasonal Sales Variation, in "Fundamental Investigations in Methods of Operations Research," *MIT Interim Technical Report* 1, pp. 12–13, July 1, 1953–Mar. 31, 1954.

*Dantzig, G. B., S. Johnson, and W. White: A Linear Programming Approach to the Chemical Equilibrium Problem, *Management Science*, vol. 5, no. 1, October, 1958.

b. The Coal Industry

Henderson, James M.: A Short-run Model for the Coal Industry, *Review of Economics and Statistics*, vol. 37, no. 4, pp. 336–346, November, 1955.

c. Commercial Airlines

Morton, George: "Application of Linear Programming Methods to Commercial Airline Operations." Paper presented to the fourteenth European meeting of the Econometric Society, Cambridge, England, Aug. 13–15, 1952; abstracted in *Econometrica*, vol. 21, no. 1, p. 193 (with discussion), January, 1953.

d. Communications Industry

Kalaba, Robert E., and Mario L. Juncosa: "Optical Design and Utilization of Communication Networks." Paper presented before a joint meeting of the Institute of Management Sciences and the Operations Research Society of America, University of California, Los Angeles, May 30, 1956; published as P-782, The RAND Corporation, 25 pp., July 13, 1956; also RM-1687, 22 pp., Apr. 23, 1956; published in *Management Science*, vol. 3, no. 1, pp. 33–44 (including references), October, 1956.

*Saaty, T. L., and G. Suzuki: A Nonlinear Programming Model in Optimum Communication Satellite Use, *SIAM Review*, vol. 7, no. 3, July, 1965.

e. Iron and Steel Industry

Fabian, Tibor: "Application of Linear Programming to Steel Production Planning." Paper presented to the seventh national meeting of the Operations Research Society of America, Los Angeles, Aug. 15–17, 1955; abstracted in *Journal of the Operations Research Society of America*, vol. 3, no. 4, p. 565, November, 1955.

*Fabian, Tibor: Blast Furnace Burdening and Production Planning, *Management Science*, vol. 14, no. 2, October, 1967.

Reinfeld, Nyles V.: Do You Want Production or Profit? *Tooling and Production*, vol. 20, no. 5, pp. 44–48, 69, August, 1954.

f. Paper Industry

*Gilmore, P. C., and R. E. Gomory: A Linear-programming Approach to the Cutting-stock Problem–Part I, *Operations Research*, vol. 9, no. 6, November–December, 1961.

*Gilmore, P. C., and R. E. Gomory: A Linear-programming Approach to the Cutting-stock Problem–Part II, *Operations Research*, vol. 11, no. 6, November–December, 1963.

*Morgan, J. I.: Survey of Operations Research, *Paper Mill News*, vol. 84, no. 11, March, 1961.

Paull, A. E.: "Linear Programming: A Key to Optimum Newsprint Production." Paper presented to the summer meeting of the Technical Section, Canadian Pulp and Paper Association Quebec, Canada, June 6–8, 1955; published in *Pulp and Paper Magazine of Canada*, vol. 57, no. 1, pp. 85–90, January, 1956, reissued, vol. 57, no. 4, pp. 145–150, March, 1956.

g. Petroleum Industry

*Aronofsky, J. S., and A. C. Williams: The Use of Linear Programming and Mathematical Models in Underground Oil Production, *Management Science*, vol. 8, no. 4, July, 1962.

*Catchpole, A. R.: The Application of Linear Programming to Integrated Supply Problems in the Oil Industry, *Operational Research Quarterly* (U.K.), vol. 13, no. 2, June, 1962.

Charnes, Abraham, William W. Cooper, and Robert Mellon: "Blending Aviation Gasolines—A Study in Programming Interdependent Activities in an Integrated Oil Company." Paper presented to the Symposium on Linear Inequalities and Programming, Washington, D.C., June 14–16, 1951, jointly sponsored by the Air Force, DCS/Comptroller, Headquarters USAF, and the National Bureau of Standards; published in Project SCOOP, Manual 10, pp. 115–146 (including references), April, 1952; also published in *Econometrica*, vol. 20, no. 2, pp. 135–159, April, 1952; abstracted in *Operational Research Quarterly*, vol. 3, no. 3, pp. 54–55, September, 1952.

Davie, J. W.: "Use of Linear Programming in Selective Blending Studies." Paper presented to the summer meeting of the Econometric Society, Montreal, Canada, Sept. 10–13, 1954; abstracted in *Econometrica*, vol. 23, no. 3, pp. 336–337, July, 1955.

*Faur, P.: Elements for Selection of Investments in the Refining Industry, *Proceedings of the Second International Conference on Operational Research*, English Universities Press, Ltd., London, and John Wiley & Sons, Inc., New York, 1960.

*Garvin, W. W., H. W. Crandell, J. B. John, and R. A. Spellman: Applications of Linear Programming in the Oil Industry, *Management Science*, vol. 3, no. 4, July, 1957.

Manne, Alan S.: "Concave Programming for Gasoline Blends," P-383, The RAND Corporation, Mar. 20, 1953. Paper presented to the 1953 annual meeting of the Operations Research Society of America, Case Institute of Technology, Cleveland, Ohio, May 15–16, 1953; abstracted in *Journal of the Operations Research Society of America*, vol. 1, no. 3, p. 148, May, 1953.

———: "Scheduling of Petroleum Refinery Operations," 181 pp., Harvard Economic Studies, vol. 48, Harvard University Press, Cambridge, Mass., 1956.

*Smith, L. W.: An Approach to Costing Joint Production Based on Mathematical Programming with an Example from Petroleum Refining, Document AD-256 624, *U.S. Government Research Reports*, Washington, D.C., February, 1961.

Symonds, Gifford H.: "A Crude Allocation Problem." A Manufacturing Technical Committee Report of Esso Standard Oil Company, Linden, N.J.; published as chap. 9 (pp. 63–74) of "Linear Programming: The Solution of Refinery Problems," Esso Standard Oil Company, New York, 1955.

———: "Linear Programming: The Solution of Refinery Problems," 74 pp., Esso Standard Oil Company, New York, 1955.

———: "Optimum Production Rates and Inventory to Meet Uncertain Seasonal Requirements." A Manufacturing Technical Committee Report of Esso

Standard Oil Company, Linden, N.J., published as chap. 6 (pp. 37–46) of "Linear Programming: The Solution of Refinery Problems," Esso Standard Oil Company, New York, 1955.

Vazsonyi, Andrew: Optimizing a Function of Additively Separated Variables Subject to a Simple Restriction, *Proceedings of the Second Symposium in Linear Programming*, vol. II, pp. 453–469 (including references) (collection of papers presented to a conference sponsored by the National Bureau of Standards and the Directorate of Management Analysis Service, DCS/Comptroller, Headquarters USAF, held in Washington, D.C., Jan. 27–29, 1955).

h. Railroad Industry

Charnes, Abraham, and M. H. Miller: "A Model for Optimal Programming of Railway Freight Train Movements." Paper presented to the meeting of the Econometric Society, New York City, Dec. 28–30, 1955; abstracted in *Econometrica*, vol. 24, no. 3, pp. 349–350, July, 1956; full article published in *Management Science*, vol. 3, no. 1, pp. 74–93 (including references), October, 1956.

Crane, Roger R.: A New Tool: Operations Research, *Modern Railroads*, vol. 9, no. 1, pp. 146–152, January, 1954.

i. Other Industries

*Androit, J., and J. Gaussens: Programme for Thermal Reactor and "Breeder Reactor" Power Stations—Fuel Economy Problem of Storage—Price of Plutonium, *Proceedings of the Second International Conference on Operational Research*, English Universities Press, Ltd., London, and John Wiley & Sons, Inc., New York, 1960.

Charnes, Abraham, William W. Cooper, and Robert O. Ferguson: Optimal Estimation of Executive Compensation by Linear Programming, *Management Science*, vol. 1, no. 2, pp. 138–151, January, 1955.

*Horowitz, J., R. Lattes, and E. Parker: Some Operational Problems Connected with Power Reactor Discharges, *Proceedings of the Second International Conference on Operational Research*, English Universities Press, Ltd., London, and John Wiley & Sons, Inc., New York, 1960.

Schwan, Harry T., and John J. Wilkinson: Linear Programming: What Can It Do? *Chemical Engineering*, August, 1956, no. 63, pp. 211–214.

4. Economic Analysis

Altschul, Eugen: "Reorientation in Economic Theory: Linear and Nonlinear Programming." Paper presented to the annual spring meeting of the Missouri Section of the Mathematical Association of America, University of Kansas City, Kansas City, Mo., Apr. 22, 1955; abstracted in *American Mathematical Monthly*, vol. 62, no. 6, p. 543, June, 1955.

Beckmann, Martin J., and Thomas Marschak: "An Activity Analysis Approach to Location Theory," 38 pp. P-649, The RAND Corporation, Apr. 5, 1955. Published in *Proceedings of the Second Symposium in Linear Programming*, vol. I, pp. 331–379 (collection of papers presented to a conference sponsored jointly by the National Bureau of Standards and the Directorate of Management Analysis Service, DCS/Comptroller, Headquarters USAF, held in Washington, D.C., Jan. 27–29, 1955).

Brown, J. A. C.: "An Experiment in Demand Analysis: The Computation of a Diet Problem on the Manchester Computer." Paper presented to the Conference on Linear Programming arranged by Ferranti, Ltd., held in London, May 4, 1954; published in *Conference on Linear Programming,* pp. 41–53, Ferranti, Ltd., London, 1954; summary in *Operations Research* (*Jorsa*), vol. 4, no. 1, p. 133, February, 1956.

Charnes, Abraham, and William W. Cooper: "An Example of Constrained Games in Industrial Economics." Presented to a meeting of the Econometric Society, Washington, D.C., Dec. 27–29, 1953; published as *ONR Research Memorandum* 12, Graduate School of Industrial Administration, Carnegie Institute of Technology, Pittsburgh, Pa.; abstracted in *Econometrica,* vol. 22, no. 4, pp. 526–527, October, 1954.

Chipman, John: Linear Programming, *The Review of Economics and Statistics,* vol. 35, no. 2, pp. 101–117, May, 1953.

Davidson, Donald, and Patrick Suppes: "Experimental Measurement of Utility by Use of a Linear Programming Model," 30 pp., *Technical Report* 3, Applied Mathematics and Statistical Laboratory, Stanford University, Apr. 6, 1956. Paper presented to the meeting of the Econometric Society, Ann Arbor, Mich., Aug. 29–Sept. 1, 1955; abstracted in *Econometrica,* vol. 24, no. 2, pp. 201–202, April, 1956.

Dorfman, Robert: "Application of Linear Programming to the Theory of the Firm, Including an Analysis of Monopolistic Firms by Nonlinear Programming," 98 pp., Bureau of Business and Economic Research, University of California, University of California Press, Berkeley, Calif., 1951. Reviewed by Erich Schneider, *Econometrica,* vol. 22, no. 1, pp. 129–130, January, 1954.

Frisch, Ragnar A. K.: "Principles of Linear Programming, with Particular Reference to the Double Gradient Form of the Logarithmic Potential Method," 219 pp., memorandum, University Institute of Economics, Oslo, Norway, Oct. 18, 1954.

Gunther, Paul: Use of Linear Programming in Capital Budgeting (letter to the editor), *Journal of the Operations Research Society of America,* vol. 3, no. 2, pp. 219–224, May, 1955.

Markowitz, Harry M.: "Portfolio Selection." Thesis submitted to the University of Chicago, Chicago, Ill., Summer, 1953; published in *Journal of Finance,* vol. 7, no. 1, pp. 77–91, March, 1952.

Martin, Alfred D., Jr.: Mathematical Programming of Portfolio Selections, *Management Science,* vol. 1, no. 2, pp. 152–156, January, 1955.

Samuelson, Paul A.: "Linear Programming and Economic Theory," *Proceedings of the Second Symposium in Linear Programming,* vol. I, pp. 251–272 (collection of papers presented to a conference sponsored by the National Bureau of Standards and the Directorate of Management Analysis Service, DCS/Comptroller, Headquarters USAF, held in Washington, D.C., Jan. 27–29, 1955); also P-685, The RAND Corporation, 17 pp., May 25, 1955.

———: "The Le Châtelier Principle in Linear Programming," 18 pp., RM-210, The RAND Corporation, Aug. 4, 1949.

Solow, Robert M.: Linear Programming: Lecture XII, "Notes from M.I.T. Summer Courses on Operations Research, June 16–July 3, 1953," pp. 116–129, Technology Press, M.I.T., Cambridge, Mass., 1953.

Whitin, Thomson M.: Classical Theory, Graham's Theory, and Linear Programming in International Trade, *The Quarterly Journal of Economics*, vol. 67, no. 4, pp. 520–544, November, 1953.

5. Military Applications

Jacobs, Walter W.: The Caterer Problem, *Naval Research Logistics Quarterly* (Office of Naval Research), vol. 1, no. 2, pp. 154–165 (including references), June, 1954.

———: Military Applications of Linear Programming, *Proceedings of the Second Symposium in Linear Programming*, vol. I, pp. 1–27 (including references) (collection of papers presented to a conference sponsored by the National Bureau of Standards and the Directorate of Management Analysis Service, DCS/Comptroller, Headquarters USAF, held in Washington, D.C., Jan. 27–29, 1955).

Joseph, Joseph A.: "The Application of Linear Programming to Weapon Selection and Target Analysis," 40 pp., *Operations Analysis Technical Memorandum* 42, Operations Analysis Division, Directorate on Operations, DCS/Operations, Headquarters USAF, Washington, D.C., Jan. 5, 1954.

Nicholson, George E., Jr., and George W. Blackwell: "Game Theory and Defense against Community Disaster," 71 pp., National Research Council, Washington, D.C., February, 1954. Reviewed in *Research Previews*, vol. 2, no. 3, pp. 1–5, May, 1954.

*Saaty, T. L., and K. W. Webb: Sensitivity and Renewals in Scheduling Aircraft Overhaul, *Proceedings of the Second International Conference on Operational Research*, English Universities Press, Ltd., London, and John Wiley & Sons, Inc., New York, 1960.

Wood, M. K., and M. A. Geisler: Development of Dynamic Models for Program Planning, chap. 12 in T. C. Koopmans (ed.), "Activity Analysis of Production and Allocation," John Wiley & Sons, Inc., New York, 1951.

6. Personnel Assignment

Dantzig, George B.: "Notes on Linear Programming: Part XIV—A Computational Procedure for a Scheduling Problem of Edie," 13 pp., RM-1290, The RAND Corporation, July 1, 1954.

7. Production Scheduling, Inventory Control, and Planning

Bellman, Richard E.: "Mathematical Aspects of Scheduling Theory," 61 pp., P-651, The RAND Corporation, May 23, 1955. Published in *Journal of the Society for Industrial and Applied Mathematics*, vol. 4, no. 3, pp. 168–205 (including references), September, 1956.

Cahn, Albert S., Jr.: The Warehouse Problem (Abstract 505), *Bulletin of the American Mathematical Society*, vol. 54, p. 1073, November, 1948.

Charnes, Abraham, and William W. Cooper: "Generalizations of the Warehousing Model," *ONR Research Memorandum* 34, Graduate School of Industrial Administration, Carnegie Institute of Technology, Pittsburg, Pa., also published in *Operational Research Quarterly*, vol. 6, no. 4, pp. 131–172 (including references), December, 1955.

———, William W. Cooper, and Donald Farr: Linear Programming and Profit Preference Scheduling for a Manufacturing Firm, *Journal of the Operations Research Society of America*, vol. 1, no. 3, pp. 114–129 (including references), May, 1953.

———, ———, and B. Mellon: A Model for Optimizing Production by Reference to Cost Surrogates, *Econometrica*, vol. 23, no. 3, pp. 307–323 (including references), July, 1955. Also published in *Proceedings of the Second Symposium in Linear Programming*, vol. I, pp. 117–150 (including references) (collection of papers presented to a conference sponsored by the National Bureau of Standards and the Directorate of Management Analysis Service, DCS/Comptroller, Headquarters USAF, held in Washington, D.C., Jan. 27–29, 1955).

*DeBoer, J., and J. VanderSloot: A Method of Cost-controlled Production Planning, *Statistica Neerlandica* (Netherlands), vol. 16, no. 1, 1962.

*Efroymson, M. A., and T. L. Ray: A Branch-bound Algorithm for Plant Location, *Operations Research*, vol. 14, no. 3, May–June, 1966.

*Fetter, R. B.: A Linear Programming Model for Long Range Capacity Planning, *Management Science*, vol. 7, no. 4, July, 1961.

Gepfert, Alan, and Charles H. Grace: Operations Research. . . . as It is Applied to Production Problems, *Tool Engineer*, vol. 36, no. 5, pp. 73–79, May, 1956.

*Gomory, R. E., and B. P. Dzielinski: Optimal Programming of Lot Sizes, Inventory and Labor Allocation, *Management Science*, vol. 11, no. 9, July, 1965.

Johnson, Selmer M.: "Optimal Two- and Three-stage Production Schedules with Setup Times Included," 10 pp., P-402, The RAND Corporation, May 5, 1953. Presented to the Econometric Society meeting, Washington, D.C., Dec. 28, 1953; published in *Naval Research Logistics Quarterly* (Office of Naval Research), vol. 1, no. 1, pp. 61–68 (including references), March, 1954.

*Kantorovich, L. V.: Mathematical Methods of Organizing and Planning Production, *Management Science*, vol. 6, no. 4, July, 1960.

*Kelley, J. E., Jr.: Critical Path Planning and Scheduling: Mathematical Basis, *Operations Research*, vol. 9, no. 3, May–June, 1961.

*Koeningsberg, E.: Some Industrial Applications of Linear Programming, *Operational Research Quarterly* (U.K.), vol. 12, no. 2, June, 1961.

Magee, John R.: Guides to Inventory Policy. Part I: Functions and Lot Sizes, *Harvard Business Review*, vol. 34, no. 1, pp. 49–60, January–February, 1956; Part II: Problems of Uncertainty, *ibid.*, no. 2, pp. 103–116, March–April, 1956; Part III: Anticipating Future Needs, *ibid.*, no. 3, pp. 57–70, May–June, 1956.

———: "Linear Programming in Production Scheduling." Paper presented to the first national meeting of the Operations Research Society of America,

Washington, D.C., Nov. 17–18, 1952; abstracted in *Journal of the Operations Research Society of America*, vol. 1, no. 2, p. 76, February, 1953.

Manne, Alan S.: "An Application of Linear Programming to the Procurement of Transport Aircraft," 2 pp. P-672A, The RAND Corporation, May 13, 1955. Presented to the second national meeting of the Institute of Management Sciences, New York City, Oct. 20–21, 1955; abstracted in *Management Science*, vol. 2, no. 2, pp. 190–191, January, 1956.

*Müller-Merbach, Heiner: The Optimum Allocation of Products to Machines by Linear Programming, *Fortschriftliche Betriebsfuhrung* (Germany), vol. 11, no. 1, February, 1962.

*Rapoport, L. A., and W. P. Drews: Mathematical Approach to Long-range Planning, *Harvard Business Review*, vol. 40, no. 3, May–June, 1962.

Salveson, Melvin E.: The Assembly Line Balancing Problem, *Proceedings of the Second Symposium in Linear Programming*, vol. I, pp. 55–101 (including references) (collection of papers presented to a conference sponsored by the National Bureau of Standards and the Directorate of Management Analysis Service, DCS/Comptroller, Headquarters USAF, held in Washington, D.C., Jan. 27–29, 1955); also published in *Transactions of the American Society of Mechanical Engineers*, vol. 77, no. 6, pp. 939–947, August, 1955; and in *Journal of Industrial Engineering*, vol. 6, no. 3, pp. 18–25 (including references), May–June, 1955.

*Smith, S. B.: Planning Transistor Production by Linear Programming, *Operations Research*, vol. 13, no. 1, January–February, 1965.

Vazsonyi, Andrew: "A Problem in Machine Shop Loading." Paper presented to the fifth national meeting of the Operations Research Society of America, Washington, D.C., Nov. 19–20, 1954; abstracted in *Journal of the Operations Research Society of America*, vol. 3, no. 1, p. 115, February, 1955.

Whitin, Thomson M.: Inventory Control Research: A Survey, *Management Science*, vol. 1, no. 1, pp. 32–40 (including references), October, 1954.

8. Structural Design

Charnes, Abraham, and Herbert J. Greenberg: "Plastic Collapse and Linear Programming. Preliminary Report." Paper presented at the summer meeting of the American Mathematical Society, September, 1951; abstracted in *Bulletin of the American Mathematical Society*, vol. 57, no. 6, p. 480, November, 1951.

Dorn, W. S., and Herbert J. Greenberg: "Linear Programming and Plastic Limit Analysis of Structures," 30 pp. (including tables), *Technical Report* 7, Carnegie Institute of Technology, Pittsburgh, Pa., August, 1955.

Heyman, Jacques: Plastic Design of Beams and Plane Frames for Minimum Material Consumption, *Quarterly of Applied Mathematics*, vol. 8, no. 4, pp. 373–381, January, 1951.

9. Traffic Analysis

Lavallee, R. Stanley: "The Application of Linear Programming to the Problem of Scheduling Traffic Signals." Paper presented at the seventh national meeting of the Operations Research Society of America, Los Angeles, Calif.,

Aug. 15–17, 1955; abstracted in *Journal of the Operations Research Society of America*, vol. 3, no. 4, p. 562, November, 1955.

*Little, J. P. C.: The Synchronization of Traffic Signals by Mixed-integer Linear Programming, *Operations Research*, vol. 14, no. 4, July–August, 1966.

10. Transportation Problems and Network Theory

*Balinski, M. L., and R. E. Quandt: On an Integer Program for a Delivery Problem, *Operations Research*, vol. 12, no. 2, March–April, 1964.

Batchelor, James H.: A Commercial Use of Linear Programming, *Proceedings of the Second Symposium in Linear Programming*, vol. I, pp. 103–116 (including references) (collection of papers presented to a conference sponsored by the National Bureau of Standards and the Directorate of Management Analysis Service, DCS/Comptroller, Headquarters USAF, held in Washington, D.C., Jan. 27–29, 1955).

Clem, William J.: "Two Techniques of Linear Programming," 61 pp. (including references), Master's Report in Industrial Engineering, Department of Industrial Engineering, Columbia University, New York, September, 1954.

Dantzig, George B., Lester R. Ford, Jr., and Delbert R. Fulkerson: "A Primal-Dual Algorithm," 16 pp., P-778, The RAND Corporation, Dec. 5, 1955. Also published as Part XXXI of "Notes on Linear Programming," 14 pp., RM-1709, The RAND Corporation, May 9, 1956.

———, and D. L. Johnson: Maximum Payloads per Unit Time Delivered through an Air Network, *Operations Research*, vol. 12, no. 2, March–April, 1964.

*——— and J. H. Ramser: The Truck Dispatching Problem, *Management Science*, vol. 6, no. 1, October, 1959.

Dwyer, Paul S.: "The Solution of the Hitchcock Transportation Problem with a Method of Reduced Matrices," University of Michigan, Ann Arbor, Mich., December, 1955.

Flood, Merrill M.: Application of Transportation Theory to Scheduling a Military Tanker Fleet, *Journal of the Operations Research Society of America*, vol. 2, no. 2, pp. 150–162, May, 1954.

———: "On the Hitchcock Distribution Problem," 26 pp., P-213, The RAND Corporation, May, 1951. Presented to the Symposium on Linear Inequalities and Programming, Washington, D.C., June 14–16, 1951, jointly sponsored by the Air Force, DCS/Comptroller, Headquarters USAF, and the National Bureau of Standards. Published in Project SCOOP, Manual 10, pp. 74–99, Apr. 1, 1952; also published in *Pacific Journal of Mathematics*, vol. 3, no. 2, pp. 369–386, June, 1953.

Ford, Lester R., Jr., and Delbert R. Fulkerson: "Notes on Linear Programming: Part XXIX—A Simple Algorithm for Finding Maximal Network Flows and an Application to the Hitchcock Problem," 21 pp. (including references), RM-1604, The RAND Corporation, Dec. 29, 1955; also P-743, The RAND Corporation.

Fulkerson, Delbert R., and George B. Dantzig: Computation of Maximal Flows in Networks, *Naval Research Logistics Quarterly* (Office of Naval Research), vol. 2, no. 4, pp. 277–283, December, 1955.

Gleyzel, Andre N.: An Algorithm for Solving the Transportation Problem

(Research Paper 2583), *Journal of Research of the National Bureau of Standards*, vol. 54, no. 4, pp. 213–216, April, 1955.

Heller, Isidor: "Least Ballast Shipping Required to Meet a Specified Shipping Program." Presented to the Symposium on Linear Inequalities and Programming, Washington, D.C., June 14–16, 1951, jointly sponsored by the Air Force, DCS/Comptroller, Headquarters USAF, and the National Bureau of Standards; published in Project SCOOP, Manual 10, pp. 164–171, Apr. 1, 1952.

Hitchcock, Frank L.: The Distribution of a Product from Several Sources to Numerous Localities, *Journal of Mathematics and Physics* (Massachusetts Institute of Technology), vol. 20, no. 3, pp. 224–230, August, 1941.

Kantorovich, L.: On the Translocation of Masses, *Comptes rendus (Doklady) de l'académie des sciences de l'URSS*, vol. 37, no. 7–8, pp. 199–201, 1942.

Koopmans, Tjalling C.: "Optimum Utilization of the Transportation System." Paper presented at the international meeting of the Econometric Society, Washington, D.C., Sept. 6–18, 1947; abstracted in *Econometrica,* vol. 16, no. 1, pp. 66–68, January, 1948; published in full in *Econometrica,* vol. 17, nos. 3 and 4, pp. 136–145 (with discussion, pp. 145–146), supplement to the July, 1949, issue.

*Lederman, J., L. Gleiberman, and J. F. Egan: Vessel Allocation by Linear Programming, *Naval Research Logistics Quarterly,* vol. 13, no. 3, September, 1966.

*Nakagawa, Shizuaki: An Application of Operations Research to Optimization of Sailing Cost, *Operations Research as a Management Science* (Japan), vol. 6, no. 2, August, 1961.

Orden, Alex: The Transshipment Problem, *Management Science,* vol. 2, no. 3, pp. 276–285, April, 1956.

11. Traveling-salesman Problem

Dantzig, George B., Delbert R. Fulkerson, and Selmer Johnson: "Solution of a Large-scale Traveling-salesman Problem," 33 pp. (with references), P-510, The RAND Corporation, Apr. 12, 1954, revised July 8, 1954. Presented to the summer meeting of the Econometric Society, Montreal, Canada, Sept. 10–13, 1954; published in *Journal of the Operations Research Society of America,* vol. 2, no. 4, pp. 393–410, November, 1954; reviewed by H. W. Kuhn in *Mathematical Reviews,* vol. 17, no. 1, p. 58, January, 1956.

Flood, Merrill N.: "The Traveling-salesman Problem," 21 pp., Seminar Paper 13, Informal Seminar in Operations Research, 1954–1955, sponsored by The Operations Research Office and held at The Johns Hopkins University, Baltimore, Md., Feb. 16, 1955. Published in *Operations Research* vol. 4, no. 1, pp. 61–75 (including references), February, 1956; also published in J. F. McCloskey, and J. M. Coppinger (eds)., "Operations Research for Management," vol. II, pp. 340–357 (including references), Johns Hopkins Press, Baltimore, 1956.

Heller, Isidor: On the Traveling Salesman's Problem, *Proceedings of the Second Symposium in Linear Programming,* vol. II, pp. 643–665 (including references) (collection of papers presented to a conference sponsored by the

National Bureau of Standards and the Directorate of Management Analysis Service, DCS/Comptroller, Headquarters USAF, held in Washington, D.C., Jan. 27–29, 1955).

*Little, J. D. C., K. G. Murty, D. W. Sweeney, and C. Karel: An Algorithm for the Traveling-salesman Problem, *Operations Research,* vol. II, no. 6, November–December, 1963.

12. Other Applications

*Aronofsky, J. S., Growing Applications of Linear Programming, *Communications of the ACM*, vol. 7, no. 6, June, 1964.

*Cohen, K. J., and F. S. Hammer: Optimal Coupon Schedules for Municipal Bonds, *Management Science*, vol. 12, no. 1, September, 1965.

Dantzig, George B., and Alan J. Hoffman: Dilworth's Theorem on Partially Ordered Sets, Paper 11 [pp. 207–214 (including references)] in "Linear Inequalities and Related Systems," *Annals of Mathematics Studies* 38, Princeton University Press, Princeton, N.J., 1956.

Ford, Lester, Jr., and Delbert R. Fulkerson: "Maximal Flow through a Network." Paper presented to the meeting of the Econometric Society, New York City, Dec. 27–30, 1955; published in *Canadian Journal of Mathematics*, vol. 8, no. 3, pp. 399–404, 1956; also RM-1400 and P-605, The RAND Corporation, 12 pp., Nov. 19, 1954.

*Freeman, R. J., D. C. Gogerty, G. W. Graves, and R. B. S. Brooks: A Mathematical Model of Supply Support for Space Operations, *Operations Research*, vol. 14, no. 1, January–February, 1966.

*Hartung, P. H.: Brand Switching and Mathematical Programming in Market Expansion, *Management Science*, vol. 11, no. 10, August, 1965.

Hoffman, Alan J., and Harold W. Kuhn: On Systems of Distinct Representatives, Paper 10 [pp. 199–206 (including references)] in "Linear Inequalities and Related Systems," *Annals of Mathematics Studies* 38, Princeton University Press, Princeton, N.J., 1956.

*Knight, U. G. W.: The Logical Design of Electrical Networks Using Linear Programming Methods, *Institute of Electrical Engineers, Proceedings,* vol. 107, no. 33, June, 1960.

*Kolesar, P. J.: Linear Programming and the Reliability of Multicomponent Systems, *Naval Research Logistics Quarterly*, vol. 14, no. 3, September, 1967.

Kruskal, Joseph B., Jr.: "On the Shortest Spanning Subtree of a Graph and the Traveling Salesman Problem," 4 pp., *Logistics Papers Issue* 11, The George Washington University Logistics Research Project, Appendix to Quarterly Progress Report 21, Nov. 16, 1954–Feb. 15, 1955, Mar. 3, 1955.

*Loucks, D. P., C. S. ReVelle, and W. R. Lynn: Linear Programming Models for Water Pollution Control, *Management Science*, vol. 4, December, 1967.

*McGuire, C. B.: Some Team Models of a Sales Organization, *Management Science*, vol. 7, no. 2, January, 1961.

Mannos, Murray: "An Application of Linear Programming to Efficiency in Operations of a System of Dams." Presented at the summer meeting of the Econometric Society, Montreal, Canada, Sept. 10–13, 1954; abstracted in *Econometrica*, vol. 23, no. 3, pp. 335–336, July, 1955.

Orden, Alex: Application of Linear Programming to Optical Filter Design (abstract), *Proceedings of the Second Symposium in Linear Programming*, vol. I, p. 185 (collection of papers presented to a conference sponsored by the National Bureau of Standards and the Directorate of Management Analysis Service, DCS/Comptroller, Headquarters USAF, held in Washington, D.C., Jan. 27–29, 1955).

*Wagner, H. M., R. J. Giglio, and R. G. Glaser: Preventive Maintenance Scheduling by Mathematical Programming, *Management Science*, vol. 10, no. 2, January, 1964.

*Wardle, P. A.: Forest Management and Operational Research: A Linear Programming Study, *Management Science*, vol. 11, no. 10, August, 1965.

*Wilson, R. C.: A Packaging Problem, *Management Science*, vol. 12, no. 4, December, 1965.

REFERENCES

1. Abadie, J.: "Nonlinear Programming," North Holland Publishing Company, Amsterdam, 1967.

1a. Abadie, J.: "On the Decomposition Principle," ORC Report 63-20, Operations Research Center, University of California, Berkeley, Calif., August, 1963.

1b. Abadie, J.: "Problèmes d'Optimisation," Tome II, Institut Blaise Pascal, Paris, June, 1965.

1c. Abadie, J., and A. C. Williams: Dual and Parametric Methods in Decomposition, pp. 149–158 of Graves and Wolfe [51*g*].

1d. Agmon, S.: The Relaxation Method for Linear Inequalities, *Canadian Journal of Mathematics*, vol. 6, 1954.

2. Allen, R. G. D.: "Mathematical Economics," The Macmillan Company, New York, 1956.

2a. Antosiewicz, H. A., and A. J. Hoffman: A Remark on the Smoothing Problem, *Management Science*, vol. 1, 1954–1955.

3. Arrow, K. J., L. Hurwicz, and H. Uzawa (eds.): "Studies in Linear and Nonlinear Programming," Stanford University Press, Stanford, Calif., 1958.

3a. Balinski, M. L.: Integer Programming: Methods, Uses, Computation, *Management Science*, vol. 12, no. 3, November, 1965.

3b. Balinski, M. L., and R. E. Gomory: A Primal Method for the Assignment and Transportation Problems, *Management Science*, vol. 10, no. 3, April, 1964.

3c. Balinski, M. L., and R. E. Quandt: On an Integer Program for a Delivery Problem, *Operations Research*, March–April, 1964.

3cc. Balintfy, J. L.: Menu Planning by Computers, *The Communications of The ACM*, vol. 7, April, 1964.

3d. Barankin, E. W., and R. Dorfman: "On Quadratic Programming," University of California Publications in Statistics 2, University of California, Berkeley, Calif., 1958.

3e. Barnett, S.: Stability of the Solution to a Linear Programming Problem, *Operational Research Quarterly*, vol. 13, no. 3, 1962.

3f. Baumol, W. J.: "Economic Theory and Operations Analysis," 2d ed., Prentice-Hall, Inc., Englewood Cliffs, N.J., 1965.

3g. Baumol, W. J., and T. Fabian: Decomposition, Pricing for Decentralization and External Economies, *Management Science*, vol. 11, no. 1, 1964.

4. Beale, E. M. L.: Cycling in the Dual Simplex Algorithm, *Naval Research Logistics Quarterly*, vol. 2, no. 4, 1955.

5. Beale, E. M. L.: An Alternative Method for Linear Programming, *Proceedings of the Cambridge Philosophical Society*, vol. 50, 1954.

5a. Beale, E. M. L.: "A Method for Solving Linear Programming Problems When Some But Not All of the Variables Must Take Integral Values," *Statistical Techniques Research Group Technical Report* 19, Princeton University, Princeton, N.J., 1958.

5b. Beale, E. M. L.: Survey of Integer Programming, *Operational Research Quarterly*, vol. 16, no. 2, June, 1965.

5c. Beale, E. M. L.: The Simplex Method Using Pseudo-basic Variables for Structured Linear Programming Problems, pp. 133–148 of Graves and Wolfe [51*g*].

5d. Beale, E. M. L.: Numerical Methods, chap. 7 in Abadie [1].

5e. Beale, E. M. L.: On Quadratic Programming, *Naval Research Logistics Quarterly*, no. 6, 1959.

6. Bellman, R.: The Theory of Dynamic Programming, chap. 11 in E. F. Beckenbach (ed.), "Modern Mathematics for the Engineer," McGraw-Hill Book Company, New York, 1956.

6a. Benders, F., Jr., A. R. Catchpole, and C. Kuiken: "Discrete Variables Optimization Problems," Koninklijke/Shell-Lab., Amsterdam, 1960.

6b. Benders, J. F.: Partitioning Procedures for Solving Mixed Variable Programming Problems, *Numerische Mathematik*, 4, pp. 238–252, 1962.

6c. Boot, J. C.: "Quadratic Programming," Rand McNally & Company, Chicago, 1964.

7. Bowman, E. H.: Production Scheduling by the Transportation Method of Linear Programming, *Operations Research*, vol. 4, 1956.

7a. Bracken, J., and G. P. McCormick: "Selected Applications of Nonlinear Programming," John Wiley & Sons, New York, 1968.

7b. Brigham, G.: A Classroom Example of Linear Programming, *Operations Research*, vol. 7, no. 4; July–August, 1959.

8. Brown, G. W.: Iterative Solution of Games by Fictitious Play, chap. 24 of Koopmans [65].

8a. Buck, R. C.: "Advanced Calculus," 2d ed., McGraw-Hill Book Company, New York, 1965.

8b. Busacker, R. G., and T. L. Saaty: "Finite Graphs and Networks," McGraw-Hill Book Company, New York, 1965.

8c. Carr, C. R., and C. W. Howe: "Quantitative Decision Procedures in Management and Economics," McGraw-Hill Book Company, New York, 1964.

8d. Catchpole, A. R.: The Application of Linear Programming to Integrated Supply Problems in the Oil Industry, *Operational Research Quarterly*, vol. 13, no. 2, 1962.

9. Charnes, A.: Optimality and Degeneracy in Linear Programming, *Econometrica*, vol. 20, 1952.

10. Charnes, A., and W. W. Cooper: The Stepping Stone Method of Explaining Linear Programming Calculations in Transportation Problems, *Management Science*, vol. 1, 1954–1955.

10a. Charnes, A., and W. W. Cooper: "Management Models and Industrial Application of Linear Programming," vols. I and II, John Wiley & Sons, Inc., New York, 1960.

11. Charnes, A., W. W. Cooper, and D. Farr: Linear Programming and Profit Preference Scheduling for a Manufacturing Firm, *Journal of the Operations Research Society of America*, vol. 1, 1953.

11a. Charnes, A., W. W. Cooper, and R. O. Ferguson: Optimal Estimation of Executive Compensation by Linear Programming, *Management Science*, vol. 1, no. 2, 1955.

12. Charnes, A., W. W. Cooper and A. Henderson: "Introduction to Linear Programming," John Wiley & Sons, Inc., New York, 1953.

13. Charnes, A., W. W. Cooper, and B. Mellon: Blending Aviation Gasolines—A Study in Programming Interdependent Activities in an Integrated Oil Company, *Econometrica*, vol. 20, 1952.

14. Charnes, A., and C. E. Lemke: "The Bounded Variables Problem," *ONR Research Memorandum* 10, Graduate School of Industrial Administration, Carnegie Institute of Technology, Pittsburgh, Pa., 1954.

15. Charnes, A., and C. E. Lemke: "Minimization of Nonlinear Separable Functions," Graduate School of Industrial Administration, Carnegie Institute of Technology, Pittsburgh, Pa., 1954.

15a. Chung, An-min: "Linear Programming," Charles E. Merrill Books, Inc., Columbus, Ohio, 1963.

16. Cooper, W. W., and A. Charnes: Linear Programming, *Scientific American*, August, 1954.

16a. Courtillot, M.: On Varying All the Parameters in a Linear-programming Problem and Sequential Solution of a Linear-programming Problem, *Operations Research*, vol. 10, no. 4, 1962.

16b. Cutler, Leola, and P. Wolfe: "Experiments in Linear Programming," RM-3402-PR, The RAND Corporation, Santa Monica, Calif., February, 1963. Also, pp. 177–200 of Graves and Wolfe [51*g*].

16c. Dantzig, G. B.: "Linear Programming and Extensions," Princeton University Press, Princeton, N.J., 1963.

17. Dantzig, G. B.: Maximization of a Linear Function of Variables Subject to Linear Inequalities, chap. 21 of Koopmans [65].

18. Dantzig, G. B.: Application of the Simplex Method to a Transportation Problem, chap. 23 of Koopmans [65].

19. Dantzig, G. B.: A Proof of the Equivalence of the Programming Problem and the Game Problem, chap. 20 of Koopmans [65].

20. Dantzig, G. B.: "Block Triangular Systems in Linear Programming," RAND Report RM-1273, The RAND Corporation, Santa Monica, Calif., 1954.

21. Dantzig, G. B.: "Computational Algorithm of the Revised Simplex Method," RAND Report RM-1266, The RAND Corporation, Santa Monica, Calif., 1953.

22. Dantzig, G. B.: "The Dual Simplex Algorithm," RAND Report RM-1270, The RAND Corporation, Santa Monica, Calif., 1954.

23. Dantzig, G. B.: "Composite Simplex–Dual Simplex Algorithm—I," RAND Report RM-1274, The RAND Corporation, Santa Monica, Calif., 1954.

24. Dantzig, G. B.: "Variables with Upper Bounds in Linear Programming," RAND Report RM-1271, The RAND Corporation, Santa Monica, Calif., 1954.

25. Dantzig, G. B.: "Linear Programming under Uncertainty," RAND Report RM-1374, The RAND Corporation, Santa Monica, Calif., 1954.

26. Dantzig, G. B.: Developments in Linear Programming, in Directorate of Management Analysis [37].

27. Dantzig, G. B.: "Discrete-variable Extremum Problems," RAND Report RM-1832, The RAND Corporation, Santa Monica, Calif., 1956.

27a. Dantzig, G. B.: "On the Significance of Solving Linear Programming Problems with Some Integer Variables," RAND Report P-1486, The RAND Corporation, Santa Monica, Calif., 1959.

27b. Dantzig, G. B.: "On Integer and Partial Integer Linear Programming Problems," RAND Report P-1410, The RAND Corporation, Santa Monica, Calif., 1958.

27c. Dantzig, G. B.: Solving Linear Programs in Integers, "Notes on Linear Programming," Part XLVII, RM-2209 (ASTIA Document AD-156047), The RAND Corporation, Santa Monica, Calif., 1958.

27d. Dantzig, G. B.: "On the Status of Multistage Linear Programming Problems," RAND Report P-1028, The RAND Corporation, Santa Monica, Calif., 1959; also in *Management Science*, vol. 6, no. 1, 1959.

27e. Dantzig, G. B.: Optimal Solution of a Dynamic Leontief Model with Substitution, *Econometrica*, vol. 23, July, 1955.

27f. Dantzig, G. B.: Compact Basis Triangularization for the Simplex Method, pp. 125–132 of Graves and Wolfe [51g].

27g. Dantzig, G. B.: "Quadratic Programming: A variant of the Wolfe-Markowitz Algorithms," Research Report No, 2, Operations Research Center, University of California, Berkeley Calif.

27h. Dantzig, G. B.: Application of Generalized Linear Programming to Control Theory, chap. 12 in Abadie [1].

28. Dantzig, G. B., L. R. Ford, Jr., and D. R. Fulkerson: "A Primal-Dual Algorithm," RAND Report RM-1709, The RAND Corporation, Santa Monica, Calif., 1956.

28a. Dantzig, G. B., and S. Johnson: "A Production Smoothing Problem," pp. 151–176 of Directorate of Management Analysis [37].

28b. Dantzig, G. B., D. R. Fulkerson, and S. M. Johnson: On a Linear Programming Combinatorial Approach to the Traveling Salesman Problem, "Linear Programming and Extensions," Part XLIX RM-2321 (ASTIA AD-212974), The RAND Corporation, Santa Monica, Calif., 1959.

28c. Dantzig, G. B., D. R. Fulkerson, and S. M. Johnson: "On a Linear Programming, Combinatorial Approach to the Traveling Salesman Problem," RAND Report P-1281, The RAND Corporation, Santa Monica, Calif., 1959; also in *Operations Research*, vol. 7, no. 1, 1959.

29. Dantzig, G. B., and W. Orchard-Hays: "Alternate Algorithm for the Revised Simplex Method," RAND Report RM-1268, The RAND Corporation, Santa Monica, Calif., 1953.

30. Dantzig, G. B., W. Orchard-Hays, and G. Waters: "Product-form Tableau for Revised Simplex Method," RAND Report RM-1268A, The RAND Corporation, Santa Monica, Calif., 1954.

31. Dantzig, G. B., and A. Orden: A Duality Theorem Based on the Simplex Method, pp. 51–55 of Directorate of Management Analysis [35].

31a. Dantzig, G. B., and A. Orden: "Duality Theorems," RAND Report RM-1265, The RAND Corporation, Santa Monica, Calif., October, 1953.

32. Dantzig, G. B., A. Orden, and P. Wolfe: "Generalized Simplex Method for Minimizing a Linear Form under Linear Inequality Restraints," RAND Report RM-1264, The RAND Corporation, Santa Monica, Calif., 1954.

32a. Dantzig, G. B., and R. M. Van Slyke: "Generalized Upper Bounded Techniques for Linear Programming," ORC Reports 64-17 and 64-18, Operations Research Center, University of California, Berkeley, Calif., 1964.

32b. Dantzig, G. B., and P. Wolfe: "A Decomposition Principle for Linear Programs," RAND Report P-1544, The RAND Corporation, Santa Monica, Calif., 1959; also in *Operations Research*, vol. 8, no. 1, 1959.

32c. Dantzig, G. B., and P. Wolfe: The Decomposition Algorithm for Linear Programs, *Econometrica*, vol. 29, no. 4, October, 1961.

32d. Demuth, O.: A Remark on the Transport Problem, Čas. pěst. mat. 86, 103–110, 1961.

32e. Dennis, J. B.: A High-speed Computer Technique for the Transportation Problem, *Journal of the Association of Computing Machinery*, vol. 5, no. 2, 1958.

32f. Dennis, J. B.: "Mathematical Programming and Electrical Networks," John Wiley & Sons, Inc., New York, 1959.

33. DiCarlo-Cottone, M.: "The Optimum Transportation Problem" (mimeograph), DCS/Comptroller, Headquarters U.S. Air Force, Washington, D.C., 1954.

34. Dickson, L. E.: "New First Course in the Theory of Equations," John Wiley & Sons, Inc., New York, 1947.

34a. Dickson, J. C., and F. P. Frederick: A Decision Rule for Improved Efficiency in Solving Linear Programming Problems with the Simplex Algorithm, *Communications*, Association for Computing Machinery, vol. 3, September, 1960.

35. Directorate of Management Analysis: "Symposium on Linear Inequalities and Programming," A. Orden and L. Goldstein (eds.), DCS/Comptroller, Headquarters U.S. Air Force, Washington, D.C., April, 1952.

36. Directorate of Management Analysis: "The Application of Linear Programming Techniques to Air Force Problems," DCS/Comptroller, Headquarters U.S. Air Force, Washington, D.C., December, 1954.

37. Directorate of Management Analysis: "Proceedings of the Second Symposium in Linear Programming," H. Antosiewicz (ed.), vols. 1 and 2, DCS/Comptroller, Headquarters U.S. Air Force, Washington, D.C., January, 1955.

37a. Doig, A. G.: "The Minimum Number of Basic Feasible Solutions to a Transport Problem," *Operational Research Quarterly*, vol. 14, no. 4, 1963.

38. Dorfman, R.: Mathematical, or "Linear," Programming, *American Economic Review*, vol. 43, December, 1953.

39. Dorfman, R.: "Application of Linear Programming to the Theory of the Firm," University of California Press, Berkeley, Calif., 1951.

40. Dorfman, R., P. A. Samuelson, and R. Solow: "Linear Programming and Economic Analysis," McGraw-Hill Book Company, New York, 1958.

40a. Dorn, W. S.: Nonlinear Programming–A Survey, *Management Science*, vol. 9, no. 2, January, 1963.

40b. Dorn, W. S.: Duality in Quadratic Programming, *Quarterly of Applied Mathematics*, 18, 1960.

40c. Dorn, W. S.: A Duality Theorem for Convex Programs, *IBM Journal of Research and Development*, vol. 4, no. 4, October, 1960.

40d. Dorn, W. S.: Linear Fractional Programming, Research Report RC 830, IBM Research Center, Yorktown Heights, N.Y., November, 1962.

40e. Duffin, R. T., E. L. Peterson, and C. Zener: "Geometric Programming," John Wiley & Sons, Inc., New York, 1967.

41. Dwyer, P. S.: Solution to the Personnel Classification Problem with the Method of Optimal Regions, *Psychometrika*, vol. 19, 1954.

41a. Eisemann, K.: The Trim Problem, *Management Science*, vol. 3, no. 3, 1957.

41b. Eisemann, K., and J. R. Lourie: "The Machine Loading Problem," IBM Applications Library, New York, 1959.

41c. Eisemann, K.: The Primal-dual Method for Bounded Variables, *Operations Research*, vol. 12, no. 1, January–February, 1964.

42. Egerváry, E.: On Combinatorial Properties of Matrices, *Matematikaés Fizikai Lapok*, vol. 38, 1931; translated as "On Combinatorial Properties of Matrices," by H. W. Kuhn, Office of Naval Research Logistics Project Report, Department of Mathematics, Princeton University, Princeton, N.J., 1953.

42a. Fenchel, W.: "Convex Cones, Sets, and Functions," Princeton University Department of Mathematics, Princeton, N.J., September, 1953.

42b. Fiacco, A. V., and G. P. McCormick: The Sequential Unconstrained Minimization Technique for Nonlinear Programming–A Primal-dual Method, *Management Science*, vol. 10, no. 4, 1964.

42c. Fiacco, A. V., and G. P. McCormick: "Nonlinear Programming," John Wiley & Sons, New York, 1968.

42d. Flood, M. M.: A Transportation Algorithm and Code, *Naval Research Logistics Quarterly*, vol. 8, no. 3, September, 1961.

43. Ford, L. R., Jr., and D. R. Fulkerson: "Solving the Transportation Problem," RAND Report RM-1736, The RAND Corporation, Santa Monica, Calif., 1956.

43a. Ford, L. R., Jr., and D. R. Fulkerson: "Flows in Networks," Princeton University Press, Princeton, N.J., 1962.

44. Ford, L. R., Jr., and D. R. Fulkerson: "A Primal Dual Algorithm for the Capacitated Hitchcock Problem," RAND Report RM-1798, The RAND Corporation, Santa Monica, Calif., 1956.

44a. Ford, L. R., and D. R. Fulkerson: "Flows in Networks," Princeton University Press, Princeton, N.J., 1962.

44b. Frisch, R.: "The Multiplex Method for Linear Programming," Memorandum, Sosialökon. Inst. University of Oslo, Norway, 1958.

44c. Fulkerson, D. R.: Flow Networks and Combinatorial Operations Research, *The American Mathematical Monthly*, vol. 73, no. 2, February, 1966.

44d. Gale, D.: "The Theory of Linear Economic Models," McGraw-Hill Book Company, New York, 1960.

45. Gale, D., H. W. Kuhn, and A. W. Tucker: Linear Programming and the Theory of Games, chap. 19 of Koopmans [65].

45a. Garvin, W. W.: "Introduction to Linear Programming," McGraw-Hill Book Company, New York, 1960.

45b. Garvin, W. W., H. W. Crandell, J. B. John, and R. A. Spellman: Applications of Linear Programming in the Oil Industry, *Management Science*, vol. 3, no. 4, July, 1957.

46. Gass, S. I.: A First Feasible Solution to the Linear Programming Problem, pp. 495–508 of Directorate of Management Analysis [37].

46a. Gass, S. I.: Recent Developments in Linear Programming, in F. Alt (ed.), "Advances in Computers," vol. II, Academic Press Inc., New York, 1962.

46b. Gass, S. I.: "The Dualplex Method for Large-scale Linear Programs," ORC Report 66-15, Operations Research Center, University of California, Berkeley, Calif., June, 1966.

47. Gass, S. I., and T. L. Saaty: Parametric Objective Function. Part II: Generalization, *Journal of the Operations Research Society of America*, vol. 3, 1955.

48. Gass, S. I., and T. L. Saaty: The Computational Algorithm for the Parametric Objective Function, *Naval Research Logistics Quarterly*, vol. 2, 1955.

49. Gerstenhaber, M., and J. E. Kelley, Jr: "Threshold Methods in Linear Programming," Remington Rand Univac, Philadelphia, Pa., 1956.

49a. Gilmore, P. C., and R. E. Gomory: A Linear-programming Approach to the Cutting-stock Problem–Part I, *Operations Research*, vol. 9, no. 6, November–December, 1961; Part II, *Operations Research*, vol. 11, no. 6, November–December, 1963.

50. Glaser, E.: "Introductory Notes on Input-output Analysis," mimeograph notes presented to the Washington section of the American Statistical Association, Washington, D.C., 1953.

51. Goldman, A. J., and A. W. Tucker: Theory of Linear Programming, pp. 53–97 of Kuhn and Tucker [68].

51a. Goldstein, L.: "The Simplex Solution of Waugh's Problem," mimeographed notes, Directorate of Management Analysis, Headquarters U.S. Air Force, Washington, D.C., 1952.

51b. Gomory, R. E.: Essentials of an Algorithm for Integer Solutions to Linear Programs, *Bulletin of the American Mathematical Society*, vol. 64, 1958.

51c. Gomory, R. E.: An Algorithm for Integer Solution to Linear Programming, pp. 269–302 of Graves and Wolfe [51*g*]. Also, Princeton-IBM Math. Research Project Technical Report no. 1, Princeton, N.J., 1958.

51d. Gomory, R. E.: "An Algorithm for the Mixed Integer Problem," RM-2597, The RAND Corporation, Santa Monica, Calif., 1960.

51e. Gomory, R. E.: All-integer Programming Algorithm, pp. 193–206 in J. F. Muth and G. L. Thompson (eds.), "Industrial Scheduling," Prentice Hall, Inc., Englewood Cliffs, N.J. First issued as Research Report RC-189, IBM Research Center, Yorktown Heights, N.Y., 1960.

51f. Gomory, R. E.: Large and Non-convex Problems in Linear Programming, *Proceedings of Symposia in Applied Mathematics*, vol. 15, American Mathematical Society, 1963.

51g. Graves, R. L., and P. Wolfe: "Recent Advances in Mathematical Programming," McGraw-Hill Book Company, New York, 1963.

52. Gross, O.: "A Class of Discrete-type Minimization Problems," RAND Report RM-1644, The RAND Corporation, Santa Monica, Calif., 1956.

52a. Gross, O.: A Simple Linear Programming Problem Explicitly Solvable in Integers, "Notes on Linear Programming," Part XXVIII, RM-1560, The RAND Corporation, Santa Monica, Calif., 1955.

52b. Hadley, G.: "Linear Algebra," Addison-Wesley Publishing Company, Inc., Reading, Mass., 1961.

52c. Hadley, G.: "Linear Programming," Addison-Wesley Publishing Company, Inc., Reading, Mass., 1962.

52d. Hadley, G.: "Nonlinear and Dynamic Programming," Addison-Wesley Publishing Company, Inc., Reading, Mass., 1964.

52e. Haley, K. B.: The Solid Transportation Problem, *Operations Research*, vol. 10, no. 4, July–August, 1962.

53. Harrison, J. O., Jr: Linear Programming and Operations Research, in J. F. McCloskey and F. N. Trefethen (eds.), "Operations Research for Management," vol. I, pp. 217–237, The Johns Hopkins Press, Baltimore, Md., 1954.

53a. Harvey, R. P.: The Decomposition Principle for Linear Programs, *International Journal of Computer Mathematics*, vol. 1, no. 1, May, 1964.

54. Heller, I., and C. B. Tompkins: An Extension of a Theorem of Dantzig, Paper 14 in Kuhn and Tucker [68].

54a. Hellerman, E.: Large-scale Linear Programs: Theory and Computation, paper presented at the Technical Association of the Pulp and Paper Industry Symposium, CEIR Inc., Bethesda, Md., March, 1966.

55. Henderson, A., and R. Schlaifer: Mathematical Programming, *Harvard Business Review*, vol. 32, May–June, 1954.

55a. Hildebrand, F. B.: "Methods of Applied Mathematics," Prentice-Hall, Inc., Englewood Cliffs, N.J., 1958.

56. Hirsch, W. M., and G. B. Dantzig: The Fixed Charge Problem, *Naval Research Logistics Quarterly*, vol. 15, no. 3, 1968.

57. Hitchcock, F. L.: Distribution of a Product from Several Sources to Numerous Localities, *Journal of Mathematical Physics*, vol. 20, 1941.

58. Hoffman, A.: "Cycling in the Simplex Algorithm," National Bureau of Standards Report, Washington, D.C., 1953.

59. Hoffman, A. J.: How to Solve a Linear Programming Problem, pp. 397–424 of Directorate of Management Analysis [37].

60. Hoffman, A. J., and W. W. Jacobs: Smooth Patterns of Production, *Management Science*, vol. 1, no. 1, 1954.

61. Hoffman, A. J., and J. G. Kruskal: Integral Boundary Points of Convex Polyhedra, Paper 13 in Kuhn and Tucker [68].

61a. Hoffman, A., M. Mannos, D. Sokolowsky, and N. Wiegmann: Computational Experience in Solving Linear Programs, *Journal of the Society for Industrial and Applied Mathematics*, vol. 1, no. 1, 1953.

61b. Hu, T. C.: Multi-commodity Network Flows, *Operations Research*, vol. 11, no. 3, 1964.

61c. Isbell, J. R., and W. H. Marlow: Attrition Games, *Naval Research Logistics Quarterly*, vol. 3, pp. 71–93, 1956.

62. Jacobs, W. W.: The Caterer Problem, *Naval Research Logistics Quarterly*, vol. 1, no. 2, 1954.

62a. Jewell, W. S.: A Classroom Example of Linear Programming, Lesson no. 2, *Operations Research*, vol. 8, no. 4, July–August, 1960.

62b. Jewell, W. S.: Optimal Flow through Networks with Gains, *Operations Research*, vol. 10, no. 4, 1962.

62c. John, F.: Extremum Problems with Inequalities as Subsidiary Conditions, "Studies and Essays," Courant Anniversary Volume, Interscience Publishers, (Division of John Wiley & Sons, Inc.), New York, 1948.

63. Joseph, J. A.: "The Application of Linear Programming to Weapon Selection and Target Analysis," *Technical Memorandum* 42, Operations Analysis Division, Headquarters U.S. Air Force, Washington, D.C., 1954.

63a. Karlin, S.: "Mathematical Methods and Theory in Games, Programming, and Economics," vol. I, Addison-Wesley Publishing Company, Inc., Reading, Mass., 1959.

64. Katzman, Irwin: Solving Feed Problems through Linear Programming, *Journal of Farm Economics*, vol. 38, 1956.

64a. Kelley, J. E.: An Application of Linear Programming to Curve Fitting, *Journal of the Society for Industrial and Applied Mathematics*, 6, pp. 15–22. 1958.

64b. Kelley, J. E., Jr.: The Cutting-plane Method for Solving Convex Programs, *Journal of the Society for Industrial and Applied Mathematics*, 8, pp. 703–712, 1960.

65. Koopmans, T. C. (ed.): "Activity Analysis of Production and Allocation," *Cowles Commission Monograph* 13, John Wiley & Sons, Inc., New York, 1951.

66. Koopmans, T. C.: Optimum Utilization of the Transportation System, *Econometrica*, vol. 17, Supplement, 1949.

67. Kuhn, H. W.: "Lectures on the Theory of Games," *Annals of Mathematics Studies* 37, Princeton University Press, Princeton, N.J., 1957. (Also Logistics Research Report sponsored by the Office of Naval Research, Project NR047-002.)

67a. Kuhn, H. W.: The Hungarian Method for the Assignment Problem, *Naval Research Logistics Quarterly*, vol. 2, nos. 1 and 2, March–June, 1955.

68. Kuhn, H. W., and A. W. Tucker: "Linear Inequalities and Related Systems," *Annals of Mathematics Studies* 38, Princeton University Press, Princeton, N.J., 1956.

69. Kuhn, H. W., and A. W. Tucker: Nonlinear Programming, in "Proceedings of the Second Berkeley Symposium on Mathematical Statistics and Probability," University of California Press, Berkeley, Calif., 1951.

69a. Kunzi, H. P., and W. Krelle: "Nonlinear Programming," Blaisdell Publishing Co., Waltham, Mass., 1966.

69b. Lageman, J. J.: A Method for Solving the Transportation Problem, *Naval Research Logistics Quarterly*, vol. 14, no. 1, March, 1967.

69c. Land, A. H., and A. G. Doig: An Automatic Method for Solving Discrete Programming Problems, *Econometrica*, vol. 28, pp. 497–520, 1960.

69d. Lawler, E. L., and D. E. Wood: Branch and Bound Methods: A Survey, *Operations Research*, vol. 14, no. 4, July–August, 1966.

70. Lemke, C. E.: The Dual Method of Solving the Linear Programming Problem, *Naval Research Logistics Quarterly*, vol. 1, no. 1, 1954.

71. Leontief, W. W.: "The Structure of the American Economy, 1919–1929," Oxford University Press, New York, 1951.

71a. Luce, R. D., and H. Raiffa: "Games and Decisions," John Wiley & Sons, Inc., New York, 1957.

72. McKinsey, J. C. C.: "Introduction to the Theory of Games," McGraw-Hill Book Company, New York, 1952.

72a. Madansky, A.: Methods of Solution of Linear Programs under Uncertainty, *Operations Research*, vol. 10, no. 4, 1962.

73. Magee, John F.: "Studies in Operations Research. I. Application of Linear Programming to Production Scheduling," Arthur D. Little, Inc., Cambridge, Mass.

73a. Manne, A. S.: "Scheduling of Refinery Operations," Harvard University Press, Cambridge, Mass., 1956.

74. Manne, A. S.: "Notes on Parametric Linear Programming," RAND Report P-468, The RAND Corporation, Santa Monica, Calif., 1953.

74a. Manne, A. S. "On the Job Shop Scheduling Problem," *Cowles Commission for Research in Economics* Contract 358(01) NR 047-066, Office of Naval Research, Washington, D.C., 1959.

74b. Manne, A. S.: "Economic Analysis for Business Decisions," McGraw-Hill Book Company, New York, 1961.

74c. Markowitz, H.: "The Elimination Form of the Inverse and Its Application to Linear Programming," RM-1470, The RAND Corporation, Santa Monica, Calif., 1955.

74d. Markowitz, H.: The Optimization of a Quadratic Function Subject to Linear Constraints, *Naval Research Logistics Quarterly*, vol. 3, nos. 1 and 2, 1956.

74e. Miller, C. E.: "The Simplex Method for Local Separable Programming," pp. 89–100 in Graves and Wolfe [51*g*].

74f. Miller, C. E., A. W. Tucker, and R. A. Zemlin: Integer Programming Formulation of Traveling Salesman Problems, *Journal of the Association of Computing Machinery*, vol. 7, no. 4, October, 1960.

74g. Moeske, P. V.: "A General Duality Theorem of Convex Programming," *Metroeconomica*, vol. 17, p. 161–170, 1965.

75. Morgenstern, O. (ed.): "Economic Activity Analysis," John Wiley & Sons, Inc., New York, 1954.

76. Motzkin, T. S.: The Multi-index Transportation Problem, *Bulletin of the American Mathematical Society*, vol. 58, no. 4, 1952.

77. Motzkin, T. S., and I. J. Schoenberg: The Relaxation Method for Linear Inequalities, *Canadian Journal of Mathematics*, vol. 6, 1954.

77a. Mueller, R. K., and L. Cooper: A Comparison of the Primal Simplex and Primal-dual Algorithms in Linear Programming, *Communications of the ACM*, vol. 18, no. 11, November, 1965.

77b. Müller-Merbach, Heiner: Das Verfahren der direkten Dekomposition in der linearen Planungsrechnung, *Ablauf- und Planungsforschung*, Heft 2, 1965.

77c. Munkres, J.: Algorithms for the Assignment and Transportation Problems, *Journal of the Society for Industrial and Applied Mathematics*, vol. 5, no. 1, 1957.

77d. Nef, W.: "Linear Algebra," McGraw-Hill Book Company, New York, 1966.

78. Neumann, J. von: A Certain Zero-sum Two-person Game Equivalent to the Optimal Assignment Problem, pp. 5–12 of *Annals of Mathematics Studies* 28, Princeton University Press, Princeton, N.J., 1953.

79. Neumann, J. von: A Numerical Method to Determine Optimum Strategy, *Naval Research Logistics Quarterly*, vol. 1, no. 2, 1954.

80. Neumann, J. von, and O. Morgenstern: "Theory of Games and Economic Behavior," Princeton University Press, Princeton, N.J., 1947.

81. Orchard-Hays, W.: "Background, Development and Extensions of the Revised Simplex Method," RAND Report RM-1433, The RAND Corporation, Santa Monica, Calif., 1954.

82. Orchard-Hays, W.: "The RAND Code for the Simplex Method," RAND Report RM-1269, The RAND Corporation, Santa Monica, Calif., 1954.

83. Orchard-Hays, W.: "A Composite Simplex Algorithm—II," RAND Report RM-1275, The RAND Corporation, Santa Monica, Calif., 1954.

83a. Orchard-Hays, W.: "Matrices, Elimination and the Simplex Method," CEIR, Inc., Bethesda, Md. 1961.

84. Orden, A.: Application of the Simplex Method to a Variety of Matrix Problems, pp. 28–50 of Directorate of Management Analysis [35].

85. Orden, A.: "A Procedure for Handling Degeneracy in the Transportation Problem," mimeograph, DCS/Comptroller, Headquarters U.S. Air Force, Washington, D.C., 1951.

85a. Orden, A.: The Transshipment Problem, *Management Science*, vol. 2, no. 3, 1956.

85aa. Owen, G.: "Game Theory," W. B. Saunders Company, Philadelphia, 1968.

85b. Prager, W.: On the Caterer Problem, *Management Science*, vol. 3, pp. 15–23, 1956.

85c. Quandt, R. E., and H. W. Kuhn: On Upper Bounds for the Number of Iterations in Solving Linear Programs, *Operations Research*, vol. 12, no. 1, 1964.

86. Raiffa, H., G. L. Thompson, and R. M. Thrall: An Algorithm for the Determination of All Solutions of a Two-person Zero-sum Game with a Finite Number of Strategies (Double Descriptive Method), pp. 100–114 of Directorate of Management Analysis [35].

86a. Rech, P.: "Decomposition and Interconnected Systems in Mathematical

Programming," ORC Report 65-31, Operations Research Center, University of California, Berkeley, Calif., September, 1965.

87. Riley, V., and S. I. Gass: "Bibliography on Linear Programming and Related Techniques," John Hopkins Press, Baltimore, 1958.

87a. Ritter, K.: "A Decomposition Method for Linear Programming Problems with Coupling Constraints and Variables," Mathematics Research Center, U.S. Army, MRC Report no. 739, University of Wisconsin, Madison, Wis., April, 1967.

87b. Rosen, J. B.: The Gradient Projection Method for Nonlinear Programming, *SIAM Journal*, Part I, vol. 8, no. 1, 1960; Part II, vol. 9, no. 4, 1961.

87c. Rosen, J. B.: Primal Partition Programming for Block Diagonal Matrices, *Numerische Mathematik* 6, pp. 250–260, 1964.

87d. Rosen, J. B.: "Optimal Control and Convex Programming," Chap. 13 in Abadie [1].

88. Saaty, T. L.: The Number of Vertices of a Polyhedron, *American Mathematical Monthly*, vol. 62, 1955.

88a. Saaty, T. L.: Coefficient Perturbation of a Constrained Extremum, *Operations Research*, vol. 7, May–June, 1959.

88b. Saaty, T. L., and J. Bram: "Nonlinear Mathematics," McGraw-Hill Book Company, New York, 1964.

89. Saaty, T. L., and S. I. Gass: The Parametric Objective Function. Part I, *Journal of the Operations Research Society of America*, vol. 2, 1954.

90. Schell, E.: Distribution of a Product by Several Properties, pp. 615–642 of Directorate of Management Analysis [37].

90a. Shetty, C. M.: Solving Linear Programming Problems with Variable Parameters, *Journal of Industrial Engineering*, vol. 10, no. 6, 1959.

90b. Simonnard, M.: "Linear Programming," Prentice-Hall Inc., Englewood Cliffs, N.J., 1966.

90c. Simonnard, M. A., and G. F. Hadley: Maximum Number of Iterations in the Transportation Problem, *Naval Research Logistics Quarterly*, vol. 6, pp. 125–129, 1959.

90d. Slater, M.: "Lagrange Multipliers Revisited," RAND Report RM-676, The RAND Corporation, Santa Monica, Calif., August, 1951.

91. Stanley, E. D., D. Honig, and L. Gainen: Linear Programming in Bid Evaluation, *Naval Research Logistics Quarterly*, vol. 1, no. 1, 1954.

91a. Stiefel, E.: Note on Jordan Elimination, Linear Programming and Tchebycheff Approximation, *Numerische Mathematik*, vol. 2, pp. 1–17, 1960.

92. Stigler, G. J.: The Cost of Subsistence, *Journal of Farm Economics*, vol. 27, 1945.

93. Suzuki, G.: "A Transportation Simplex Algorithm for Machine Computation Based on the Generalized Simplex Method," Report 959, The David W. Taylor Model Basin, Washington, D.C., 1955.

93a. Symonds, G. W.: "Linear Programming: The Solution of Refinery Problems," Esso Standard Oil Company, N.Y., 1955.

93b. Tintner, G.: Stochastic Linear Programming with Applications to Agricultural Economics, pp. 197–228 of Directorate of Management Analysis [37].

94. Tompkins, C.: Projection Methods in Calculation, pp. 425–447 of Directorate of Management Analysis [37].

94a. Torng, H. C.: Optimization of Discrete Control Systems through Linear Programming, *Journal of the Franklin Institute*, vol. 278, no. 1, July, 1964.

94b. Trauth, C. A., Jr., and R. E. Woolsey: "Practical Aspects of Integer Linear Programming," monograph Sc-R-66-925, Sandia Corp., Albuquerque, N.M., August, 1966.

95. Tucker, A. W.: Linear Programming, pp. 651–657 of "The Quality Control Conference Papers, 1953," American Society for Quality Control, Inc., New York.

96. Tucker, A. W.: Linear and Non-linear Programming, *Operations Research*, vol. 5, April, 1957.

97. Vajda, S.: "The Theory of Games and Linear Programming," John Wiley & Sons, Inc., New York, 1956.

97a. Vajda, S.: "Mathematical Programming," Addison-Wesley Publishing Company, Inc., Reading, Mass., 1961.

97b. Vajda, S.: "Readings in Mathematical Programming," John Wiley & Sons, Inc., New York, 1962.

97c. Valentine, F. A.: "Convex Sets," McGraw-Hill Book Company, New York, 1964.

97d. Van de Panne, C., and A. Whinston: The Simplex and the Dual Method for Quadratic Programming, *Operational Research Quarterly*, vol. 15, no. 4, 1964.

97e. Van de Panne, C., and A. Whinston: Simplicial Methods for Quadratic Programming, *Naval Research Logistics Quarterly*, vol. 11, no. 4, December, 1964.

97f. Votaw, D. F., and A. Orden: The Personnel Assignment Problem, pp. 155–163 of Directorate of Management Analysis [35].

98. Wagner, H. M.: "A Linear Programming Solution to Dynamic Leontief Type Models," RAND Report RM-1343, The RAND Corporation, Santa Monica, Calif., 1954.

99. Wagner, H. M.: A Comparison of the Original and Revised Simplex Methods, *Operations Research*, vol. 5, no. 3, 1957.

99a. Wagner, H. M.: On the Distribution of Solutions in Linear Programming Problems, *Journal of the American Statistical Association*, vol. 53, pp. 161–163, 1958.

99b. Wagner, H. M.: An Integer Linear-programming Model for Machine Scheduling, *Naval Research Logistics Quarterly*, vol. 6, no. 2, 1959.

99c. Wagner, H. M.: Linear Programming Techniques for Regression Analysis, *American Statistical Association Journal*, March, 1959.

99d. Wagner, H. M.: The Dual Simplex Algorithm for Bounded Variables, *Naval Logistics Research Quarterly*, vol. 5, no. 3, September, 1958.

99e. Wagner, H. M., and T. M. Whitin: Dynamic Version of the Economic Lot Size Model, *Management Science*, vol. 5, no. 1, 1958.

99f. Ward, L. E., Jr.: Linear Programming and Approximation Problems, *The American Mathematical Monthly*, January, 1961.

100. Waugh, F. V.: The Minimum-cost Dairy Feed, *Journal of Farm Economics*, August, 1951.

100a. Webb, K. W.: "The Mathematical Theory of Sensitivity," Paper presented at the eighteenth national meeting of the Operations Research Society of America, Detroit, Mich., Oct. 12, 1960.

100b. Whalen, B. H.: On Minimal-time and Minimal-fuel Problems, *Institute of Radio Engineers Transactions on Automatic Control*, p. 46, July, 1962.

100c. Widder, D. V.: "Advanced Calculus," Prentice-Hall, Inc., Englewood Cliffs, N.J., 1947.

100cc. Wilde, D. J., and C. S. Beightler: "Foundations of Optimization," Prentice-Hall, Inc., Englewood Cliffs, N.J., 1967.

100d. Williams, A. C.: "Transportation and Transportation-like Problems," *Electronic Computer Center Report* ECC 60.2, Socony Mobil Oil Co., New York, 1960.

101. Williams, J.: "The Compleat Strategyst," McGraw-Hill Book Company, New York, 1954.

102. Wolfe, P.: "The Simplex Method for Quadratic Programming," RAND Report P-1295, The RAND Corporation, Santa Monica, Calif., October, 1957.

103. Wolfe, P.: "A Technique for Resolving Degeneracy in Linear Programming," RAND Report RM-2995-PR, The RAND Corporation, Santa Monica, Calif., May, 1962.

104. Wolfe, P.: "Recent Developments in Non-linear Programming," R-401-PR, The RAND Corporation, Santa Monica, Calif., May, 1962.

105. Wolfe, P.: Some Simplex-like Non-linear Programming Procedures, *Operations Research*, vol. 10, no. 4, July–August, 1962.

105a. Wolfe, P.: The Simplex Method for Quadratic Programming, *Econometrica*, vol. 27, no. 3, July, 1959.

105b. Wolfe, P.: Recent Developments in Nonlinear Programming, pp. 155–187 in F. Alt (ed.), "Advances in Computers," vol. 3, Academic Press Inc., New York, 1962.

105c. Wolfe, P.: Methods of Nonlinear Programming, pp. 66–85 of Graves and Wolfe [51*g*].

105d. Wolfe, P.: "Foundations of Nonlinear Programming," RAND Report RM-4669-PR, The RAND Corporation, Santa Monica, Calif., August, 1965.

105e. Wolfe, P.: The Composite Simplex Algorithm, *SIAM Review*, vol. 7, no. 1, 1965.

105f. Wolfe, P.: Methods of Nonlinear Programming, chap. 6 in Abadie [1].

106. Yudin, D. B., and E. G. Gol'shtein: "Linear Programming," Israel Program for Scientific Translations Ltd., Jerusalem, 1965.

107. Zadeh, L. A.: An Optimal Control and Linear Programming, *Institute of Radio Engineers Transactions on Automatic Control*, pp. 45–46, July, 1962.

108. Zoutendijk, G.: "Methods of Feasible Directions," Elsevier Publishing Company, Amsterdam, 1960.

109. Zukhovitskiy, S. I., and L. I. Avdeyeva: "Linear and Convex Programming," W. B. Saunders Company, Philadelphia, 1966.

INDEX

INDEX